AF814063

Generalized Quasilinearization for Nonlinear Problems

Mathematics and Its Applications

Managing Editor:

M. HAZEWINKEL
Centre for Mathematics and Computer Science, Amsterdam, The Netherlands

Volume 440

Generalized Quasilinearization for Nonlinear Problems

by

V. Lakshmikantham

Florida Institute of Technology,
Division of Applied Mathematics,
Melbourne, Florida, U.S.A.

and

A.S. Vatsala

University of Southwestern Louisiana,
College of Sciences,
Department of Mathematics,
Lafayette, Louisiana, U.S.A.

KLUWER ACADEMIC PUBLISHERS
DORDRECHT / BOSTON / LONDON

A C.I.P. Catalogue record for this book is available from the Library of Congress.

ISBN 0-7923-5038-3

Published by Kluwer Academic Publishers,
P.O. Box 17, 3300 AA Dordrecht, The Netherlands.

Sold and distributed in North, Central and South America
by Kluwer Academic Publishers,
101 Philip Drive, Norwell, MA 02061, U.S.A.

In all other countries, sold and distributed
by Kluwer Academic Publishers,
P.O. Box 322, 3300 AH Dordrecht, The Netherlands.

Printed on acid-free paper

All Rights Reserved
©1998 Kluwer Academic Publishers
No part of the material protected by this copyright notice may be reproduced or
utilized in any form or by any means, electronic or mechanical,
including photocopying, recording or by any information storage and
retrieval system, without written permission from the copyright owner

Printed in the Netherlands.

Contents

Preface

The problems of modern society are complex, interdisciplinary and nonlinear. Nonlinear problems are therefore abundant in several diverse disciplines. Since explicit analytic solutions of nonlinear problems in terms of familiar, well-trained functions of analysis are rarely possible, one needs to exploit various approximate methods. There do exist a number of powerful procedures for obtaining approximate solutions of nonlinear problems such as, Newton-Raphson method, Galerkins method, expansion methods, dynamic programming, iterative techniques, truncation methods, method of upper and lower bounds and Chapligin method, to name a few.

Let us turn to the fruitful idea of Chapligin, see [27] (vol I), for obtaining approximate solutions of a nonlinear differential equation

$$u' = f(t, u), \quad u(0) = u_0. \tag{*}$$

Let f_1, f_2 be such that the solutions of $u' = f_1(t, u)$, $u(0) = u_0$, and $u' = f_2(t, u)$, $u(0) = u_0$ are comparatively simple to solve, such as linear equations, and lower order equations. Suppose that we have

$$f_1(t, u) \le f(t, u) \le f_2(t, u), \quad \text{for} \quad \text{all } (t, u).$$

Then the solutions of the simpler differential equations can be employed to bound the solutions of original equation (*). This technique brings into play the theories of differential and integral inequalities and monotone operators. A systematic way of obtaining these bounding functions is furnished by the method known as quasilinearization, which combines linear approximation techniques with capabilities of the digital computer in various adroit fashions. The approximations are constructed to yield rapid convergence and monotonicity as well. The origin of quasilinearization lies in the theory of dynamic programming. See Bellman [2], Bellman and Kalaba [3], Lee [43]. In many important cases, one can express

$$f(t,u) = \max_{v} g(t,u,v)$$

where the function g is linear in u for every choice of the function v. It then follows that for any v, we get

$$u(t) \geq w(t)$$

in some interval, where w is the solution of $w' = g(t,w,v)$, $w(0) = u_0$. By choosing the initial approximation $v(t)$ carefully, one can generate useful lower bounds for the solution of (*). The most important application of this technique which is called quasilinearization, is that $g(t,u,v)$ is linear in u. This can be achieved if f is convex in u, in which case, one can define the function

$$g(t,u,v) = f(t,v) + f_u(t,v)(u-v),$$

where f_u is the partial derivative of f. Of course, if f is concave in u, then one can get upper bounds of the solution of (*) as a dual result.

It is well known that method of upper and lower solutions together with monotone iterative technique offers monotone sequences which converge to the solution of the original problem (*). If we utilize the technique of lower and upper solutions coupled with the method of quasilinearization and employ the idea of Newton-Fourier, it is possible to construct concurrently lower and upper bounding monotone sequences whose elements are the solutions of the corresponding linear problems. Of course, both sequences converge rapidly. Moreover, this unification provides a mechanism to enlarge the class of nonlinear problems considerably to which the method is applicable. For example, it is not necessary to impose convexity assumption on f, since one can allow much weaker assumption. Also, the problem when f admits a decomposition $f = f_1 + f_2$ where f_1 is convex and f_2 is concave can be handled. In fact, there are several possibilities that can be investigated with this unified methodology. We therefore call this technique, the method of generalized quasilinearization. Furthermore, these ideas can be extended, generalized and refined to other types of nonlinear problems.

In this monograph, we shall develop precisely these ideas and describe the theory of the generalized quasilinearization. We present, in Chapter 1, a systematic development of the method of generalized quasilinearization to first order initial value problem, and to systems with initial conditions, bringing out clearly its various ramifications. In chapter 2, we continue to investigate first order differential equations relative to periodic and anti-periodic boundary value problems and higher order of convergence etc.

Chapter 3 is devoted to the second order boundary value problems. Available results are discussed in detail for simpler as well as general boundary value problems etc.

In Chapter 4, we provide extensions to a variety of differential equations such as, delay differential equations, Stochastic differential equations, Integro-differential equations, differential equations in a Banach space, and dynamic systems on a time scale. The study of equations on time scales (on measure chains) contains both the theory of difference equations and differential equations, which is a new area of investigation. We only provide typical results in each case to bring out the principal ideas.

Some important features of the monograph are as follows.

(1) It provides a systematic development of the method of generalized quasilinearization indicating the notions and technical difficulties that are encountered in the unified approach.

(2) It enhances considerably the usefulness of the method of quasilinearization which has proved to be very effective in several areas of investigation and in applications.

(3) It includes the well known monotone iterative technique as a special case.

We do hope that the timely recognition of this unified approach, namely, the method of generalized quasilinearization will provide an impetus for further advances which are fruitful in applications to real world problems.

We wish to express our thanks to Mrs. Donn Miller-Kermani for her excellent typing of this manuscript.

V. LAKSHMIKANTHAM AND A. S. VATSALA

1 FIRST ORDER DIFFERENTIAL EQUATIONS

1.0 INTRODUCTION

This chapter introduces the theory of the method of lower and upper solutions, monotone iterative technique and the method of generalized quasilinearization relative to the first order differential equations with initial conditions, and consequently forms the basis for the remaining chapters where essential ideas are extended to other types of dynamic systems.

The basic idea of the original method of quasilinearization developed by Bellman and Kalaba is to provide an explicit analytic representation for the solution of nonlinear differential equations, which yields pointwise lower estimates for the solution of the problem whenever the function involved is convex. The most important applications of this method has been to obtain a sequence of lower bounds which are the solutions of linear differential equations that converge quadratically. As a result, the method has been popular in applied areas. See Bellman [2, 3]. However, the convexity assumption that is demanded by the method of quasilinearization has been a stumbling block for further development of the theory. Recently, this method has been generalized, refined and extended in several directions so as to be applicable to a much larger class of nonlinear problems by not demanding convexity property. Moreover, other possibilities have been explored which make the method of generalized quasilinearization universally useful in applications.

In Section 1.1, we shall prove necessary results concerning the method of lower and upper solutions, including an existence result in a closed set generated by lower and upper solutions. We shall describe, in Section 1.2, the basic idea of the well known method of quasilinearization. When one utilizes the method of lower and upper solutions and employ the trick of Newton-Fourier, the method of quasilinearization can be refined so as to obtain simultaneously lower and upper bounds in terms of monotone sequences that converge to the solution quadratically. A dual result when the function is concave is also indicated in this set up.

We shall discuss, in Section 1.3, several possible extensions of the method of quasilinearization when the function in question admits a decomposition into a difference of two convex or concave parts. This splitting permits us to use lower and upper solutions in a natural form or coupled forms which generate several different problems of interest.

We shall generalize the method of quasilinearization in Section 1.4 by not demanding convexity assumption which, without a new idea, leads to a disadvantage of solving nonlinear differential equations for generating monotone sequences, although much simpler than the given differential equation. This section describes the development of this aspect of generalization which utilizes a multistage process first, and then discovers a simple algorithmic approach that directly offers the construction of monotone sequences which are the solutions of linear differential equations. Furthermore, the condition imposed to relax convexity assumption, appears as a needed condition at the first sight but as noted in Remark 1.3.1, the required condition always holds since we are working in a compact set. The final result in this development provides the direct approach avoiding the multistage process and this important idea is followed in later sections.

Further splitting of the function involved is described in Section 1.5 showing the limitations of how far we can go in this method of generalized quasilinearization. Several results are offered to demonstrate the possible choices with corresponding outcomes.

1.1 METHOD OF UPPER AND LOWER SOLUTIONS

An interesting and fruitful technique for proving existence results for nonlinear problems is the method of upper and lower solutions. The basic idea is to modify the given problem suitably into a simpler problem and then employ Leray-Schauder theory of topological degree or known existence results of the modified problem, together with the theory of inequalities, to establish the existence results of the given problem. Moreover, this technique yields, as a by-product, existence of solutions in a closed set, namely, the ordered interval, generated by upper and lower solutions.

We shall devote this section to study the method of upper and lower solutions which plays a prominent part in our later discussions. We shall begin with the scalar initial value problems (IVP for short) where we shall bring out the important ideas involved in the method of upper and lower solutions.

Consider the IVP

$$x' = f(t,x), \quad x(0) = x_0, \tag{1.1.1}$$

where $f \in C[J \times R, R]$ and $J = [0,T]$. A function $v \in C^1[J,R]$ is said to be a lower solution of (1.1.1) if

$$v' \leq f(t,v), \quad v(0) \leq x_0,$$

and an upper solution of (1.1.1) if the inequalities are reversed.

A fundamental result concerning upper and lower solutions is the following theorem.See [21].

Theorem 1.1.1 *Let v, $w \in C^1[J,R]$ be lower and upper solutions respectively and suppose that*

$$f(t,x) - f(t,y) \leq L(x-y),$$

whenever $x \geq y$ for some $L > 0$. Then $v(0) \leq w(0)$ implies $v(t) \leq w(t)$ on J.

Sometimes, it may be convenient to consider coupled upper and lower solutions of (1.1.1) defined by

$$v' \leq f(t,w), \quad v(0) \leq x_0,$$

$$w' \geq f(t,v), \quad w(0) \geq x_0,$$

in which case, we would also need a result corresponding to Theorem 1.1.1. In fact, we require in our later discussions, a variety of such results which we collect in the following result.

Theorem 1.1.2 *Let v, $w \in C^1[J,R]$ and $F \in C[J \times R^2, R]$. Suppose further that any one of the following conditions hold:*

(H_1) $v' \leq F(t,v,v)$, $w' \geq F(t,w,w)$,

$$F(t,x_1,y_1) - F(t,x_2,y_2) \leq L[(x_1-x_2)+(y_1-y_2)], \ L \geq 0,$$

whenever $x_1 \geq x_2,\ y_1 \geq y_2$;

(H_2) $v' \leq F(t,v,w),\ w' \geq F(t,w,v)$,

$$\begin{aligned} F(t,x_1,y) - F(t,x_2,y) &\leq L(x_1 - x_2),\ L \geq 0, \\ F(t,x,y_1) - F(t,x,y_2) &\geq -L(y_1 - y_2), \end{aligned}$$

whenever $x_1 \geq x_2,\ y_1 \geq y_2$;

(H_3) $v' \leq F(t,w,v),\ w' \geq F(t,v,w)$,

$$\begin{aligned} F(t,x,y_1) - F(t,x,y_2) &\leq L(y_1 - y_2),\ L \geq 0, \\ F(t,x_1,y) - F(t,x_2,y) &\geq -L(x_1 - x_2), \end{aligned}$$

whenever $x_1 \geq x_2,\ y_1 \geq y_2$;

(H_4) $v' \leq F(t,w,w),\ w' \geq F(t,v,v)$,

$$F(t,x_1,y_1) - F(t,x_2,y_2) \geq -L\left[(x_1 - x_2) + (y_1 - y_2)\right],\ L \geq 0,$$

whenever $x_1 \geq x_2,\ y_1 \geq y_2$.

Then $v(0) \leq w(0)$ *implies* $v(t) \leq w(t)$ *on* J.

Proof Let us first prove the conclusion relative to (H_1) which includes Theorem 1.1.1 as a special case. To do this, let $w_0 = w + \varepsilon e^{3Lt}$ for small $\varepsilon > 0$ so that $w_0(t) > w(t)$ and $w_0(0) > v(0)$. Then, using (H_1), we have

$$\begin{aligned} w_0' &\geq F(t,w,w) + 3L\varepsilon e^{3Lt} \\ &\geq F(t,w_0,w_0) - 2L\varepsilon e^{3Lt} + 3L\varepsilon e^{3Lt} \\ &> F(t,w_0,w_0). \end{aligned}$$

We shall show that $v(t) < w_0(t)$ on J, which proves the conclusion as $\varepsilon \to 0$. If this is false, there would exist a $t_0 \in (0,T]$ such that

$$v(t_0) = w_0(t_0) \text{ and } v(t) < w_0(t),\ 0 \leq t < t_0.$$

This implies $v'(t_0) \geq w_0'(t_0)$ which yields a contradiction

$$F(t_0, v(t_0), v(t_0)) \geq v'(t_0) \geq w_0'(t_0) > F(t_0, w_0(t_0), w_0(t_0)).$$

Hence the claim of the theorem is true when (H_1) holds.

We shall next show that the claim of the theorem is valid when (H_2) holds. The conclusion relative to (H_3) and (H_4) can be proved using similar arguments.

To prove the result when (H_2) holds, we set

$$w_0 = w + \varepsilon e^{3Lt} \text{ and } v_0 = v - \varepsilon e^{3Lt}$$

for small $\varepsilon > 0$ so that $w_0 > w$, $v_0 < v$ and $w_0(0) > v_0(0)$. Then, using (H_2), we get

$$\begin{aligned} v_0' &\leq F(t, v, w) - 3L\varepsilon e^{3Lt} \\ &\leq F(t, v, w_0) + L\varepsilon e^{3Lt} - 3L\varepsilon e^{3Lt} \\ &\leq F(t, v_0, w_0) + 2L\varepsilon e^{3Lt} - 3L\varepsilon e^{3Lt} \\ &< F(t, v_0, w_0). \end{aligned}$$

Similarly,

$$\begin{aligned} w_0' &\geq F(t, w, v) + 3L\varepsilon e^{3Lt} \\ &\geq F(t, w_0, v) - L\varepsilon e^{3Lt} + 3L\varepsilon e^{3Lt} \\ &\geq F(t, w_0, v_0) - 2L\varepsilon e^{3Lt} + 3L\varepsilon e^{3Lt} \\ &> F(t, w_0, v_0). \end{aligned}$$

It is enough to show $v_0(t) < w_0(t)$ on J, which proves the claim of the theorem as $\varepsilon \longrightarrow 0$. Suppose that $v_0(t) < w_0(t)$ on J is not true. Then there would exist a $t_0 \in (0, T]$ such that $v_0(t_0) = w_0(t_0)$ and $v_0(t) < w_0(t)$ for $0 \leq t < t_0$ and consequently $v_0'(t_0) \geq w_0'(t_0)$. As a result, we arrive at the contradiction

$$F(t_0, v_0(t_0), w_0(t_0)) > v_0'(t_0) \geq w_0'(t_0) > F(t_0, w_0(t_0), v_0(t_0)),$$

proving the theorem.

If we know the existence of upper and lower solutions v, w of (1.1.1) such that $v(t) \leq w(t)$ on J then we can prove the following result :

Theorem 1.1.3 *Let v, $w \in C^1[J, R]$ be lower and upper solutions of* (1.1.1) *such that $v(t) \leq w(t)$ on J, and let $f \in C[\Omega, R]$ where $\Omega = [(t, x) : v(t) \leq x \leq w(t),\ t \in J]$, then there exists a solution $x(t)$ of* (1.1.1) *satisfying $v(t) \leq x(t) \leq w(t)$ on J.*

We shall not prove this theorem since we shall consider below an extension of this result to finite systems of differential equations which leads to more general results. For this purpose, we need to use vectorial inequalities which are understood to mean the same inequalities hold between their corresponding components. Also, when we deal with systems of inequalities, a certain property known as quasimonotone nondecreasing property is necessary which we define below.

A function $f \in C[R^n, R^n]$ is said to possess quasimonotone nondecreasing property if for each i such that $1 \leq i \leq n$, $u \leq v$ and $u_i = v_i$ implies $f_i(t,u) \leq f_i(t,v)$.

Consider now the differential system

$$x' = f(t,x), \quad x(0) = x_0, \tag{1.1.2}$$

where $f \in C[J \times R^n, R^n]$. The following result is an extension of Theorem 1.1.1 for systems of inequalities.

Theorem 1.1.4 *Let v, $w \in C^1[J \times R^n, R^n]$ be lower and upper solutions of (1.1.2) respectively. Assume that $f \in C[J \times R^n, R^n]$, $f(t,x)$ is quasimonotone nondecreasing in x for each $t \in J$ and for each i,*

$$f_i(t,x) - f_i(t,y) \leq L \sum_{j=1}^{n} (x_j - y_j), \quad L \geq 0, \tag{1.1.3}$$

where $x \geq y$. Then $v(0) \leq w(0)$ implies $v(t) \leq w(t)$.

One can extend Theorem 1.1.2 to the case of systems so that Theorem 1.1.4 would be a special case. The next result is precisely in this direction.

Theorem 1.1.5 *Let v, $w \in C^1[J, R^n]]$and $F \in C\left[J \times R^{2n}, R^n\right]$. Suppose that one of the following conditions hold:*

(H_1) $v' \leq F(t,v,v)$, $w' \geq F(t,w,w)$, $F(t,x,y)$ is quasimonotone nondecreasing in x and y for each $t \in J$ and for each i,

$$F_i(t,x_1,y_1) - F_i(t,x_2,y_2) \leq L \sum_{j=1}^{n} \left[(x_{1j} - x_{2j}) + (y_{1j} - y_{2j})\right],$$

whenever $x_1 \geq x_2$, $y_1 \geq y_2$;

(H_2) $v' \leq F(t,v,w)$, $w' \geq F(t,w,v)$, $F(t,x,y)$ is quasimonotone nondecreasing in x for each (t,y) and nonincreasing in y for each (t,x) and for each i,

$$\begin{aligned} F_i(t,x_1,y) - F_i(t,x_2,y) &\leq L \sum_{j=1}^{n} (x_{1j} - x_{2j}), \ L \geq 0, \text{ for each } (t,y), \\ F_i(t,x,y_1) - F_i(t,x,y_2) &\geq -L \sum_{j=1}^{n} (y_{1j} - y_{2j}), \text{ for each } (t,x), \end{aligned}$$

whenever $x_1 \geq x_2$, $y_1 \geq y_2$;

(H_3) $v' \leq F(t,w,v)$, $w' \geq F(t,v,w)$, $F(t,x,y)$ *is quasimonotone nonincreasing in* x *for each* (t,y) *and nondecreasing in* y *for each* (t,x) *and for each* i,

$$\begin{aligned} F_i(t,x_1,y) - F_i(t,x_2,y) &\geq -L\sum_{j=1}^{n}(x_{1j} - x_{2j}),\ L \geq 0,\ \textit{for each } (t,y), \\ F_i(t,x,y_1) - F_i(t,x,y_2) &\leq L\sum_{j=1}^{n}(y_{1j} - y_{2j}),\ \textit{for each } (t,x), \end{aligned}$$

whenever $x_1 \geq x_2$, $y_1 \geq y_2$;

(H_4) $v' \leq F(t,w,w)$, $w' \geq F(t,v,v)$, $F(t,x,y)$ *is quasimonotone nonincreasing in* x *and* y *for each* $t \in J$ *and for each* i,

$$F_i(t,x_1,y_1) - F_i(t,x_2,y_2) \geq -L\sum_{j=1}^{n}\left[(x_{1j} - x_{2j}) + (y_{1j} - y_{2j})\right],\ L \geq 0$$

whenever $x_1 \geq x_2$, $y_1 \geq y_2$.

Then $v(0) \leq w(0)$ *implies* $v(t) \leq w(t)$ *on* J.

Proof Suppose that (H_1) holds. Let $w_0 = w + \varepsilon e^{(2n+1)Lt}$ for some $\varepsilon > 0$ and $\varepsilon = (\varepsilon, \varepsilon, \cdots, \varepsilon)$. Then $w_0(0) > v(0)$ and $w_0 > w$, and consequently, using (H_1), we obtain, for each i,

$$\begin{aligned} w'_{0,i} &= w'_i + (2n+1)L\varepsilon e^{(2n+1)Lt} \\ &\geq F_i(t,w,w) + (2n+1)L\varepsilon e^{(2n+1)Lt} \\ &\geq F_i(t,w_0,w_0) - 2nL\varepsilon e^{(2n+1)Lt} + (2n+1)L\varepsilon e^{(2n+1)Lt} \\ &> F_i(t,w_0,w_0). \end{aligned}$$

It is enough to show that $v(t) < w_0(t)$ on J so that as $\varepsilon \to 0$, we arrive at the conclusion of the theorem. Suppose that $v(t) < w_0(t)$ on J is false. Then there would exist an index j, $1 \leq j \leq n$ and a $t_0 \in (0,T]$ such that

$$v_j(t_0) = w_{0j}(t_0),\ v_j(t) \leq w_{0j}(t),\ 0 \leq t \leq t_0 \text{ and } v_i(t_0) \leq w_i(t_0) \text{ for } i \neq j.$$

It then follows that $v'_j(t_0) \geq w'_{0j}(t_0)$, which in view of ($H_1$) yields

$$F_j(t_0, v(t_0), v(t_0)) \geq v'_j(t_0) \geq w'_{0j}(t_0) > F_i(t_0, w_0(t_0), w_0(t_0)).$$

This contradiction proves the claim. Note that we have used the quasimonotonicity of F in x and y to get the foregoing inequality.

It is now easy to construct the proofs of the conclusions relative to (H_2), (H_3) and (H_4) following the proof of Theorem 1.1.2 and the proof given above. We omit the details. The proof of this theorem is therefore complete.

Next we shall present a theorem which extends the result of Theorem 1.1.3 to systems of differential equations.

Theorem 1.1.6 *Let v, $w \in C^1[J, R^n]$ be lower and upper solutions of* (1.1.2) *respectively such that $v(t) \leq w(t)$ on J. Suppose that $f \in C[\Omega, R^n]$, where, as before*

$$\Omega = [(t,x) : v(t) \leq x \leq w(t),\ t \in J],$$

and $f(t,x)$ is quasimonotone nondecreasing in x for each $t \in J$. Then there exists a solution $x(t)$ of (1.1.2) satisfying $v(t) \leq x(t) \leq w(t)$ on J provided $v(0) \leq x_0 \leq w(0)$.

The validity of Theorem 1.1.6 is also true without demanding $f(t,x)$ to be quasimonotone nondecreasing in x, provided we strengthen the notion of upper and lower solutions of (1.1.2) as follows:

For each i, $1 \leq i \leq n$,

$$\begin{aligned} v_i' &\leq f_i(t,\sigma) \text{ for all } \sigma \text{ such that } v_i(t) = \sigma_i \text{ and } v(t) \leq \sigma \leq w(t), \\ w_i' &\geq f_i(t,\sigma) \text{ for all } \sigma \text{ such that } w_i(t) = \sigma_i \text{ and } v(t) \leq \sigma \leq w(t). \end{aligned} \tag{1.1.4}$$

We shall prove the following existence result which includes Theorem 1.1.6 as a special case.

Theorem 1.1.7 *Let v, $w \in C^1[J, R^n]$ with $v(t) \leq w(t)$ satisfy the relations* (1.1.4) *and $f \in C[\Omega, R^n]$. Then the conclusion of Theorem* 1.1.6 *is valid.*

Proof Consider $P : J \times R^n \to R^n$ defined by

$$P_i(t,x) = \max\{v_i(t),\ \min[x_i, w_i(t)]\}, \text{ for each } i.$$

Then $f(t, P(t,x))$ defines a continuous extension of f to $J \times R^n$ which is also bounded since f is bounded on Ω. Therefore, $x' = f(t, P(t,x))$ has a solution x on J with $x(0) = x_0$. Let us show that $v(t) \leq x(t) \leq w(t)$ and therefore a solution of (1.1.2). For $\varepsilon > 0$ and $e = (1, 1, \cdots, 1)$, consider $w_\varepsilon(t) = w(t) + \varepsilon(1 + t)e$ and $v_\varepsilon(t) = v(t) - \varepsilon(1+t)e$. We have $v_\varepsilon(0) < x_0 < w_\varepsilon(0)$. Suppose that $t_1 \in J$ is such that $v_\varepsilon(t) < x(t) < w_\varepsilon(t)$ in $[0, t_1)$ but $x_i(t_1) = w_{\varepsilon i}(t_1)$. Then we have $v(t_1) \leq P(t_1, x(t_1)) \leq w(t_1)$ and $P_i(t_1, x(t_1)) = w_i(t_1)$, hence

$$w_i'(t_1) \geq f_i(t_1, P(t_1, x(t_1))) = x_i'(t_1),$$

which implies $x_i'(t_1) < w_{\varepsilon i}'(t_1)$, contradicting $x_i(t_1) = w_{\varepsilon i}(t_1)$. Therefore $v_\varepsilon(t) < x(t) < w_\varepsilon(t)$ in J. Now $\varepsilon \to 0$ yields $v(t) \le x(t) \le w(t)$. This completes the proof.

One could also have an arbitrary coupling relative to lower and upper solutions of (1.1.2). To define this let for each i, $1 \le i \le n$, p_i, q_i represent two nonnegative integers such that $p_i + q_i = n - 1$ so that we can split the system and define coupled lower and upper quasi solutions of (1.1.2) as follows:

$$\begin{aligned} v_i' &\le f_i\left(t, v_i, [v]_{p_i}, [w]_{q_i}\right), \\ w_i' &\ge f_i\left(t, w_i, [w]_{p_i}, [v]_{q_i}\right). \end{aligned} \tag{1.1.5}$$

A function $f(t,x)$ is said to possess a mixed quasimonotone property if for each i, $f_i\left(t, x_i, [x]_{p_i}, [x]_{q_i}\right)$ is monotone nondecreasing in $[x]_{p_i}$ and monotone nonincreasing in $[x]_{q_i}$. Observe that if f satisfies a mixed quasimonotone property, then (1.1.4) holds for coupled lower and upper quasi solutions (v,w) defined by (1.1.5). Consequently one has the following result as well, which is also a special case of Theorem 1.1.7.

Theorem 1.1.8 *Let v, $w \in C[J, R^n]$ be coupled lower and upper solutions of (1.1.2) and $f \in C[\Omega, R^n]$. If $f(t,x)$ possesses a mixed quasimonotone property, then there exists a solution $x(t)$ of (1.1.2) such that $v(t) \le x(t) \le w(t)$ on J.*

Remark 1.1.1 It is easy to note from the proofs that v, w need not be C^1 and it is enough to define various lower and upper solutions v, w in terms of Dini derivatives. For example, it is enough to suppose that v, $w \in C[J, R^n]$ and

$$D_- v \le f(t,v), \qquad D_- w \ge f(t,w),$$

where $D_- v = \liminf\limits_{h \to 0^-} \frac{1}{h}\left[v(t+h) - v(t)\right]$. See [27] for details.

Sometimes we need the following corollary of Theorem 1.1.4 which can be considered as the generalized Gronwall Lemma.

Corollary 1.1.1 *Let $v \in C^1[J, R^n]$ and $v' \le Av + \sigma$, where $A = (a_{ij})$ is an $n \times n$ matrix satisfying $a_{ij} \ge 0$, $i \ne j$ and $\sigma \in C[J, R^n]$. Then we have*

$$v(t) \le v(0)e^{At} + \int_0^t e^{A(t-s)}\sigma(s)ds, \quad t \in J.$$

1.2 METHOD OF QUASILINEARIZATION

The basic idea of the well known method of quasilinearization is very simple. Consider the IVP

$$x' = f(t,x), \quad x(0) = x_0, \tag{1.2.1}$$

where $f \in C^2[I \times \Omega, R]$ and $f_{xx}(t,x) > 0$ on $I \times \Omega$, I and Ω being open sets in R containing $(0, x_0)$. We define the function

$$g(t,x,y) = f(t,y) + f_x(t,y)(x-y),$$

which is linear in x for each (t,y) so that we can write

$$x' = f(t,x) = \max_y g(t,x,y). \tag{1.2.2}$$

It then follows that, for any function $y \in C[I,R]$, we have

$$x' \geq g(t,x,y)$$

and by comparison theorem we get

$$x(t) \geq w(t,y), \quad t \in I_0, \tag{1.2.3}$$

where $w(t,y)$ is the solution of

$$w' = g(t,w,y(t)), \quad w(0) = x_0 \tag{1.2.4}$$

and $x(t)$ is the solution of (1.2.1) existing on some interval $I_0 = [0,a] \subset I$. Choosing $y(t)$ in an adroit fashion, one can obtain useful lower bounds for solution of (1.2.1) on I_0. The most important application of this technique, which is called quasilinearization, is that $g(t,x,y)$ is linear in x.

Returning to (1.2.3), we find that (1.2.3) is valid for all function $y(t)$ and that the equality holds for the function $y(t) = x(t)$. We can therefore write

$$x(t) = \max_y w(t,y), \tag{1.2.5}$$

which provides an explicit analytic representation for the solution of (1.2.1). Its values lie not so much in the explicit representation as in the pointwise estimates that can be obtained from (1.2.3) by simple choices of $y(t)$.

We observe that the function which maximizes (1.2.2), in view of the definition of $g(t,x,y)$ is the solution $x(t)$ itself. This suggests that we employ a method of successive approximations in which $y(t)$ is chosen at each stage to be a lower estimate of the desired solution $x(t)$ of (1.2.1). Let $y_0(t)$ to be an initial approximation which is a lower solution of (1.2.1), namely,

$$y_0' \leq f(t,y_0), \quad y_0(0) \leq x_0.$$

Then we can construct the sequence of iterates $\{y_n(t)\}$ as the solutions of the linear differential equations

$$y_n' = f(t, y_{n-1}) + f_x(t, y_{n-1})(y_n - y_{n-1}), \quad y_n(0) = x_0$$

and show under the assumptions, that the sequence $\{y_n(t)\}$ is monotone nondecreasing and satisfies

$$y_0(t) \leq y_1(t) \leq y_2(t) \leq \cdots \leq y_n(t) \leq x(t) \text{ on } I_0,$$

where $x(t)$ is the solution of (1.2.1) on I_0. The standard arguments (see [3, 21]), then show that $y_n(t)$ converges uniformly to $x(t)$ on I_0. Moreover, one can prove that the convergence of $y_n(t)$ to $x(t)$ is quadratic, that is

$$\max_{I_0} |x(t) - y_n(t)| \leq k \max_{I_0} |x(t) - y_{n-1}(t)|^2,$$

for some constant k.

If, on the other hand, $f(t, x)$ is uniformly concave and the initial function y_0 satisfies $y_0' \geq f(t, y_0)$, $y_0(0) \geq x_0$, then one can prove a dual result where one can derive upper bounds for the solution $x(t)$ of (1.2.1). Furthermore, we can construct the sequence $\{y_n(t)\}$ that converge to the unique solution $x(t)$ of (1.2.1) on some interval.

We note that lower and upper solutions of (1.2.1) are implicitly involved in the method of quasilinearization. Also, it is not easy, in general to obtain upper sequences bounding the solution, when we assume $f(t, x)$ to be convex, since a direct application of the method of quasilinearization yields upper sequence whose elements are the solutions of nonlinear problems. A dual situation occurs when $f(t, x)$ is concave. However, if we utilize the technique of lower and upper solutions coupled with the method of quasilinearization, and employ the idea of Newton-Fourier it is possible to construct concurrently upper and lower bounding monotone sequences, whose elements are solutions of linear initial value problems. Of course, both converge quadratically to the unique solution of (1.2.1). This approach also offers existence of solution of (1.2.1) in the closed set generated by lower and upper solutions of (1.2.1). Let us prove the following result which exploits the foregoing idea.

Theorem 1.2.1 *Assume that*

(A_1) α_0, $\beta_0 \in C^1[J, R]$ *are lower and upper solutions of (1.2.1) such that* $\alpha_0(t) \leq \beta_0(t)$ *on* $J = [0, T]$;

(A_2) $f \in C[\Omega, R]$, $f_x(t, x)$, $f_{xx}(t, x) \geq 0$ *exists and are continuous on* Ω, *where* $\Omega = \{(t, x) : \alpha_0(t) \leq x \leq \beta_0(t),\ t \in J\}$.

Then there exist monotone sequences $\{\alpha_n(t)\}$, $\{\beta_n(t)\}$ which converge uniformly to the unique solution $x(t)$ of (1.2.1) on J and the convergence is quadratic.

Proof Observe that (A_2) implies that for any $\alpha_0(t) \le y \le x \le \beta_0(t)$

$$f(t,x) \ge f(t,y) + f_x(t,y)(x-y) \tag{1.2.6}$$

and

$$f(t,x) - f(t,y) \le L(x-y),$$

for some constant $L \ge 0$. Consider the IVPs

$$\begin{aligned} \alpha_1' &= g(t,\alpha_1;\alpha_0), \quad \alpha_1(0) = x_0, && (1.2.7)\\ \beta_1' &= G(t,\beta_1;\alpha_0,\beta_0), \quad \beta_1(0) = x_0, && (1.2.8)\end{aligned}$$

where $\alpha_0(0) \le x_0 \le \beta_0(0)$,

$$\begin{aligned} g(t,\alpha_1;\alpha_0) &= f(t,\alpha_0) + f_x(t,\alpha_0)(\alpha_1 - \alpha_0),\\ G(t,\beta_1;\alpha_0,\beta_0) &= f(t,\beta_0) + f_x(t,\alpha_0)(\beta_1 - \beta_0).\end{aligned}$$

Let $p = \alpha_0 - \alpha_1$ so that $p(0) \le 0$. Then in view of (1.2.7),

$$p' \le f(t,\alpha_0) - g(t,\alpha_1;\alpha_0) = f_x(t,\alpha_0)p$$

which implies that $p(t) \le 0$ on J, showing that $\alpha_0 \le \alpha_1$ on J. Similarly letting $q = \beta_1 - \beta_0$, we arrive using (1.2.8) at $q(0) \le 0$ and

$$q' \le G(t,\beta_1;\alpha_0,\beta_0) - f(t,\beta_0) = f_x(t,\beta_0)q.$$

This proves that $\beta_1 \le \beta_0$ on J. Next we let $p = \alpha_1 - \beta_1$ and note that $p(0) = 0$. Because of (1.2.7) and (1.2.8) and the fact that $f_x(t,x)$ is nondecreasing in x, we find using mean value theorem, that

$$\begin{aligned} p' &= f_x(t,\xi)(\alpha_0 - \beta_0) + f_x(t,\alpha_0)\left[\alpha_1 - \alpha_0 + \beta_0 - \beta_1\right]\\ &\le f_x(t,\alpha_0)p,\end{aligned}$$

where $\alpha_0(t) \le \xi \le \beta_0(t)$. As a result, we get $\alpha_1 \le \beta_1$ on J and therefore, it follows that

$$\alpha_0 \le \alpha_1 \le \beta_1 \le \beta_0 \text{ on } J.$$

This process can be continued successively to obtain monotone sequences $\{\alpha_n(t)\}$, $\{\beta_n(t)\}$ satisfying

$$\alpha_0 \le \alpha_1 \le \cdots \le \alpha_n \le \beta_n \le \cdots \le \beta_1 \le \beta_0 \text{ on } J,$$

where α_n, β_n are the solutions of the linear IVPs

$$\begin{aligned}
\alpha_n' &= f(t, \alpha_{n-1}) + f_x(t, \alpha_{n-1})(\alpha_n - \alpha_{n-1}), \quad \alpha_n(0) = x_0, \\
\beta_n' &= f(t, \beta_{n-1}) + f_x(t, \alpha_{n-1})(\beta_n - \beta_{n-1}), \quad \beta_n(0) = x_0,
\end{aligned}$$

on J. Employing standard arguments, it is now easy to prove that the sequences $\{\alpha_n(t)\}$, $\{\beta_n(t)\}$ converge uniformly and monotonically to the unique solution $x(t)$ of (1.2.1) on J.

We shall next show the convergence of α_n, β_n to $x(t)$ the unique solution of (1.2.1) is quadratic. For this purpose, consider $p_{n+1} = x - \alpha_{n+1} \geq 0$ and $q_{n+1} = \beta_{n+1} - x \geq 0$ and note that $p_{n+1}(0) = q_{n+1}(0) = 0$ for $n \geq 1$. Now

$$\begin{aligned}
p_{n+1}' &= f(t, x) - [f(t, \alpha_n) + f_x(t, \alpha_n)(\alpha_{n+1} - \alpha_n)] \\
&= f_x(t, \xi)p_n - f_x(t, \alpha_n)(-p_{n+1} + p_n) \\
&\leq [f_x(t, x) - f_x(t, \alpha_n)]\, p_n + f_x(t, \alpha_n)p_{n+1} \\
&= f_{xx}(t, \sigma)p_n^2 + f_x(t, \alpha_n)p_{n+1}
\end{aligned}$$

where $\alpha_n(t) < \xi$, $\sigma < x(t)$. Thus we get

$$p_{n+1}' \leq Mp_{n+1} + Np_n^2,$$

where $|f_x(t, x)| \leq M$, $|f_{xx}(t, x)| \leq N$ on Ω. Consequently, applying Gronwall's Lemma, treating Np_n^2 as a forcing term, it follows that

$$0 \leq x(t) - \alpha_{n+1}(t) \leq N \int_0^t [x(s) - \alpha_n(s)]^2 e^{M(t-s)} ds, \quad t \in J,$$

which yields the desired result

$$\max_J |x(t) - \alpha_{n+1}(t)| \leq \frac{Ne^{Mt}}{M} \max_J |x(t) - \alpha_n(t)|^2 .$$

Similarly, we find that

$$\begin{aligned}
q_{n+1}' &= f(t, \beta_{n-1}) + f_x(t, \beta_{n-1})(\beta_n - \beta_{n-1}) - f(t, x) \\
&= f_x(t, \xi)q_n + f_x(t, \alpha_n)(q_{n+1} - q_n) \\
&\leq [f_x(t, \beta_n) - f_x(t, \alpha_n)]\, q_n + f_x(t, \alpha_n)q_{n+1} \\
&= f_{xx}(t, \sigma)\, [q_n + p_n]\, q_n + f_x(t, \alpha_n)q_{n+1} \\
&\leq Mq_{n+1} + \frac{3N}{2}q_n^2 + \frac{N}{2}p_n^2,
\end{aligned}$$

where $x(t) < \xi$, $\sigma < \beta_n(t)$, N, M being the same constant as earlier. This yields, using Gronwall's Lemma, as before

$$\max_J |\beta_{n+1} - x(t)| \leq \frac{3}{2}\frac{Ne^{Mt}}{M} \max_J |\beta_n - x(t)|^2 + \frac{1}{2}\frac{Ne^{Mt}}{M} \max_J |x(t) - \alpha_n(t)|^2 .$$

The proof is therefore complete.

One can prove a dual result for the case when $f(t,x)$ is concave. This we state as a problem.

Problem 1.2.1 Assume (A_1) of Theorem 1.2.1 holds. Suppose that $f \in C[\Omega, R]$ and $f_x(t,x)$, $f_{xx}(t,x) \leq 0$ exist and are continuous on Ω. Then there exist monotone sequences $\{\alpha_n(t)\}$, $\{\beta_n(t)\}$ which converge uniformly to the unique solution $x(t)$ of (1.2.1) on J and the convergence is quadratic.

Remark 1.2.1 It is important to realize that in Theorem 1.2.1 the construction of $\{\beta_n\}$, we use an idea of Newton-Fourier's method so that one can have β_n as the solution of linear IVPs.

One can weaken the conditions of Theorem 1.2.1 and obtain superlinear convergence rather than quadratic. This we state as another problem.

Problem 1.2.2 Let (A_1) of Theorem 1.2.1 holds. Suppose that $f \in C[\Omega, R]$, $f_x(t,x)$ exists, continuous on Ω and satisfies $f(t,x) \geq f(t,y) + f_x(t,x)(x-y)$, if $\alpha_0(t) \leq y \leq x \leq \beta_0(t)$, and $|f_x(t,x) - f_x(t,y)| \leq L|x-y|^\gamma$, $0 < \gamma < 1$ for (t,x), uniformly to the unique solution $x(t)$ of (1.2.1) on J and the convergence is superlinear. If $\gamma = 1$, the convergence is quadratic.

Remark 1.2.2 Sometimes it may not be easy to find both lower and upper solutions required in Theorem 1.2.1 at the same time. In this case, one can utilize the solution $x(t)$ as the missing lower or upper solutions and obtain only corresponding monotone sequence.

As an example, consider the Ricatti equation

$$x' = x^2 + a(t), \quad x(0) = x_0.$$

Here it is easy to find a lower solution $\alpha_0(t)$. Since $x^2 \geq 0$, $\alpha_0(t)$ may be chosen as $\alpha_0(t) = \alpha_0(0) + \int_0^t a(s)ds$ with $\alpha_0(0) \leq x_0$. It is not easy to choose an upper solution in this case. Thus letting $\alpha_0'(t) = a(t)$, one can obtain only the nondecreasing sequence $\alpha_n(t)$ which converges to $x(t)$ on the interval of existence of $x(t)$ of (1.2.10).

As we shall see later in Section 1.4 that convexity requirement of $f(t,x)$ can be weakened in which case finding lower and upper solutions becomes comparatively easier in several cases.

1.3 EXTENSIONS

We have seen in Section 1.2 that when $f(t,x)$ in (1.2.1) is convex, one can provide lower and upper bounds simultaneously that converge quadratically to the unique solution of the IVP (1.2.1) and a similar conclusion is valid when $f(t,x)$ is concave. The question, whether one can prove corresponding results when $f(t,x)$ admits a decomposition into a difference of two convex or concave functions or equivalently f admits a splitting of convex and concave parts, is natural. In this section, we shall discuss this problem and show that we can obtain a variety of results with the same conclusion each of which is interesting in itself. Naturally, these results include other possible cases when f is either convex or concave and whether lower and upper solutions are of natural form or of coupled form. We state these as corollaries.

Let α_0, $\beta_0 \in C^1[J,R]$ with $\alpha_0(t) \le \beta_0(t)$ on J and define, as before

$$\Omega = [(t,u) : \alpha_0(t) \le u \le \beta_0(t), \quad t \in J].$$

We shall consider the IVP

$$u' = f(t,u) + g(t,u), \quad u(0) = u_0 \tag{1.3.1}$$

where f, $g \in C[\Omega, R]$. We shall discuss four different kinds of problems. Let us begin with the following result where lower and upper solutions are of natural type.

Theorem 1.3.1 *Assume that*

(A_1) α_0, $\beta_0 \in C^1[J,R]$, $\alpha_0(t) \le \beta_0(t)$ *on* J, *and*

$$\begin{aligned} \alpha_0' &\le f(t,\alpha_0) + g(t,\alpha_0) \\ \beta_0' &\ge f(t,\beta_0) + g(t,\beta_0), \quad t \in J = [0,T]; \end{aligned}$$

(A_2) f, $g \in C[\Omega,R]$, f_u, g_u, f_{uu}, g_{uu} *exist, are continuous and satisfy*

$$f_{uu}(t,u) \ge 0, \quad g_{uu}(t,u) \le 0 \quad \text{for } (t,u) \in \Omega.$$

Then there exist monotone sequences $\{\alpha_n(t)\}$, $\{\beta_n(t)\}$ *which converge uniformly to the unique solution of* (1.3.1) *and the convergence is quadratic.*

Proof We note that the assumptions $f_{uu}(t,u) \ge 0$, $g_{uu}(t,u) \le 0$ yield the following inequalities

$$f(t,u) \ge f(t,v) + f_u(t,v)(u-v), \quad \text{and} \tag{1.3.2}$$

$$g(t,u) \ge g(t,v) + g_u(t,u)(u-v) \tag{1.3.3}$$

for $u \geq v$. It is also clear that f and g satisfy, for any $\alpha_0(t) \leq u_2 \leq u_1 \leq \beta_0(t)$,

$$-L(u_1 - u_2) \leq f(t, u_1) - f(t, u_2) \leq L(u_1 - u_2), \tag{1.3.4}$$

$$-L(u_1 - u_2) \leq g(t, u_1) - g(t, u_2) \leq L(u_1 - u_2), \tag{1.3.5}$$

for some $L > 0$.

Consider the IVPs

$$\begin{aligned} u' = F(t, \alpha_0, \beta_0; u) = \;& f(t, \alpha_0) + f_u(t, \alpha_0)(u - \alpha_0) \\ & + g(t, \alpha_0) + g_u(t, \beta_0)(u - \alpha_0), \qquad (1.3.6) \\ & u(0) = u_0, \end{aligned}$$

$$\begin{aligned} v' = G(t, \alpha_0, \beta_0; v) = \;& f(t, \beta_0) + f_u(t, \alpha_0)(v - \beta_0) \\ & + g(t, \beta_0) + g_u(t, \beta_0)(v - \beta_0), \qquad (1.3.7) \\ & v(0) = v_0, \end{aligned}$$

$$(1.3.8)$$

where $\alpha_0(0) \leq u_0 \leq \beta_0(0)$. The inequalities (1.3.2), (1.3.3) and (A_1) imply

$$\begin{aligned} \alpha_0' \leq f(t, \alpha_0) + g(t, \alpha_0) \;&\equiv\; F(t, \alpha_0, \beta_0; \alpha_0) \\ \beta_0' \geq f(t, \beta_0) + g(t, \beta_0) \;&\geq\; f(t, \alpha_0) + f_u(t, \alpha_0)(\beta_0 - \alpha_0) \\ &\quad + g(t, \alpha_0) + g_u(t, \beta_0)(\beta_0 - \alpha_0) \\ &\equiv\; F(t, \alpha_0, \beta_0; \beta_0). \end{aligned}$$

Hence by Theorem 1.1.3, and the fact that equation (1.3.6) is linear, it follows that there exists, a unique solution $\alpha_1(t)$ of (1.3.6) such that $\alpha_0(t) \leq \alpha_1(t) \leq \beta_0(t)$ on J.

Similarly, using (1.3.2), (1.3.3) and (A_1), we obtain

$$\begin{aligned} \alpha_0' \leq f(t, \alpha_0) + g(t, \alpha_0) \;&\leq\; f(t, \beta_0) + f_u(t, \alpha_0)(\alpha_0 - \beta_0) \\ &\quad + g(t, \beta_0) + g_u(t, \beta_0)(\alpha_0 - \beta_0) \\ &\equiv\; G(t, \alpha_0, \beta_0; \alpha_0) \\ \beta_0' \geq f(t, \beta_0) + g(t, \beta_0) \;&\equiv\; G(t, \alpha_0, \beta_0; \beta_0), \end{aligned}$$

and therefore, as before, there exists a unique solution $\beta_1(t)$ of (1.3.7) such that $\alpha_0(t) \leq \beta_1(t) \leq \beta_0(t)$ on J.

Now since $\alpha_1' = F(t, \alpha_0, \beta_0; \alpha_1)$, we get using (1.3.6)

$$\begin{aligned} \alpha_1' &= f(t, \alpha_0) + f_u(t, \alpha_0)(\alpha_1 - \alpha_0) + g(t, \alpha_0) + g_u(t, \beta_0)(\alpha_1 - \alpha_0) \\ &\leq f(t, \alpha_1) + g(t, \alpha_1) + g_u(t, \alpha_1)(\alpha_0 - \alpha_1) + g_u(t, \beta_0)(\alpha_1 - \alpha_0) \\ &\leq f(t, \alpha_1) + g(t, \alpha_1) + [g_u(t, \beta_0) - g_u(t, \alpha_1)]\,(\alpha_1 - \alpha_0) \\ &\leq f(t, \alpha_1) + g(t, \alpha_1) \end{aligned}$$

because of the fact $g_u(t,u)$ is nonincreasing and $\alpha_1 \leq \beta_0$. Similarly, using (1.3.7), since $\beta_1' = G(t,\alpha_0,\beta_0;\beta_1)$, we obtain

$$\begin{aligned}
\beta_1' &= f(t,\beta_0) + f_u(t,\alpha_0)(\beta_1 - \beta_0) + g(t,\beta_0) + g_u(t,\beta_0)(\beta_1 - \beta_0) \\
&\geq f(t,\beta_1) + f_u(t,\beta_1)(\beta_0 - \beta_1) + f_u(t,\alpha_0)(\beta_1 - \beta_0) + g(t,\beta_1) \\
&= f(t,\beta_1) + [-f_u(t,\beta_1) + f_u(t,\alpha_0)](\beta_1 - \beta_0) + g(t,\beta_1) \\
&\geq f(t,\beta_1) + g(t,\beta_1)
\end{aligned}$$

because of the fact $f_u(t,u)$ is nondecreasing in u and $\alpha_0 \leq \beta_1$. It then follows from Theorem 1.1.2, (1.3.4) and (1.3.5) that $\alpha_1(t) \leq \beta_1(t)$ on J. As a result we get

$$\alpha_0(t) \leq \alpha_1(t) \leq \beta_1(t) \leq \beta_0(t) \text{ on } J. \tag{1.3.9}$$

Next we consider the IVPs

$$u' = F(t,\alpha_1,\beta_1;u), \quad u(0) = u_0, \tag{1.3.10}$$
$$v' = G(t,\alpha_1,\beta_1;v), \quad v(0) = v_0. \tag{1.3.11}$$

Observe that

$$\begin{aligned}
\alpha_1' &\leq f(t,\alpha_1) + g(t,\alpha_1) \equiv F(t,\alpha_1,\beta_1;\alpha_1) \\
\beta_1' &\geq f(t,\beta_1) + g(t,\beta_1) \\
&\geq f(t,\alpha_1) + f_u(t,\alpha_1)(\beta_1 - \alpha_1) + g(t,\alpha_1) + g_u(t,\alpha_1)(\beta_1 - \alpha_1) \\
&\equiv F(t,\beta_1,\alpha_1;\beta_1)
\end{aligned}$$

in view of (1.3.2) and (1.3.3).

Consequently by Theorem 1.1.2 and Theorem 1.1.3 and the fact (1.3.10) and (1.3.11) are linear, we obtain, as before, a unique solution $\alpha_2(t)$ of (1.3.10) exists such that

$$\alpha_1(t) \leq \alpha_2(t) \leq \beta_1(t) \leq \beta_0(t) \text{ on } J.$$

Similarly, since

$$\begin{aligned}
\alpha_1' &\leq f(t,\alpha_1) + g(t,\alpha_1) \\
&\leq f(t,\beta_1) + f_u(t,\alpha_1)(\alpha_1 - \beta_1) + g(t,\beta_1) + g_u(t,\alpha_1)(\alpha_1 - \beta_1) \\
&\equiv G(t,\alpha_1,\beta_1;\alpha_1) \\
\beta_1' &\geq f(t,\beta_1) + g(t,\beta_1) \equiv G(t,\alpha_1,\beta_1;\beta_1),
\end{aligned}$$

we find that there exists unique solution $\beta_2(t)$ of (1.3.11) satisfying

$$\alpha_1(t) \leq \beta_2(t) \leq \beta_1(t) \text{ on } J.$$

In view of the fact

$$\begin{aligned} \alpha_2' &= F(t, \alpha_1, \beta_1; \alpha_2), \\ \beta_2' &= G(t, \alpha_1, \beta_1; \beta_2), \end{aligned}$$

we have as before, using (1.3.2) and (1.3.3)

$$\begin{aligned} \alpha_2' &\le f(t, \alpha_2) + g(t, \alpha_2) \\ \beta_2' &\ge f(t, \beta_2) + g(t, \beta_2), \end{aligned}$$

which yield by Theorem 1.1.2, (1.3.4) and 1.3.5) that $\alpha_2(t) \le \beta_2(t)$ on J. It therefore follows because of (1.3.9)

$$\alpha_0(t) \le \alpha_1(t) \le \alpha_2(t) \le \beta_2(t) \le \beta_1(t) \le \beta_0(t) \text{ on } J.$$

This process can be continued successively to arrive at

$$\alpha_0(t) \le \alpha_1(t) \le \cdots \le \alpha_n(t) \le \beta_n(t) \le \cdots \le \beta_1(t) \le \beta_0(t) \text{ on } J,$$

where the elements of the monotone sequences $\{\alpha_n(t)\}$, $\{\beta_n(t)\}$ are the unique solutions of the linear IVP's

$$\begin{aligned} \alpha_{n+1}' &= F(t, \alpha_n, \beta_n; \alpha_{n+1}), \quad \alpha_{n+1}(0) = u_0, \\ \beta_{n+1}' &= G(t, \alpha_n, \beta_n; \beta_{n+1}), \quad \beta_{n+1}(0) = v_0. \end{aligned}$$

Employing the standard argument as before, it is easy to conclude that the sequences $\{\alpha_n(t)\}$, $\{\beta_n(t)\}$ converge uniformly to the unique solution of (1.3.1) on J.

We shall now show that the convergence of the sequences $\{\alpha_n(t)\}$, $\{\beta_n(t)\}$ to the unique solution $u(t)$ of (1.3.1) is quadratic. To do this, consider

$$\begin{aligned} p_n(t) &= u(t) - \alpha_n(t) \ge 0, \\ q_n(t) &= \beta_n(t) - u(t) \ge 0, \end{aligned}$$

and note that $p_n(0) = q_n(0) = 0$.

Using the definition of α_n, β_n and the mean value theorem together with (A_2), we obtain successively,

$$\begin{aligned} p_n' &= f(t, u) + g(t, u) - [f(t, \alpha_{n-1}) + f_u(t, \alpha_{n-1})(\alpha_n - \alpha_{n-1}) \\ &\quad + g(t, \alpha_{n-1}) + g_u(t, \beta_{n-1})(\alpha_n - \alpha_{n-1})] \\ &= f_u(t, \xi) p_{n-1} + g_u(t, \sigma) p_{n-1} + f_u(t, \alpha_{n-1})(-p_{n-1} + p_n) \\ &\quad + g_u(t, \beta_{n-1})(-p_{n-1} + p_n) \\ &\le [f_u(t, u) - f_u(t, \alpha_{n-1})] p_{n-1} - [g_u(t, \beta_{n-1}) - g_u(t, \alpha_{n-1})] p_{n-1} \\ &\quad + [f_u(t, \alpha_{n-1}) + g_u(t, \beta_{n-1})] p_n \\ &= f_{uu}(t, \xi_1) p_{n-1}^2 - g_{uu}(t, \sigma_1) [\beta_{n-1} - \alpha_{n-1}] p_{n-1} \\ &\quad + [f_u(t, \alpha_{n-1}) + g_u(t, \beta_{n-1})] p_n \end{aligned}$$

where $\alpha_{n-1} < \xi$, $\sigma < u$ and $\alpha_{n-1} < \xi_1 < u$, $\alpha_{n-1} < \sigma_1 < \beta_{n-1}$.

But

$$\begin{aligned} -g_{uu}(t,\sigma_1)\left[\beta_{n-1} - \alpha_{n-1}\right] p_{n-1} &\leq N_2\left[q_{n-1} + p_{n-1}\right] p_{n-1} \\ &\leq N_2\left[p_{n-1}^2 + p_{n-1}q_{n-1}\right] \\ &\leq \frac{3}{2}N_2 p_{n-1}^2 + \frac{1}{2}N_2 q_{n-1}^2. \end{aligned}$$

Thus

$$p_n' \leq M p_n + \left(N_1 + \frac{3}{2}N_2\right) p_{n-1}^2 + \frac{1}{2}N_2 q_{n-1}^2,$$

where

$$\begin{aligned} &|f_u(t,u)| \,|\leq M_1,\ |g_u(t,u)| \leq M_2,\ |f_{uu}(t,u)| \leq N_1, \\ &|g_{uu}(t,u)| \leq N_2 \text{ and } M = M_1 + M_2. \end{aligned} \tag{1.3.12}$$

Now Gronwall's inequality implies

$$0 < p_n(t) \leq \int_0^t e^{M(t-s)} \left[(N_1 + \frac{3}{2}N_2)p_{n-1}^2(s) + \frac{1}{2}N_2 q_{n-1}^2(s)\right] ds \text{ on } J.$$

This yields the desired result,

$$\begin{aligned} \max_J |u(t) - \alpha_n(t)| \leq \tfrac{e^{Mt}}{M} &\left[\left(N_1 + \tfrac{3}{2}N_2\right) \max_J |u(t) - \alpha_{n-1}(t)|^2 \right. \\ &\left. + \tfrac{1}{2}N_2 \max_J |\beta_{n-1}(t) - u(t)|^2\right]. \end{aligned} \tag{1.3.13}$$

Similarly,

$$\begin{aligned} q_n' &= f(t,\beta_{n-1}) + f_u(t,\alpha_{n-1})(\beta_n - \beta_{n-1}) + g(t,\beta_{n-1}) \\ &\quad + g_u(t,\beta_{n-1})(\beta_n - \beta_{n-1}) - f(t,u) - g(t,u) \\ &= f_u(t,\xi)q_{n-1} + g_u(t,\sigma)q_{n-1} + f_u(t,\alpha_{n-1})(q_n - q_{n-1}) \\ &\quad + g_u(t,\beta_{n-1})(q_n - q_{n-1}) \\ &\leq \left[f_u(t,\beta_{n-1}) - f_u(t,\alpha_{n-1})\right] q_{n-1} + \left[g_u(t,\sigma) - g_u(t,\beta_{n-1})\right] q_{n-1} \\ &\quad + \left[f_u(t,\alpha_{n-1}) + g_u(t,\beta_{n-1})\right] q_n \\ &= f_{uu}(t,\xi_1)\left[\beta_{n-1} - \alpha_{n-1}\right] q_{n-1} - g_{uu}(t,\sigma_1) q_{n-1}^2 \\ &\quad + \left[f_u(t,\alpha_{n-1}) + g_u(t,\beta_{n-1})\right] q_n \end{aligned}$$

where $u < \xi$, $\sigma < \beta_{n-1}$ and $\alpha_{n-1} < \xi_1 < \beta_{n-1}$, $u < \sigma_1 < \beta_{n-1}$.

But

$$\begin{aligned} f_{uu}(t,\xi_1)\left[q_{n-1}+p_{n-1}\right]q_{n-1} &\leq N_1\left[q_{n-1}^2+p_{n-1}q_{n-1}\right] \\ &\leq \frac{3}{2}N_1 q_{n-1}^2+\frac{1}{2}N_1 p_{n-1}^2. \end{aligned}$$

Hence

$$q_n' \leq Mq_n + \left(\frac{3}{2}N_1+N_2\right)q_{n-1}^2+\frac{1}{2}N_1p_{n-1}^2,$$

where the constants are as in (1.3.12).

Using Gronwall's inequality again, we get

$$0 < q_n(t) \leq \int_0^t e^{M(t-s)}\left[(\frac{3}{2}N_1+N_2)q_{n-1}^2(s)+\frac{1}{2}N_1p_{n-1}^2(s)\right]ds \text{ on } J.$$

This yields the desired result,

$$\begin{aligned} \max_J|\beta_n(t)-u(t)| \leq \tfrac{e^{Mt}}{M}&\left[\left(\tfrac{3}{2}N_1+N_2\right)\max_J|\beta_{n-1}(t)-u(t)|^2\right. \\ &\left.+\tfrac{1}{2}N_1\max_J|\alpha_{n-1}(t)-u(t)|^2\right], \end{aligned} \tag{1.3.14}$$

and the proof is complete.

The next result assumes the existence of upper and lower solutions completely opposite to that supposed in Theorem 1.3.1 and the result requires extra conditions to obtain the same conclusion.

Theorem 1.3.2 *Assume that*

(B_1) $\alpha_0, \beta_0 \in C^1[J,R]$, $\alpha_0(t) \leq \beta_0(t)$ *on* J, *and*

$$\begin{aligned} \alpha_0' &\leq f(t,\beta_0)+g(t,\beta_0) \\ \beta_0' &\geq f(t,\alpha_0)+g(t,\alpha_0); \end{aligned}$$

(B_2) (A_2) *holds;*

(B_3) $f_u(t,u) \leq 0$ *and* $g_u(t,u) \leq 0$ *on* Ω.

Then the conclusion of Theorem 1.3.1 *remains valid.*

Proof The assumption $f_{uu}(t,u) \geq 0$ and $g_{uu}(t,u) \leq 0$ yield the following inequalities which will be useful in the present situation.

$$f(t,u) \leq f(t,v)+f_u(t,u)(u-v) \text{ for } u \geq v \tag{1.3.15}$$

$$g(t,u) \leq g(t,v)+g_u(t,v)(u-v). \tag{1.3.16}$$

As before, $f$and $g$satisfy for any $\alpha_0(t) \leq u_2 \leq u_1 \leq \beta_0(t)$, $t \in J$,

$$L(u_1 - u_2) \geq f(t, u_1) - f(t, u_2) \geq -L(u_1 - u_2), \text{for some } L > 0 \quad (1.3.17)$$

$$L(u_1 - u_2) \geq g(t, u_1) - g(t, u_2) \geq -L(u_1 - u_2). \quad (1.3.18)$$

Consider the coupled IVP

$$\begin{aligned} u' &= F(t, \alpha_0, \beta_0; v) \\ &= f(t, \beta_0) + f_u(t, \beta_0)(v - \beta_0) \qquad (1.3.19) \\ &\quad +g(t, \beta_0) + g_u(t, \beta_0)(v - \beta_0), \quad u(0) = u_0 \\ v' &= G(t, \alpha_0, \beta_0; u) \\ &= f(t, \alpha_0) + f_u(t, \alpha_0)(u - \alpha_0) \qquad (1.3.20) \\ &\quad +g(t, \alpha_0) + g_u(t, \alpha_0)(u - \alpha_0), \quad v(0) = u_0, \end{aligned}$$

where $\alpha_0(0) \leq u_0 \leq \beta_0(0)$.

The inequalities (1.3.15), (1.3.16) and (B_1) imply successively,

$$\begin{aligned} \alpha_0' \leq f(t, \beta_0) + g(t, \beta_0) &\equiv F(t, \alpha_0, \beta_0; \beta_0) \\ \beta_0' \geq f(t, \alpha_0) + g(t, \alpha_0) &\geq f(t, \beta_0) + f_u(t, \beta_0)(\alpha_0 - \beta_0) \\ &\quad +g(t, \beta_0) + g_u(t, \alpha_0)(\alpha_0 - \beta_0) \\ &\equiv F(t, \alpha_0, \beta_0; \alpha_0) \\ \alpha_0' \leq f(t, \beta_0) + g(t, \beta_0) &\leq f(t, \alpha_0) + f_u(t, \beta_0)(\beta_0 - \alpha_0) \\ &\quad +g(t, \alpha_0) + g_u(t, \alpha_0)(\beta_0 - \alpha_0) \\ &\equiv G(t, \alpha_0, \beta_0; \beta_0) \\ \beta_0' \geq f(t, \alpha_0) + g(t, \alpha_0) &\equiv G(t, \alpha_0, \beta_0; \alpha_0). \end{aligned}$$

Since $f_u(t, u) \leq 0$, $g_u(t, u) \leq 0$, it is clear that $F(t, \alpha_0, \beta_0; v)$ and $G(t, \alpha_0, \beta_0; u)$ are nonincreasing in v and u respectively. It therefore follows from Theorem 1.1.6 there exist unique solutions (α_1, β_1) of (1.3.19) and 1.3.20), such that $\alpha_0 \leq \alpha_1$, $\beta_1 \leq \beta_0$ on J. Now since

$$\begin{aligned} \alpha_1' &= F(t, \alpha_0, \beta_0; \beta_1) \\ \beta_1' &= G(t, \alpha_0, \beta_0; \alpha_1), \end{aligned}$$

we obtain using (1.3.15) and (1.3.16),

$$\begin{aligned} \alpha_1' &= f(t, \beta_0) + f_u(t, \beta_0)(\beta_1 - \beta_0) + g(t, \beta_0) + g_u(t, \alpha_0)(\beta_1 - \beta_0) \\ &\leq f(t, \beta_1) + f_u(t, \beta_0)(\beta_0 - \beta_1) + f_u(t, \beta_0)(\beta_1 - \beta_0) \\ &\quad +g(t, \beta_1) + g_u(t, \beta_1)(\beta_0 - \beta_1) + g_u(t, \alpha_0)(\beta_1 - \beta_0) \\ &= f(t, \beta_1) + g(t, \beta_1) + [g_u(t, \beta_1) - g_u(t, \alpha_0)]\,(\beta_0 - \beta_1) \\ &\leq f(t, \beta_1) + g(t, \beta_1), \end{aligned}$$

$$\begin{aligned}
\beta_1' &= f(t,\alpha_0) + f_u(t,\beta_0)(\alpha_1 - \alpha_0) + g(t,\alpha_0) + g_u(t,\alpha_0)(\alpha_1 - \alpha_0) \\
&\geq f(t,\alpha_1) + f_u(t,\alpha_1)(\alpha_0 - \alpha_1) + f_u(t,\beta_0)(\alpha_1 - \alpha_0) \\
&\quad + g(t,\alpha_1) + g_u(t,\alpha_0)(\alpha_0 - \alpha_1) + g_u(t,\alpha_0)(\alpha_1 - \alpha_0) \\
&= f(t,\alpha_1) + g(t,\alpha_1) + [f_u(t,\beta_0) - f_u(t,\alpha_1)](\alpha_1 - \alpha_0) \\
&\geq f(t,\alpha_1) + g(t,\alpha_1),
\end{aligned}$$

because of the fact that f_u is nondecreasing and g_u is nonincreasing in u. Hence by Theorem 1.1.2 with (H_4), $\alpha_1(t) \leq \beta_1(t)$ on J and as a result, we have

$$\alpha_0(t) \leq \alpha_1(t) \leq \beta_1(t) \leq \beta_0(t) \text{ on } J. \tag{1.3.21}$$

Consider next the coupled IVP

$$u' = F(t,\alpha_1,\beta_1;v), \quad u(0) = u_0, \tag{1.3.22}$$

$$v' = G(t,\alpha_1,\beta_1;u), \quad v(0) = u_0. \tag{1.3.23}$$

We note that (1.3.10) implies relative to (1.3.22) and (1.3.23),

$$\begin{aligned}
\alpha_1' \leq f(t,\beta_1) + g(t,\beta_1) &\equiv F(t,\alpha_1,\beta_1;\beta_1) \\
\beta_1' \geq f(t,\alpha_1) + g(t,\alpha_1) &\geq f(t,\beta_1) + f_u(t,\beta_1)(\alpha_1 - \beta_1) \\
&\quad + g(t,\beta_1) + g_u(t,\alpha_1)(\alpha_1 - \beta_1) \\
&\equiv F(t,\alpha_1,\beta_1;\alpha_1) \\
\alpha_1' \leq f(t,\beta_1) + g(t,\beta_1) &\leq f(t,\alpha_1) + f_u(t,\beta_1)(\beta_1 - \alpha_1) \\
&\quad + g(t,\alpha_1) + g_u(t,\alpha_1)(\beta_1 - \alpha_1) \\
&\equiv G(t,\alpha_1,\beta_1;\beta_1) \\
\beta_1' \geq f(t,\alpha_1) + g(t,\alpha_1) &\equiv G(t,\alpha_1,\beta_1;\alpha_1).
\end{aligned}$$

It therefore follows by Theorem 1.1.3 that there exists unique solution (α_2, β_2) of (1.3.22) and (1.3.23) such that $\alpha_1 \leq \alpha_2$, $\beta_2 \leq \beta_1$ on J. Also, we can show, as before, that

$$\begin{aligned}
\alpha_2' &\leq f(t,\beta_2) + g(t,\beta_2), \\
\beta_2' &\geq f(t,\alpha_2) + g(t,\alpha_2),
\end{aligned}$$

so that by Theorem 1.1.2 with (H_4) and (1.3.11) it follows that $\alpha_2(t) \leq \beta_2(t)$ on J. Consequently we see that

$$\alpha_0 \leq \alpha_1 \leq \alpha_2 \leq \beta_2 \leq \beta_1 \leq \beta_0 \text{ on } J.$$

Continuing this process successively, we obtain

$$\alpha_0 \leq \alpha_1 \leq \alpha_2 \leq \cdots \leq \alpha_n \leq \beta_n \leq \cdots \leq \beta_2 \leq \beta_1 \leq \beta_0 \text{ on } J, \tag{1.3.24}$$

where the elements of the monotone sequences $\{\alpha_n(t)\}$, $\{\beta_n(t)\}$ are the solutions of the linear coupled IVP

$$\alpha'_{n+1} = F(t, \alpha_n, \beta_n; \beta_{n+1}), \quad \alpha_{n+1}(0) = u_0, \tag{1.3.25}$$
$$\beta'_{n+1} = G(t, \alpha_n, \beta_n; \alpha_{n+1}), \quad \beta_{n+1}(0) = u_0. \tag{1.3.26}$$

Employing the standard arguments, it is easy to conclude that the sequences $\{\alpha_n(t)\}$, $\{\beta_n(t)\}$ converge uniformly to the unique solution of (1.3.1) on J, in view of (1.3.4), (1.3.5), (1.3.2) and (1.3.3).

To show the quadratic convergence, we set

$$\begin{aligned} p_n(t) &= u(t) - \alpha_n(t) \\ q_n(t) &= \beta_n(t) - u(t), \end{aligned}$$

where $u(t)$ is the unique solution of (1.3.1) on J. Then using the definition of α_n, β_n and mean value theorem together with (B_2), we have

$$\begin{aligned} p'_n &= f(t,u) + g(t,u) - [f(t,\beta_{n-1}) + f_u(t,\beta_{n-1})(\beta_n - \beta_{n-1}) \\ &\quad + g(t,\beta_{n-1}) + g_u(t,\alpha_{n-1})(\beta_n - \beta_{n-1})] \\ &= -f_u(t,\xi)q_{n-1} - g_u(t,\sigma)q_{n-1} - f_u(t,\beta_{n-1})\,[q_n - q_{n-1}] \\ &\quad - g_u(t,\alpha_{n-1})\,[q_n - q_{n-1}] \\ &\leq [f_u(t,\beta_{n-1}) - f_u(t,u)]\,q_{n-1} + [g_u(t,\alpha_{n-1}) - g_u(t,\beta_{n-1})]\,q_{n-1} + Mq_n \\ &= f_{uu}(t,\xi_1)q_{n-1}^2 - g_{uu}(t,\sigma_1)\,[q_{n-1} + p_{n-1}]\,q_{n-1} + Mq_n \\ &\leq \left(N_1 + \frac{3}{2}N_2\right) q_{n-1}^2 + \frac{N_2}{2}p_{n-1}^2 + Mq_n \end{aligned}$$

which results in the differential inequality

$$p'_n \leq Mq_n + \left(N_1 + \frac{3}{2}N_2\right) q_{n-1}^2 + \frac{N_2}{2}p_{n-1}^2, \tag{1.3.27}$$

where $u < \xi$, $\sigma < \beta_{n-1}$, $\alpha_{n-1} < \sigma_1 < \beta_{n-1}$, $u < \xi_1 < \beta_{n-1}$ and the constants are same as given in (1.3.12).

Similarly,

$$\begin{aligned} q'_n &= f(t,\alpha_{n-1}) + f_u(t,\beta_{n-1})(\alpha_n - \alpha_{n-1}) + g(t,\alpha_{n-1}) \\ &\quad + g_u(t,\alpha_{n-1})(\alpha_n - \alpha_{n-1}) - f(t,u) - g(t,u) \\ &= f_u(t,\xi)(\alpha_{n-1} - u) + g_u(t,\sigma)(\alpha_{n-1} - u) \\ &\quad + f_u(t,\beta_{n-1})(\alpha_n - \alpha_{n-1}) + g_u(t,\alpha_{n-1})(\alpha_n - \alpha_{n-1}) \\ &= -f_u(t,\xi)p_{n-1} - g_u(t,\sigma)p_{n-1} + f_u(t,\beta_{n-1})p_{n-1} \end{aligned}$$

$$\begin{aligned}
& -f_u(t,\beta_{n-1})p_n + g_u(t,\alpha_{n-1})p_{n-1} - g_u(t,\alpha_{n-1})p_n \\
= \; & [-f_u(t,\xi) + f_u(t,\beta_{n-1})]\, p_{n-1} + [-g_u(t,\sigma) + g_u(t,\alpha_{n-1})]\, p_{n-1} \\
& -f_u(t,\beta_{n-1})p_n - g_u(t,\alpha_{n-1})p_n \\
\leq \; & [f_u(t,\beta_{n-1}) - f_u(t,\alpha_{n-1})]\, p_{n-1} + [g_u(t,\alpha_{n-1}) - g_u(t,u)]\, p_{n-1} \\
& - [f_u(t,\beta_{n-1}) + g_u(t,\alpha_{n-1})]\, p_n \\
= \; & f_{uu}(t,\xi_1)(\beta_{n-1} - \alpha_{n-1})p_{n-1} - g_{uu}(t,\sigma)p_{n-1}^2 \\
& - [f_u(t,\beta_{n-1}) + g_u(t,\alpha_{n-1})]\, p_n
\end{aligned}$$

where $\alpha_{n-1} < \xi, \sigma < u$, $\alpha_{n-1} < \xi_1 < \beta_{n-1}$ and $\alpha_{n-1} < \sigma_1 < u$. But

$$\begin{aligned}
f_{uu}(t,\xi_1)\,(\beta_{n-1} - \alpha_{n-1}) &= f_{uu}(t,\xi_1)\,[q_{n-1} + p_{n-1}]\, p_{n-1} \\
&\leq N_1 \left[\frac{3}{2}p_{n-1}^2 + \frac{1}{2}q_{n-1}^2\right].
\end{aligned}$$

Hence,

$$q_n' \leq Mp_n + \left(N_2 + \frac{3}{2}N_1\right)p_{n-1}^2 + \left(\frac{N_1}{2}\right)q_{n-1}^2, \tag{1.3.28}$$

M, N_1, N_2 being the same constants as before. We can write the system (1.3.27) and (1.3.28) in vectorial form

$$r_n' \leq Ar_n + Br_{n-1}^2,$$

where

$$r_n = \begin{pmatrix} p_n \\ q_n \end{pmatrix}, \; A = \begin{pmatrix} 0 & M \\ M & 0 \end{pmatrix} \text{ and } B = \begin{pmatrix} \frac{1}{2}N_2 & N_1 + \frac{3}{2}N_2 \\ N_2 + \frac{3}{2}N_1 & \frac{1}{2}N_1 \end{pmatrix}.$$

An application of Corollary 1.1.1, treating Br_{n-1}^2 as a forcing term, gives

$$0 \leq r_n(t) \leq \int_0^t e^{A(t-s)}Br_{n-1}^2(s)ds, \; t \in J,$$

which, in turn, yields

$$\max_J |r_n(t)| \leq A^{-1}e^{AT}B \max_j |r_{n-1}(t)|^2, \tag{1.3.29}$$

completing the proof.

We shall utilize coupled upper and lower solutions in the next two results.

Theorem 1.3.3 *Assume that*

(C_1) $\alpha_0,\ \beta_0 \in C^1[J,R]$, $\alpha_0(t) \leq \beta_0(t)$ *on* J *and*

$$\begin{aligned} \alpha_0' &\leq f(t,\alpha_0) + g(t,\beta_0) \\ \beta_0' &\geq f(t,\beta_0) + g(t,\alpha_0), \qquad t \in J = [0,T] \end{aligned}$$

(C_2) (A_2) *holds.*

(C_3) $g_u(t,u) \leq 0$ *on* Ω.

Then the conclusion of Theorem 1.3.1 *holds.*

Proof The assumption $f_{uu}(t,u) \geq 0$ and $g_{uu}(t,u) \leq 0$ yield the following inequalities that are useful in the present situation,

$$f(t,u) \geq f(t,v) + f_u(t,v)(u-v), \quad \text{for } u \geq v. \tag{1.3.30}$$

$$g(t,u) \leq g(t,v) + g_u(t,v)(u-v). \tag{1.3.31}$$

It is also clear that f and g satisfy for any $\alpha_0(t) \leq u_2 \leq u_1 \leq \beta_0(t)$, $t \in J$,

$$-L(u_1-u_2) \leq f(t,u_1) - f(t,u_2) \leq L(u_1-u_2) \text{ for some } L > 0, \tag{1.3.32}$$

$$-L(u_1-u_2) \leq g(t,u_1) - g(t,u_2) \leq L(u_1-u_2). \tag{1.3.33}$$

Consider the coupled IVP

$$\begin{aligned} u' &= F(t,\alpha_0,\beta_0;u,v) \\ &\equiv f(t,\alpha_0) + f_u(t,\alpha_0)(u-\alpha_0) + g(t,\beta_0) \\ &\quad + g_u(t,\alpha_0)(v-\beta_0), \qquad u(0) = u_0 \end{aligned} \tag{1.3.34}$$

$$\begin{aligned} v' &= G(t,\alpha_0,\beta_0;v,u) \\ &\equiv f(t,\beta_0) + f_u(t,\alpha_0)(v-\beta_0) + g(t,\alpha_0) \\ &\quad + g_u(t,\alpha_0)(u-\alpha_0), \qquad v(0) = u_0, \end{aligned} \tag{1.3.35}$$

where $\alpha_0(0) \leq u_0 \leq \beta_0(0)$.

$$\begin{aligned}
\alpha_0' \le f(t,\alpha_0) + g(t,\beta_0) &\le f(t,\beta_0) + f_u(t,\alpha_0)(\alpha_0 - \beta_0) \\
&\quad + g(t,\alpha_0) + g_u(t,\alpha_0)(\beta_0 - \alpha_0) \qquad (1.3.38) \\
&\equiv G(t,\alpha_0,\beta_0;\alpha_0,\beta_0), \\
\beta_0' \ge f(t,\beta_0) + g(t,\alpha_0) &\equiv G(t,\alpha_0,\beta_0;\beta_0,\alpha_0). \qquad (1.3.39)
\end{aligned}$$

Since $g_u \le 0$, it is clear that $F(t,\alpha_0,\beta_0;u,v)$, $G(t,\alpha_0,\beta_0;v,u)$ are nonincreasing in v, u respectively. Hence by Theorem 1.1.8 there exists coupled unique solution (α_1,β_1) of (1.3.34) and (1.3.35) such that $\alpha_0 \le \alpha_1$, $\beta_1 \le \beta_0$ on J.

Now since

$$\begin{aligned}
\alpha_1' &= F(t,\alpha_0,\beta_0;\alpha_1,\beta_1) \\
\beta_1' &= G(t,\alpha_0,\beta_0;\beta_1,\alpha_1),
\end{aligned}$$

we obtain using (1.3.30) and (1.3.31)

$$\begin{aligned}
\alpha_1' &\le f(t,\alpha_1) + g(t,\beta_1) + [g_u(t,\beta_1) - g_u(t,\alpha_0)(\beta_0 - \beta_1)] \\
&\le f(t,\alpha_1) + g(t,\beta_1),
\end{aligned}$$

using the fact g_u is nonincreasing in u and

$$\begin{aligned}
\beta_1' &\ge f(t,\beta_1) + [f_u(t,\beta_1) - f_u(t,\alpha_0)]\,(\beta_0 - \beta_1) + g(t,\alpha_1) \\
&\ge f(t,\beta_1) + g(t,\alpha_1)
\end{aligned}$$

using the fact f_u is nondecreasing in u.

Hence by Theorem 1.1.2 with (H_2), $\alpha_1(t) \le \beta_1(t)$ on J, and as a result we have

$$\alpha_0 \le \alpha_1 \le \beta_1 \le \beta_0 \text{ on } J.$$

Consider next the coupled IVP

$$u' = F(t,\alpha_1,\beta_1;u,v), \qquad u(0) = u_0, \qquad (1.3.40)$$

$$v' = G(t,\alpha_1,\beta_1;v,u), \qquad v(0) = u_0. \qquad (1.3.41)$$

We note that (1.3.30) implies relative to (1.3.40),

$$\begin{aligned}
\alpha_1' \le f(t,\alpha_1) + g(t,\beta_1) &\equiv F(t,\alpha_1,\beta_1;\alpha_1,\beta_1), \\
\beta_1' \ge f(t,\beta_1) + g(t,\alpha_1) &\ge f(t,\alpha_1) + f_u(t,\alpha_1)(\beta_1 - \alpha_1) \\
&\quad + g(t,\beta_1) + g_u(t,\alpha_1)(\alpha_1 - \beta_1) \\
&\equiv F(t,\alpha_1,\beta_1;\beta_1,\alpha_1),
\end{aligned}$$

$$\begin{aligned}
\alpha_1' \le f(t,\alpha_1) + g(t,\beta_1) &\le f(t,\beta_1) + f_u(t,\alpha_1)(\alpha_1 - \beta_1) \\
&\quad + g(t,\alpha_1) + g_u(t,\alpha_1)(\beta_1 - \alpha_1) \\
&\equiv G(t,\alpha_1,\beta_1;\alpha_1,\beta_1), \\
\beta_1' \ge f(t,\beta_1) + g(t,\alpha_1) &\ge G(t,\alpha_1,\beta_1;\beta_1,\alpha_1).
\end{aligned}$$

It therefore follows by Theorem 1.1.7 that there exist unique solution (α_2, β_2) of (1.3.40) such that $\alpha_1 \leq \alpha_2$, $\beta_2 \leq \beta_1$ on J.

In view of the fact

$$\begin{aligned} \alpha_2' &= F(t, \alpha_1, \beta_1; \alpha_2, \beta_2) \\ \beta_2' &= G(t, \alpha_1, \beta_1; \beta_2, \alpha_2), \end{aligned}$$

we obtain arguing as before,

$$\begin{aligned} \alpha_2' &\leq f(t, \alpha_2) + g(t, \beta_2) \\ \beta_2' &\geq f(t, \beta_2) + g(t, \alpha_2), \end{aligned}$$

which yield by Theorem 1.1.7 with (H_2) and (1.3.32), $\alpha_2(t) \leq \beta_2(t)$ on J. Therefore we have

$$\alpha_0 \leq \alpha_1 \leq \alpha_2 \leq \beta_2 \leq \beta_1 \leq \beta_0 \text{ on } J.$$

This process can be continued successively to arrive at

$$\alpha_0 \leq \alpha_1 \leq \cdots \leq \alpha_n \leq \beta_n \leq \cdots \leq \beta_1 \leq \beta_0 \text{ on } J, \tag{1.3.42}$$

where the element of the monotone sequences $\{\alpha_n(t)\}$, $\{\beta_n(t)\}$ are unique solutions of the linear IVP's

$$\alpha_{n+1}' = F(t, \alpha_n, \beta_n; \alpha_{n+1}, \beta_{n+1}); \qquad \alpha_{n+1}(0) = u_0, \tag{1.3.43}$$

$$\beta_{n+1}' = G(t, \alpha_n, \beta_n; \beta_{n+1}, \alpha_{n+1}); \qquad \beta_{n+1}(0) = u_0. \tag{1.3.44}$$

Employing the standard arguments, it is easy to conclude that the sequences $\{\alpha_n(t)\}$, $\{\beta_n(t)\}$ converge uniformly to the unique solution of (1.3.1) on J, in view of (1.3.4) and (1.3.5.

To show the quadratic convergence, we set

$$\begin{aligned} p_n(t) &= u(t) - \alpha_n(t), & p_n(0) = 0, \\ q_n(t) &= \beta_n(t) - u(t), & q_n(0) = 0, \end{aligned}$$

where $u(t)$ is the unique solution of (1.3.1) on J. Then using the definitions of α_n, β_n and mean value theorem together with (C_2), we see that

$$\begin{aligned} p_n' &= f(t, u) + g(t, u) - [f(t, \alpha_{n-1}) + f_u(t, \alpha_{n-1})(\alpha_n - \alpha_{n-1}) \\ &\quad + g(t, \beta_{n-1}) + g_u(t, \alpha_{n-1})(\beta_n - \beta_{n-1})] \\ &= f_u(t, \xi)p_{n-1} - g_u(t, \sigma)q_{n-1} - f_u(t, \alpha_{n-1})\,[-p_n + p_{n-1}] \\ &\quad - g_u(t, \alpha_{n-1})\,[q_n - q_{n-1}] \end{aligned}$$

$$
\begin{aligned}
&\leq [f_u(t,u) - f_u(t,\alpha_{n-1})]\, p_{n-1} + f_u(t,\alpha_{n-1})p_n \\
&\quad + [g_u(t,\alpha_{n-1}) - g_u(t,\beta_{n-1})]\, q_{n-1} - g_u(t,\alpha_{n-1})q_n \\
&= f_{uu}(t,\xi_1)p_{n-1}^2 + f_u(t,\alpha_{n-1})p_n + g_{uu}(t,\sigma_1)\,[\alpha_{n-1} - \beta_{n-1}]\, q_{n-1} \\
&\quad - g_u(t,\alpha_{n-1})q_n \\
&\leq N_1 p_{n-1}^2 - g_{uu}(t,\sigma_1)\,[\beta_{n-1} - \alpha_{n-1}]\, q_{n-1} + M_1 p_n + M_2 q_n \\
&= N_1 p_{n-1}^2 + N_2\left[q_{n-1}^2 + p_{n-1}q_{n-1}\right] + M_1 p_n + M_2 q_n
\end{aligned}
$$

$$
p_n' \leq \left(N_1 + \frac{1}{2}N_2\right) p_{n-1}^2 + \frac{3}{2}N_2 q_{n-1}^2 + M_1 p_n + M_2 q_n \tag{1.3.45}
$$

where $\alpha_{n-1} < \xi < u$, $u < \sigma$, $\beta_{n-1} < \xi_1 < u$, $\alpha_{n-1} < \sigma_1 < \beta_{n-1}$ and the constants are same as in (1.3.12).

Similarly

$$
\begin{aligned}
q_n' &= f(t,\beta_{n-1}) + f_u(t,\alpha_{n-1})(\beta_n - \beta_{n-1}) + g(t,\alpha_{n-1}) \\
&\quad + g_u(t,\alpha_{n-1})(\alpha_n - \alpha_{n-1}) - f(t,u) - g(t,u) \\
&= f_u(t,\xi)q_{n-1} - g_u(t,\sigma)p_{n-1} + f_u(t,\alpha_{n-1})(q_n - q_{n-1}) \\
&\quad + g_u(t,\alpha_{n-1})(-p_n + p_{n-1}) \\
&\leq [f_u(t,\beta_{n-1}) - f_u(t,\alpha_{n-1})]\, q_{n-1} + [g_u(t,\alpha_{n-1}) - g_u(t,u)]\, p_{n-1} \\
&\quad + f_u(t,\alpha_{n-1})q_n - g_u(t,\alpha_{n-1})p_n \\
&= f_{uu}(t,\xi_1)(\beta_{n-1} - \alpha_{n-1})q_{n-1} + g_{uu}(t,\sigma_1)\,(\alpha_{n-1} - u)\, p_{n-1} \\
&\quad + f_u(t,\alpha_{n-1})q_n - g_u(t,\alpha_{n-1})p_n \\
&\leq N_1\,(q_{n-1} + p_{n-1})\, q_{n-1} + N_2 p_{n-1}^2 + M_1 q_n + M_2 p_n
\end{aligned}
$$

which reduces to

$$
q_n' \leq \frac{3}{2}N_1 q_{n-1}^2 + \left(\frac{1}{2}N_1 + N_2\right) p_{n-1}^2 + M_1 q_n + M_2 p_n \tag{1.3.46}
$$

where $u < \xi < \beta_{n-1}$, $\alpha_{n-1} < \sigma < u$, $\beta_{n-1} < \xi_1 < \alpha_{n-1}$, and $\alpha_{n-1} < \sigma < u$.

Now proceeding as in Theorem 1.3.2, we can obtain an estimate similar to (1.3.29) where the matrices A and B take the form,

$$
A = \begin{pmatrix} M_1 & M_2 \\ M_2 & M_1 \end{pmatrix}, \quad B = \begin{pmatrix} N_1 + \frac{1}{2}N_2 & \frac{3}{2}N_2 \\ \frac{3}{2}N_1 & \frac{1}{2}N_1 + N_2 \end{pmatrix}, \tag{1.3.47}
$$

in view of (1.3.45) and (1.3.46), provided $M_1 \neq M_2$. If $M_1 = M_2$, one can get a simpler estimate as in Theorem 1.3.1. Hence the Theorem is proved.

Theorem 1.3.4 *Assume that*

(D_1) *α_0,$\beta_0 \in C^1[J, R]$, $\alpha_0(t) \leq \beta_0(t)$ on J, and*

$$\alpha_0' \leq f(t, \beta_0) + g(t, \alpha_0), \quad \beta_0' \geq f(t, \alpha_0) + g(t, \beta_0);$$

(D_2) *(A_2) is satisfied;*

(D_3) *$f_u(t, u) \leq 0$ on Ω.*

Then the conclusion of Theorem 1.3.1 is true.

Proof Since the proof can be constructed on the basis of the proof of previous theorems, we shall only indicate the needed modifications. In place of (1.3.30) and (1.3.31), we require to use

$$\begin{aligned} f(t, u) &\leq f(t, v) + f_u(t, u)(u - v), && (1.3.48) \\ g(t, u) &\geq g(t, v) + g_u(t, u)(u - v), \quad u \geq v. && (1.3.49) \end{aligned}$$

The coupled IVP (1.3.34) and (1.3.35) changes to

$$\begin{aligned} u' &= F(t, \alpha_0, \beta_0; v, u) \\ &\equiv f(t, \beta_0) + f_u(t, \beta_0)(v - \beta_0) + g(t, \alpha_0) && (1.3.50) \\ &\quad + g_u(t, \beta_0)(u - \alpha_0), \qquad u(0) = u_0, \\ v' &= G(t, \alpha_0, \beta_0; v, u) \\ &\equiv f(t, \alpha_0) + f_u(t, \beta_0)(u - \alpha_0) + g(t, \beta_0) && (1.3.51) \\ &\quad + g_u(t, \beta_0)(v - \beta_0), \qquad v(0) = u_0, \end{aligned}$$

where $\alpha_0(0) \leq u_0 \leq \beta_0(0)$. The rest of the proof is similar with suitable modifications in arguments. We omit the details.

The foregoing theorems contain the following results which are new because we get simultaneously lower and upper bounds by means of monotone sequences that converge to the unique solution of (1.3.1) quadratically. These we state as corollaries.

Corollary 1.3.1 *Assume that*

(a) *α_0, $\beta_0 \in C^1[J, R]$, $\alpha_0(t) \leq \beta_0(t)$ on J and*
$\alpha_0' \leq f(t, \alpha_0)$, $\beta_0' \geq f(t, \beta_0)$ on J;

(b) *$f \in C[\omega, R]$, f_u, f_{uu} exist and are continuous on Ω;*

(c) *$f_{uu}(t, u) \geq 0$ on Ω.*

Then there exist monotone sequences $\{\alpha_n(t)\}$, $\{\beta_n(t)\}$ *which converge uniformly to the unique solution* $u(t)$ *of* (1.3.1) *on* J, *and the convergence is quadratic.*

Corollary 1.3.2 *Suppose that in the assumptions of Corollary* 1.3.1, *we replace* (c) *by*

(c^*) $f_{uu}(t,u) \le 0$ *on* Ω.

Then the conclusion of Corollary 1.3.1 *remains valid.*

Corollary 1.3.3 *Assume that*

(*i*) $\alpha_0, \beta_0 \in C^1[J,R]$, $\alpha_0(t) \le \beta_0(t)$ *on* J *and*
$\alpha_0' \le f(t,\beta_0)$, $\beta_0' \ge f(t,\alpha_0)$ *on* J;

(*ii*) *condition* (b) *holds;*

(*iii*) *condition* (c) *holds;*

(*iv*) $f_u(t,u) \le 0$ *on* Ω.

Then the conclusion of Corollary 1.3.1 *is true.*

Corollary 1.3.4 *Suppose that, in the assumptions of Corollary* 1.3.3, *we replace* (c) *by* (c^*). *Then the conclusion of Corollary* 1.3.1 *is valid.*

1.4 GENERALIZATIONS

We shall generalize, in this section, the method of quasilinearization by not demanding $f(t,x)$ to be convex but imposing a less restrictive assumption, namely, $f(t,x)+\phi(t,x)$ to be convex for some convex function $\phi(t,x)$. Nonetheless, the conclusion is the same as that of the method of quasilinearization.

Let us consider the IVP

$$x' = f(t,x), \qquad x(0) = x_0, \tag{1.4.1}$$

where $f \in C[J \times R, R]$ and $J = [0,T]$. We begin with the following result.

Theorem 1.4.1 *Assume that:*

(A_1) $\alpha_0, \beta_0 \in C^1[J,R]$ *are lower and upper solutions of (1.4.1) such that* $\alpha_0(t) \le \beta_0(t)$ *on* J;

(A_2) *$f \in C[\Omega, R]$, $f_x(t,x)$, $f_{xx}(t,x)$ exist and are continuous satisfying $f_{xx}(t,x) + \phi_{xx}(t,x) \geq 0$ on Ω, where $\phi \in C[\Omega, R]$ and $\phi_x(t,x)$, $\phi_{xx}(t,x)$ exist, are continuous and $\phi_{xx}(t,x) > 0$ on Ω.*

Then there exist monotone sequences $\{\alpha_n(t)\}$, $\{\beta_n(t)\}$ which converge uniformly to the unique solution of (1.4.1) *and the convergence is quadratic.*

Proof In view of (A_2), it is clear that for $\alpha_0(t) \leq x_2 \leq x_1 \leq \beta_0(t)$,

$$f(t,x_1) - f(t,x_2) \leq L_1(x_1 - x_2), \qquad \text{for some } L_1 > 0. \tag{1.4.2}$$

Also, the assumption $f_{xx}(t,x) + \phi_{xx}(t,x) \geq 0$ yields the inequality

$$f(t,x) \geq f(t,y) + [f_x(t,y) + \phi_x(t,y)](x - y) - [\phi(t,x) - \phi(t,y)] \tag{1.4.3}$$

for $x \geq y$, x, $y \in \Omega$.

Now, consider the IVPs

$$u' = f(t,\alpha_0) + [f_x(t,\alpha_0) + \phi_x(t,\alpha_0)](u - \alpha_0) - [\phi(t,u) - \phi(t,\alpha_0)], \tag{1.4.4}$$
$$v' = f(t,\beta_0) + [f_x(t,\alpha_0) + \phi_x(t,\alpha_0)](v - \beta_0) - [\phi(t,v) - \phi(t,\beta_0)], \tag{1.4.5}$$

which we shall denote for convenience in the following form

$$\begin{aligned} u' &= g(t,u;\alpha_0) \equiv F(t,\alpha_0) + F_x(t,\alpha_0)(u - \alpha_0) - \phi(t,u), \\ v' &= G(t,v;\alpha_0,\beta_0) \equiv F(t,\beta_0) + F_x(t,\alpha_0)(v - \beta_0) - \phi(t,v), \end{aligned}$$

where $F(t,x) = f(t,x) + \phi(t,x)$, $u(0) = x_0 = v(0)$ and $\alpha_0(0) \leq x_0 \leq \beta_0(0)$.

The assumption (A_1) and inequality (1.4.3) imply

$$\alpha_0' \leq f(t,\alpha_0) \equiv g(t,\alpha_0;\alpha_0), \tag{1.4.6}$$

$$\begin{aligned} \beta_0' &\geq f(t,\beta_0) \geq f(t,\alpha_0) \\ &\quad + [f_x(t,\alpha_0) + \phi_x(t,\alpha_0)](\beta_0 - \alpha_0) - [\phi(t,\beta_0) - \phi(t,\alpha_0)] \\ &\equiv g(t,\beta_0;\alpha_0). \end{aligned} \tag{1.4.7}$$

Also observe that because of (1.4.3) and the mean value theorem, the function $g(t,u;\alpha_0)$ satisfies

$$g(t,u_1;\alpha_0) - g(t,u_2;\alpha_0) \leq L_2(u_1 - u_2), \tag{1.4.8}$$

for some $L_2 > 0$ whenever $\alpha_0(t) \leq u_2 \leq u_1 \leq \beta_0(t)$. Hence, by Theorem 1.1.3, there exists a unique solution $\alpha_1(t)$ of (1.4.5) such that $\alpha_0(t) \leq \alpha_1(t) \leq \beta_0(t)$ on J.

Similarly, using (1.4.3), we get

$$\begin{aligned} \alpha_0' &\le f(t,\alpha_0) \le f(t,\beta_0) - [f_x(t,\alpha_0) + \phi_x(t,\alpha_0)](\beta_0 - \alpha_0) \\ &\quad + [\phi(t,\beta_0) - \phi(t,\alpha_0)] \\ &\equiv G(t,\alpha_0;\alpha_0,\beta_0), \end{aligned} \tag{1.4.9}$$

and

$$\beta_0' \ge f(t,\beta_0) \equiv G(t,\beta_0;\alpha_0,\beta_0). \tag{1.4.10}$$

This function $G(t,u;\alpha_0,\beta_0)$ also satisfies

$$G(t,u_1;\alpha_0,\beta_0) - G(t,u_2;\alpha_0,\beta_0) \le L_3(u_1 - u_2) \tag{1.4.11}$$

for some $L_3 > 0$ whenever $\alpha_0(t) \le u_2 \le u_1 \le \beta_0(t)$, $t \in J$. Hence, by Theorem 1.1.3, there exists a unique solution $\beta_1(t)$ of (1.4.5) such that $\alpha_0(t) \le \beta_1(t) \le \beta_0(t)$ on J.

Now, since $\alpha_1' = g(t,\alpha_1;\alpha_0)$, using (1.4.3) we get

$$\begin{aligned} \alpha_1' &= f(t,\alpha_0) + [f_x(t,\alpha_0) + \phi_x(t,\alpha_0)](\alpha_1 - \alpha_0) - [\phi(t,\alpha_1) - \phi(t,\alpha_0)] \\ &\le f(t,\alpha_1) \end{aligned}$$

and similarly, since $\beta_1' = G(t,\beta_1;\alpha_0,\beta_0)$, using (1.4.3) again,

$$\begin{aligned} \beta_1' &= f(t,\beta_0) + [f_x(t,\alpha_0) + \phi_x(t,\alpha_0)](\beta_1 - \beta_0) - [\phi(t,\beta_1) - \phi(t,\beta_0)] \\ &\ge f(t,\beta_1). \end{aligned}$$

Thus, it follows from (1.4.2) and Theorem 1.1.1 that $\alpha_1(t) \le \beta_1(t)$ on J and, consequently, we have

$$\alpha_0(t) \le \alpha_1(t) \le \beta_1(t) \le \beta_0(t) \qquad \text{on } J. \tag{1.4.12}$$

Next, consider the IVPs

$$u' = g(t,u;\alpha_1), \qquad u(0) = x_0, \tag{1.4.13}$$

$$v' = G(t,v;\alpha_1,\beta_1), \qquad v(0) = x_0. \tag{1.4.14}$$

Observe that

$$\alpha_1' \le f(t,\alpha_1) \equiv g(t,\alpha_1;\alpha_1), \tag{1.4.15}$$

and using (1.4.3) and (1.4.12), it follows that

$$\begin{aligned} \beta_1' &\ge f(t,\beta_1) \ge f(t,\alpha_1) + [f_x(t,\alpha_1) + \phi_x(t,\alpha_1)](\beta_1 - \alpha_1) \\ &\quad - [\phi(t,\beta_1) - \phi(t,\alpha_1)] \\ &\equiv g(t,\beta_1;\alpha_1). \end{aligned} \tag{1.4.16}$$

Thus, by (1.4.8) and Theorem 1.1.3, there exists a unique solution $\alpha_2(t)$ of (1.4.13) such that $\alpha_1(t) \le \alpha_2(t) \le \beta_1(t)$ on J.

Similarly, again using (1.4.3) and (1.4.12), we have

$$\begin{aligned} \alpha_1' &\le f(t,\alpha_1) \le f(t,\beta_1) - [f_x(t,\alpha_1) + \phi_x(t,\alpha_1)](\beta_1 - \alpha_1) \\ &\quad + [\phi(t,\beta_1) - \phi(t,\alpha_1)] \\ &\equiv G(t,\alpha_1;\alpha_1,\beta_1) \end{aligned} \tag{1.4.17}$$

and

$$\beta_1' \ge f(t,\beta_1) \equiv G(t,\beta_1;\alpha_1,\beta_1). \tag{1.4.18}$$

Hence, in view of (1.4.11) and Theorem 1.1.3, there exist $\alpha_2(t)$, $\beta_2(t)$ which are the unique solutions of (1.4.13) and (1.4.14) such that

$$\alpha_1(t) \le \alpha_2(t), \beta_2(t) \le \beta_1(t) \qquad \text{on } J.$$

Furthermore, using (1.4.3), we obtain

$$\alpha_2' = g(t,\alpha_2;\alpha_1) \le f(t,\alpha_2)$$

and, using (1.4.3) and mean value theorem,

$$\begin{aligned} \beta_2' &= G(t,\beta_2;\alpha_1,\beta_1) \\ &= f(t,\beta_1) + [f_x(t,\alpha_1) + \phi_x(t,\alpha_1)](\beta_2 - \beta_1) - [\phi(t,\beta_2) - \phi(t,\beta_1)] \\ &\ge f(t,\beta_2) + [f_x(t,\beta_2) + \phi_x(t,\beta_2) - f_x(t,\alpha_1) - \phi_x(t,\alpha_1)](\beta_1 - \beta_2) \\ &\ge f(t,\beta_2) + [F_x(t,\beta_2) - F_x(t,\alpha_1)](\beta_1 - \beta_2) \\ &= f(t,\beta_2) + F_{xx}(t,\sigma)(\beta_2 - \alpha_1)(\beta_1 - \beta_2) \\ &\ge f(t,\beta_2), \end{aligned}$$

where $\alpha_1 \le \sigma \le \beta_1$ and $F(t,u) = f(t,u) + \phi(t,u)$. Therefore, by Theorem 1.1.1 and (1.4.2), we have $\alpha_2(t) \le \beta_2(t)$ on J and, consequently,

$$\alpha_0(t) \le \alpha_1(t) \le \alpha_2(t) \le \beta_2(t) \le \beta_1(t) \le \beta_0(t) \quad \text{on } J.$$

This process can be continued successively to obtain

$$\alpha_0(t) \le \alpha_1(t) \le \cdots \le \alpha_n(t) \le \beta_n(t) \le \cdots \le \beta_1(t) \le \beta_0(t) \qquad \text{on } J,$$

where $\{\alpha_n(t)\}$, $\{\beta_n(t)\}$ are the unique solutions of the following initial value problems

$$\alpha_{n+1}' = g(t,\alpha_{n+1};\alpha_n), \qquad \alpha_{n+1}(0) = x_0, \tag{1.4.19}$$

$$\beta_{n+1}' = G(t,\beta_{n+1};\alpha_n,\beta_n), \qquad \beta_{n+1}(0) = x_0. \tag{1.4.20}$$

Employing standard techniques, it can be shown that the sequences $\{\alpha_n(t)\}$, $\{\beta_n(t)\}$ converge uniformly and monotonically to the unique solution $x(t)$ of (1.4.1) on J.

We shall now prove that the convergence of the sequences $\{\alpha_n(t)\}$, $\{\beta_n(t)\}$ to the unique solution $x(t)$ of (1.4.1) is quadratic. For this purpose, consider

$$\begin{aligned} p_{n+1}(t) &= x(t) - \alpha_{n+1}(t) \geq 0, \\ q_{n+1}(t) &= \beta_{n+1}(t) - x(t) \geq 0, \end{aligned}$$

and note that $p_{n+1}(0) = q_{n+1}(0) = 0$. Using $F(t,x) = f(t,x) + \phi(t,x)$ and the mean value theorem, we obtain

$$\begin{aligned} p'_{n+1} &= x' - \alpha'_{n+1} \\ &= F(t,x) - F(t,\alpha_n) - F_x(t,\alpha_n)(\alpha_{n+1} - \alpha_n) - [\phi(t,x) - \phi(t,\alpha_{n+1})] \\ &= F_x(t,\xi)(x - \alpha_n) - F_x(t,\alpha_n)(x - \alpha_n) + F_x(t,\alpha_n)(x - \alpha_{n+1}) \\ &\quad -\phi_x(t,\sigma)(x - \alpha_{n+1}) \\ &= F_{xx}(t,\eta)(\xi - \alpha_n)(x - \alpha_n) + [F_x(t,\alpha_n) - \phi_x(t,\sigma)](x - \alpha_{n+1}) \\ &\leq F_{xx}(t,\eta)(x - \alpha_n)^2 + [F_x(t,\alpha_n) - \phi_x(t,\alpha_n)](x - \alpha_{n+1}), \end{aligned}$$

where $\alpha_n \leq \eta \leq \xi < x$, $\alpha_{n+1} \leq \sigma \leq x$. Thus,

$$p'_{n+1} \leq (N + L)p_n^2 + Mp_{n+1},$$

where

$$|f_{xx}(t,x)| \leq N, \qquad |\phi_{xx}(t,x)| \leq L \qquad \text{and } |f_x(t,x)| \leq M. \tag{1.4.21}$$

As a result, treating $(N + L)p_n^2$ as the forcing term and applying Gronwall's inequality, it follows that

$$0 \leq p_{n+1}(t) \leq (N + L) \int_0^t e^{M(t-s)} [x(s) - \alpha_n(s)]^2 \, ds \qquad \text{on } J$$

which yields the desired result

$$\max_J |x(t) - \alpha_{n+1}(t)| \leq (N + L)\frac{e^{MT}}{M} \max_J |x(t) - \alpha_n(t)|^2.$$

similarly,

$$\begin{aligned} q'_{n+1} &= \beta'_{n+1} - x' \\ &= f(t,\beta_n) + [f_x(t,\alpha_n) + \phi_x(t,\alpha_n)](\beta_{n+1} - \beta_n) \end{aligned}$$

$$
\begin{aligned}
& -[\phi(t,\beta_{n+1})-\phi(t,\beta_n)]-f(t,x)\\
= \;& F(t,\beta_n)-F(t,x)+F_x(t,\alpha_n)(\beta_{n+1}-\beta_n)-\phi(t,\beta_{n+1})+\phi(t,x)\\
= \;& F_x(t,\xi)(\beta_n-x)+F_x(t,\alpha_n)(\beta_{n+1}-x)+F_x(t,\alpha_n)(x-\beta_n)\\
& -\phi_x(t,\sigma)(\beta_{n+1}-x)\\
\le \;& [F_x(t,\beta_n)-F_x(t,\alpha_n)](\beta_n-x)+[F_x(t,\alpha_n)-\phi_x(t,\sigma)](\beta_{n+1}-x)\\
= \;& F_{xx}(t,\eta)(\beta_n-\alpha_n)(\beta_n-x)+[F_x(t,\alpha_n)-\phi_x(t,\sigma)](\beta_{n+1}-x)\\
= \;& F_{xx}(t,\eta)\left[(\beta_n-x)^2+(x-\alpha_n)(\beta_n-x)\right]\\
& +[F_x(t,\alpha_n)-\phi_x(t,\sigma)](\beta_{n+1}-x),
\end{aligned}
$$

where $x \le \xi \le \beta_n$, $\alpha_n \le \eta \le \beta_n$, and $\alpha_n \le x \le \sigma \le \beta_{n+1}$. Hence, we get

$$
\begin{aligned}
q'_{n+1} &\le (N+L)\left[q_n^2+p_nq_n\right]+Mq_{n+1}\\
&\le (N+L)\left[q_n^2+\frac{1}{2}(p_n^2+q_n^2)\right]+Mq_{n+1}\\
&= \frac{3}{2}(N+L)q_n^2+\frac{1}{2}(N+L)p_n^2+Mq_{n+1},
\end{aligned}
$$

where the constants M, N, L are as in (1.4.21). Consequently, treating $\frac{3}{2}(N+L)q_n^2+\frac{1}{2}(N+L)p_n^2$ as the forcing term and applying Gronwall's inequality, we get

$$
0 \le q_{n+1}(t) \le \int_0^t e^{M(t-s)}\left[\frac{3}{2}(N+L)q_n^2(s)+\frac{1}{2}(N+L)p_n^2(s)\right]ds
$$

which yields the desired results

$$
\begin{aligned}
\max_J|\beta_{n+1}(t)-x(t)| \le \;& \frac{e^{MT}}{M}\left[\frac{3}{2}(N+L)\max_J|\beta_n(t)-x(t)|^2\right.\\
& \left.+\frac{1}{2}(N+L)\max_J|x(t)-\alpha_n(t)|^2\right].
\end{aligned}
$$

This completes the proof of the theorem.

We note that members of the sequences $\{\alpha_n(t)\}$, $\{\beta_n(t)\}$ given by (1.4.19) and (1.4.20) are unique solutions of nonlinear differential equations and we have a disadvantage in practice to obtain them. The question naturally would be whether under the same general conditions imposed in Theorem 1.4.1, is it possible to obtain the sequences $\{\alpha_n(t)\}$, $\{\beta_n(t)\}$ whose elements are solutions of linear differential equations so that their construction becomes easier? The answer is positive and to do this, we simply need to utilize the properties of the functions g and G which in the proof of Theorem 1.4.1 have not been employed. It is easy to see, in view of the assumption on $\phi(t,u)$ that

$$g_{uu}(t,u;\eta) < 0, \qquad G_{uu}(t,u;\eta,\sigma) < 0 \qquad \text{in } \Omega, \tag{1.4.22}$$

which implies

$$\begin{aligned} g(t,u;\eta) &\geq g(t,v;\eta) + g_x(t,u;\eta)(u-v)\text{if } u \geq v, & (1.4.23)\\ G(t,u;\eta,\sigma) &\geq G(t,v;\eta,\sigma) + G_u(t,u;\eta,\sigma)(u-v)\text{if } u \geq v, & (1.4.24)\end{aligned}$$

for every η, σ in Ω and u, $v \in \Omega$.

We proceed as follows: given α_0, β_0, define α_1, β_1 by

$$\begin{aligned} \alpha_1' &= g(t,\alpha_0;\alpha_0) + g_u(t,\beta_0;\alpha_0)(\alpha_1 - \alpha_0), \alpha_1(0) = x_0, & (1.4.25)\\ \beta_1' &= G(t,\beta_0;\alpha_0,\beta_0) + G_u(t,\beta_0;\alpha_0,\beta_0)(\beta_1 - \beta_0), & \\ \beta_1(0) &= x_0. & (1.4.26)\end{aligned}$$

Consider the linear IVP (1.4.25) written as

$$u' = h(t,u;\alpha_0,\beta_0) \equiv g(t,\alpha_0;\alpha_0) + g_u(t,\beta_0;\alpha_0)(u-\alpha_0) \qquad u(0) = x_0. \tag{1.4.27}$$

Now, from (1.4.6),

$$\alpha_0' \leq g(t,\alpha_0;\alpha_0) \equiv h(t,\alpha_0;\alpha_0,\beta_0).$$

Now, from (1.4.7) using (1.4.23) and 1.4.24)

$$\begin{aligned} \beta_0' &\geq g(t,\beta_0;\alpha_0)\\ &\geq g(t,\alpha_0;\alpha_0) + g_u(t,\beta_0;\alpha_0)(\beta_0 - \alpha_0)\\ &\equiv h(t,\beta_0;\alpha_0,\beta_0).\end{aligned}$$

Hence, by Theorem 1.1.3 and the fact that (1.4.27) is linear, it follows that there exists a unique solution $\alpha_1(t)$ of (1.4.27) such that $\alpha_0(t) \leq \alpha_1(t) \leq \beta_0(t)$ on J.

Next, we show that $\alpha_1(t)$ is a lower solution for (1.4.1), using (1.4.22)-(1.4.25), (1.4.3) and the mean value theorem, we get

$$\begin{aligned} \alpha_1' &= g(t,\alpha_0;\alpha_0) + g_u(t,\beta_0;\alpha_0)(\alpha_1 - \alpha_0)\\ &\leq g(t,\alpha_1;\alpha_0) - g_u(t,\alpha_1;\alpha_0)(\alpha_1 - \alpha_0) + g_u(t,\beta_0;\alpha_0)(\alpha_1 - \alpha_0)\\ &= g(t,\alpha_1;\alpha_0) + g_{uu}(t,\xi;\alpha_0)(\beta_0 - \alpha_1)(\alpha_1 - \alpha_0)\\ &\leq g(t,\alpha_1;\alpha_0)\\ &= f(t,\alpha_0) + [f_x(t,\alpha_0) + \phi_x(t,\alpha_0)](\alpha_1 - \alpha_0) - [\phi(t,\alpha_1) - \phi(t,\alpha_0)]\\ &\leq f(t,\alpha_1).\end{aligned}$$

Consider the linear IVP

$$\begin{aligned} v' &= G(t,\beta_0;\alpha_0,\beta_0) + G_u(t,\beta_0;\alpha_0,\beta_0)(v-\beta_0) \\ &\equiv H(t,v;\alpha_0,\beta_0), \\ v(0) &= x_0. \end{aligned} \tag{1.4.28}$$

Using (1.4.9), and (1.4.24), we get

$$\begin{aligned} \alpha_0' &\le G(t,\alpha_0;\alpha_0,\beta_0) \\ &\le G(t,\beta_0;\alpha_0,\beta_0) - G_u(t,\beta_0;\alpha_0,\beta_0)(\beta_0-\alpha_0) \\ &\equiv H(t,\alpha_0;\alpha_0,\beta_0). \end{aligned}$$

Similarly, using (1.4.10),

$$\beta_0' \ge G(t,\beta_0;\alpha_0,\beta_0) \equiv H(t,\beta_0;\alpha_0,\beta_0).$$

Hence, as before, there exists unique solution $\beta_1(t)$of (1.4.28) such that $\alpha_0(t) \le \beta_1(t) \le \beta_0(t)$ on J. Now, we shall show that $\beta_1(t)$ is an upper solution of (1.4.1). Using (1.4.24), (1.4.3), (1.4.5) and the mean value theorem, we have

$$\begin{aligned} \beta_1' &= G(t,\beta_0;\alpha_0,\beta_0) + G_u(t,\beta_0;\alpha_0,\beta_0)(\beta_1-\beta_0) \\ &\ge G(t,\beta_1;\alpha_0,\beta_0) \\ &= f(t,\beta_0) + [f_x(t,\alpha_0)+\phi_x(t,\alpha_0)]\,(\beta_1-\beta_0) - [\phi(t,\beta_1)-\phi(t,\beta_0)] \\ &\ge f(t,\beta_1) + [f_x(t,\beta_1)+\phi_x(t,\beta_1)]\,(\beta_0-\beta_1) - [\phi(t,\beta_0)-\phi(t,\beta_1)] \\ &\quad + [f_x(t,\alpha_0)+\phi_x(t,\alpha_0)]\,(\beta_1-\beta_0) - [\phi(t,\beta_1)-\phi(t,\beta_0)] \\ &= f(t,\beta_1) + [F_x(t,\beta_1)-F_x(t,\alpha_0)]\,(\beta_0-\beta_1) \\ &= f(t,\beta_1) + F_{xx}(t,\xi)(\beta_1-\alpha_0)(\beta_0-\beta_1) \\ &\ge f(t,\beta_1), \end{aligned}$$

where $\alpha_0 \le \xi \le \beta_1$. Thus, we have

$$\alpha_1' \le f(t,\alpha_1), \qquad \beta_1' \ge f(t,\beta_1), \qquad \alpha_1(0) = \beta_1(0),$$

which implies, by Theorem 1.1.1, that

$$\alpha_1(t) \le \beta_1(t) \text{ on } J \tag{1.4.29}$$

and consequently,

$$\alpha_0(t) \le \alpha_1(t) \le \beta_1(t) \le \beta_0(t) \qquad \text{on } J. \tag{1.4.30}$$

Next, consider the linear IVP

$$\begin{aligned} u' &= g(t,\alpha_1;\alpha_1) + g_u(t,\beta_1;\alpha_1)(u-\alpha_1) \equiv h(t,u;\alpha_1,\beta_1), \\ u(0) &= x_0. \end{aligned} \tag{1.4.31}$$

Using (1.4.3), (1.4.4) and (1.4.23),

$$\alpha_1' \le f(t,\alpha_1) \equiv h(t,\alpha_1;\alpha_1,\beta_1).$$

and

$$\begin{aligned} \beta_1' &\ge f(t,\beta_1) \ge g(t,\beta_1;\alpha_1) \\ &\ge g(t,\alpha_1;\alpha_1) + g_u(t,\beta_1;\alpha_1)(\beta_1-\alpha_1) \\ &\equiv h(t,\beta_1;\alpha_1,\beta_1). \end{aligned}$$

Here, we have α_1, β_1 as lower and upper solutions of the linear problem (1.4.31) with $\alpha_1(0) = \beta_1(0)$. Consequently, by Theorem 1.1.3 there exists a unique solution $\alpha_2(t)$ of (1.4.31) such that $\alpha_1(t) \le \alpha_2(t) \le \beta_1(t)$ on J.

Now, consider the linear IVP

$$\begin{aligned} v' &= G(t,\beta_1;\alpha_1,\beta_1) + G_u(t,\beta_1;\alpha_1,\beta_1)(v-\beta_1) \\ &\equiv H(t,v;\alpha_1,\beta_1), \\ v(0) &= x_0. \end{aligned} \tag{1.4.32}$$

Using (1.4.17), (1.4.24), and (1.4.26), it follows that

$$\begin{aligned} \alpha_1' &\le G(t,\alpha_1;\alpha_1,\beta_1) \\ &\le G(t,\beta_1;\alpha_1,\beta_1) - G_u(t,\beta_1;\alpha_1,\beta_1)(\beta_1-\alpha_1) \\ &\equiv H(t,\alpha_1;\alpha_1,\beta_1). \end{aligned}$$

Similarly, from (1.4.18),

$$\beta_1' \ge G(t,\beta_1;\alpha_1,\beta_1) \equiv H(t,\beta_1;\alpha_1,\beta_1).$$

Thus, as before, by Theorem 1.1.3, there exists a unique solution $\beta_2(t)$ of (1.4.32) such that $\alpha_1(t) \le \beta_2(t) \le \beta_1(t)$ on J. Next, we will show that $\alpha_2(t)$, $\beta_2(t)$ are lower and upper solutions of (1.4.1). In view of the fact

$$\alpha_2' = g(t,\alpha_1;\alpha_1) + g_u(t,\beta_1;\alpha_1)(\alpha_2-\alpha_1),$$

we obtain using (1.4.31), (1.4.23), (1.4.24), (1.4.3) and the mean value theorem

$$\begin{aligned} \alpha_2' &\le g(t,\alpha_2;\alpha_1) - g_u(t,\alpha_2;\alpha_1)(\alpha_2-\alpha_1) + g_u(t,\beta_1;\alpha_1)(\alpha_2-\alpha_1) \\ &\le g(t,\alpha_2;\alpha_1) + g_{uu}(t,\xi;\alpha_1)(\beta_1-\alpha_2)(\alpha_2-\alpha_1) \\ &\le g(t,\alpha_2;\alpha_1) \\ &= f(t,\alpha_1) + [f_x(t,\alpha_1)+\phi_x(t,\alpha_1)](\alpha_2-\alpha_1) - [\phi(t,\alpha_2)-\phi(t,\alpha_1)] \\ &\le f(t,\alpha_2). \end{aligned}$$

Similarly, in view of the fact

$$\beta_2' = G(t,\beta_1;\alpha_1,\beta_1) + G_u(t,\beta_1;\alpha_1,\beta_1)(\beta_2 - \beta_1)$$

we get, using (1.4.32), (1.4.26), (1.4.3) and the mean value theorem

$$\begin{aligned}
\beta_2' &= G(t,\beta_2;\alpha_1,\beta_1)\\
&\equiv f(t,\beta_1) + [f_x(t,\alpha_1) + \phi_x(t,\alpha_1)]\,(\beta_2 - \beta_1) + [\phi(t,\beta_1) - \phi(t,\beta_2)]\\
&\geq f(t,\beta_2) + [f_x(t,\beta_2) + \phi_x(t,\beta_2)]\,(\beta_1 - \beta_2) - [\phi(t,\beta_1) - \phi(t,\beta_2)]\\
&\quad + [f_x(t,\alpha_1) + \phi_x(t,\alpha_1)]\,(\beta_2 - \beta_1) + [\phi(t,\beta_1) - \phi(t,\beta_2)]\\
&= f(t,\beta_2) + [F_x(t,\beta_2) - F_x(t,\alpha_1)]\,(\beta_1 - \beta_2)\\
&= f(t,\beta_2) + F_{xx}(t,\xi)(\beta_2 - \alpha_1)(\beta_1 - \beta_2)\\
&\geq f(t,\beta_2).
\end{aligned}$$

Hence, as before, by Theorem 1.1.1, we have $\alpha_2(t) \leq \beta_2(t)$ on J and, consequently, $\alpha_0 \leq \alpha_1 \leq \alpha_2 \leq \beta_2 \leq \beta_1 \leq \beta_0$ on J.

This process can be continued successively to obtain

$$\alpha_0(t) \leq \alpha_1(t) \leq \cdots \leq \alpha_n(t) \leq \beta_n(t) \leq \cdots \leq \beta_1(t) \leq \beta_0(t) \text{ on } J,$$

where $\{\alpha_n(t)\}$, $\{\beta_n(t)\}$ are unique solutions of the linear IVPs

$$\alpha_{n+1}' = g(t,\alpha_n;\alpha_n) + g_u(t,\beta_n;\alpha_n)(\alpha_{n+1} - \alpha_n), \qquad \alpha_{n+1}(0) = x_0, \tag{1.4.33}$$

$$\beta_{n+1}' = G(t,\beta_n;\alpha_n,\beta_n) + G_u(t,\beta_n;\alpha_n,\beta_n)(\beta_{n+1} - \beta_n), \qquad \beta_{n+1}(0) = x_0. \tag{1.4.34}$$

As before, using standard techniques, it can be shown that $\{\alpha_n(t)\}$, $\{\beta_n(t)\}$ converge monotonically and uniformly to the unique solution $x(t)$ of (1.4.1) on J.

Next, we shall show that the convergence is quadratic. To obtain this, consider

$$p_{n+1}(t) = x(t) - \alpha_{n+1}(t) \geq 0, \qquad q_{n+1}(t) = \beta_{n+1}(t) - x(t) \geq 0.$$

Note that $p_{n+1}(0) = q_{n+1}(0) = 0$. Using $F(t,x) = f(t,x) + \phi(t,x)$, we obtain

$$\begin{aligned}
p_{n+1}' &= x' - \alpha_{n+1}'\\
&= f(t,x) - [g(t,\alpha_n;\alpha_n) + g_u(t,\beta_n;\alpha_n)(\alpha_{n+1} - \alpha_n)]\\
&= f(t,x) - \{f(t,\alpha_n) + [f_x(t,\alpha_n) + \phi_x(t,\alpha_n) - \phi_x(t,\beta_n)]\,(\alpha_{n+1} - \alpha_n)\}\\
&= F(t,x) - F(t,\alpha_n) - [F_x(t,\alpha_n) - \phi_x(t,\beta_n)]\,(\alpha_{n+1} - x + x - \alpha_n)
\end{aligned}$$

$$
\begin{aligned}
&\quad - [\phi(t,x) - \phi(t,\alpha_n)] \\
&= [F_x(t,\xi) - F_x(t,\alpha_n)](x-\alpha_n) - [F_x(t,\alpha_n) - \phi_x(t,\beta_n)](\alpha_{n+1}-x) \\
&\quad + [\phi_x(t,\beta_n) - \phi_x(t,\sigma)](x-\alpha_{n+1}) \\
&\leq [F_x(t,x) - F_x(t,\alpha_n)](x-\alpha_n) - [F_x(t,\alpha_n) - \phi_x(t,\beta_n)](\alpha_{n+1}-x) \\
&\quad + [\phi_x(t,\beta_n) - \phi_x(t,\sigma)](x-\alpha_n) \\
&= F_{xx}(t,\eta)(x-\alpha_n)^2 - [F_x(t,\alpha_n) - \phi_x(t,\beta_n)](\alpha_{n+1}-x) \\
&\quad + \phi_{xx}(\nu)(\beta_n-\alpha_n)(x-\alpha_n) \\
&= F_{xx}(t,\eta)(x-\alpha_n)^2 - [F_x(t,\alpha_n) - \phi_x(t,\beta_n)](\alpha_{n+1}-x) \\
&\quad + \phi_{xx}(\nu)\left[(\beta_n - x)(x-\alpha_n) + (x-\alpha_n)^2\right]
\end{aligned}
$$

where $\alpha_n \leq \sigma,\ \eta,\ \xi \leq x$, $\alpha_n \leq \nu \leq \beta_n$. Hence, we have

$$
\begin{aligned}
p'_{n+1} &\leq (N+L)p_n^2 + Mp_{n+1} + L(q_n p_n + p_n^2) \\
&\leq (N + \frac{5}{2}L)p_n^2 + \frac{1}{2}Lq_n^2 + Mp_{n+1}.
\end{aligned}
$$

Hence, using $(N + \frac{5}{2}L)p_n^2 + \frac{1}{2}Lq_n^2$ as the forcing term and applying Gronwall's inequality, we get

$$
0 \leq p_{n+1}(t) \leq \int_0^t e^{M(t-s)} \left[(N + \frac{5}{2}L)p_n^2(s) + \frac{1}{2}Lq_n^2(s)\right] ds \qquad \text{on } J,
$$

which yields the desired estimate

$$
\max_J |x(t) - \alpha_{n+1}(t)|
$$

$$
\leq \frac{e^{MT}}{M}\left[(N + \frac{5}{2}L)\max_J |x(t) - \alpha_n(t)|^2 + \frac{1}{2}L\max_J |\beta_n(t) - x(t)|^2\right].
$$

Similarly,

$$
\begin{aligned}
q'_{n+1} &= \beta'_{n+1} - x' \\
&= G(t,\beta_n;\alpha_n,\beta_n) + G_u(t,\beta_n;\alpha_n,\beta_n)(\beta_{n+1} - \beta_n) - f(t,x) \\
&= f(t,\beta_n) + [f_x(t,\alpha_n) + \phi_x(t,\alpha_n) - \phi_x(t,\beta_n)](\beta_{n+1}-\beta_n) - f(t,x) \\
&= F(t,\beta_n) - F(t,x) + [F_x(t,\alpha_n) - \phi_x(t,\beta_n)](\beta_{n+1}-\beta_n) \\
&\quad - [\phi(t,\beta_n) - \phi(t,x)] \\
&= F_x(t,\eta)(\beta_n - x) + F_x(t,\alpha_n)(x - \beta_n) \\
&\quad + [F_x(t,\alpha_n) - \phi_x(t,\beta_n)](\beta_{n+1} - x) + [\phi_x(t,\beta_n) - \phi_x(t,\sigma)](\beta_n - x) \\
&\leq [F_x(t,\beta_n) - F_x(t,\alpha_n)](\beta_n - x) + [F_x(t,\alpha_n) - \phi_x(t,\beta_n)](\beta_{n+1} - x)
\end{aligned}
$$

$$
\begin{aligned}
& +[\phi_x(t,\beta_n)-\phi_x(t,\sigma)](\beta_n-x)\\
= \;& F_{xx}(t,\xi)\left[(\beta_n-x)^2+(x-\alpha_n)(\beta_n-x)\right]\\
& +[F_x(t,\alpha_n)-\phi_x(t,\beta_n)](\beta_{n+1}-x)+\phi_{xx}(\nu)(\beta_n-x)^2,
\end{aligned}
$$

where $x \le \eta$, σ, $\nu \le \beta_n$, $\alpha_n \le \xi \le \beta_n$. Consequently, we have

$$
\begin{aligned}
q'_{n+1} &\le (N+L)\left[q_n^2+p_nq_n\right]+Lq_n^2+Mq_{n+1}\\
&\le (N+L)\left[\frac{3}{2}q_n^2+\frac{1}{2}p_n^2\right]+Lq_n^2+Mq_{n+1}\\
&= (\frac{3}{2}N+\frac{5}{2}L)q_n^2+\frac{1}{2}(N+L)p_n^2+Mq_{n+1},
\end{aligned}
$$

where the constants M, N, L are as in (1.4.21). Using $(\frac{3}{2}N+\frac{5}{2}L)q_n^2+\frac{1}{2}(N+L)p_n^2$ as the forcing term, an application of Gronwall's inequality yields

$$
0 \le q_{n+1}(t) \le \int_0^t e^{M(t-s)}\left[(\frac{3}{2}N+\frac{5}{2}L)q_n^2(s)+\frac{1}{2}(N+L)p_n^2(s)\right]ds
$$

on J and, hence,

$$
\begin{aligned}
\max_J|\beta_{n+1}(t)-x(t)| \le\; & \frac{e^{MT}}{M}\left[(\frac{3}{2}N+\frac{5}{2}L)\max_J|\beta_n(t)-x(t)|^2\right.\\
&\left.+\frac{1}{2}(N+L)\max_J|x(t)-\alpha_n(t)|^2\right].
\end{aligned}
$$

The proof is complete.

The above result generalizes the method of quasilinearization under less restrictive assumptions. The question is whether it is possible to avoid the multistage process and develop a simple algorithmic approach which directly offers the construction of monotone sequences that are the solutions of linear differential equations. The answer is affirmative, and the next result yields a technique which does this precisely.

Theorem 1.4.2 *Assume that*

(A_1) α_0, $\beta_0 \in C^1[J,R]$ *such that*

$$
\alpha_0' \le f(t,\alpha_0),\ \beta_0' \ge f(t,\beta_0) \text{ and } \alpha_0(t) \le \beta_0(t),\ t \in J;
$$

(A_2) f, $\phi \in C^{0,2}[\Omega,R]$, $f_{xx}(t,x)+\phi_{xx}(t,x) \ge 0$ *on* Ω *and* $\phi_{xx}(t,x) > 0$ *on* Ω *where* $\Omega = [(t,x) : \alpha_0(t) \le x \le \beta_0(t),\ t \in J]$.

Then there exists monotone sequences $\{\alpha_n(t)\}$, $\{\beta_n(t)\}$ which converge uniformly to the unique solution of (1.4.1) *and the convergence is quadratic.*

Proof In view of (A_2), we see that whenever $\alpha_0(t) \le x_2 \le x_1 \le \beta_0(t)$,

$$f(t, x_1) - f(t, x_2) \le L(x_1 - x_2) \tag{1.4.35}$$

for some $L \ge 0$, and

$$f(t, x) \ge F(t, y) + F_x(t, y)(x - y) - \phi(t, x), \tag{1.4.36}$$

where $F(t, y) = f(t, y) + \phi(t, y)$.

Let α_1, β_1 be the solutions of linear IVPs

$$\alpha_1' = f(t, \alpha_0) + [F_x(t, \alpha_0) - \phi_x(t, \beta_0)] (\alpha_1 - \alpha_0), \quad \alpha_1(0) = x_0, \tag{1.4.37}$$

and

$$\beta_1' = f(t, \beta_0) + [F_x(t, \alpha_0) - \phi_x(t, \beta_0)] (\beta_1 - \beta_0), \quad \beta_1(0) = x_0, \tag{1.4.38}$$

where $\alpha_0(0) \le x_0 \le \beta_0(0)$.

Set $p = \alpha_0 - \alpha_1$ so that

$$\begin{aligned} p' &= \alpha_0' - \alpha_1' \\ &\le f(t, \alpha_0) - [f(t, \alpha_0) + (F_x(t, \alpha_0) - \phi_x(t, \beta_0)) (\alpha_1 - \alpha_0)] \\ &= [F_x(t, \alpha_0) - \phi_x(t, \beta_0)] \, p. \end{aligned}$$

Since $p(0) \le 0$, we get $\alpha_0(t) \le \alpha_1(t)$ on J. Next we let $p = \alpha_1 - \beta_0$ and note that $p(0) \le 0$. Also,

$$p' = \alpha_1' - \beta_0' \le f(t, \alpha_0) + [F_x(t, \alpha_0) - \phi_x(t, \beta_0)] (\alpha_1 - \alpha_0) - f(t, \beta_0).$$

But since $\beta_0 \ge \alpha_0$, using (1.4.36) and (A_2), we have

$$f(t, \beta_0) \ge f(t, \alpha_0) + F_x(t, \alpha_0)(\beta_0 - \alpha_0) - [\phi(t, \beta_0) - \phi(t, \alpha_0)].$$

Now by the mean value theorem, $\phi(t, \beta_0) - \phi(t, \alpha_0) = \phi_x(t, \xi)(\beta_0 - \alpha_0)$ where $\alpha_0 < \xi < \beta_0$ and because $\phi_x(t, x)$ is increasing in x, it follows that $\phi(t, \beta_0) - \phi(t, \alpha_0) \le \phi_x(t, \beta_0)(\beta_0 - \alpha_0)$. Thus we get

$$f(t, \beta_0) \ge f(t, \alpha_0) + [F_x(t, \alpha_0) - \phi_x(t, \beta_0)] (\beta_0 - \alpha_0)$$

which in turn yields

$$p' \le [F_x(t, \alpha_0) - \phi_x(t, \beta_0)] \, p.$$

We therefore obtain $p(t) \leq 0$ on J proving that $\alpha_1(t) \leq \beta_0(t)$ on J. As a result, we have $\alpha_0(t) \leq \alpha_1(t) \leq \beta_0(t)$ on J.

Similarly, we can prove that $\alpha_0(t) \leq \beta_1(t) \leq \beta_0(t)$ on J. We need to show that $\alpha_1(t) \leq \beta_1(t)$ on J so that it follows

$$\alpha_0(t) \leq \alpha_1(t) \leq \beta_1(t) \leq \beta_0(t) \quad \text{on } J. \tag{1.4.39}$$

Since $\alpha_0 \leq \alpha_1 \leq \beta_0$, using (1.4.36) and (1.4.37), we see that

$$\begin{aligned}
\alpha_1' &\leq f(t,\alpha_1) + \phi(t,\alpha_1) - \phi(t,\alpha_0) - \phi_x(t,\beta_0)(\alpha_1 - \alpha_0) \\
&\leq f(t,\alpha_1) + \phi_x(t,\beta_0)(\alpha_1 - \alpha_0) - \phi_x(t,\beta_0)(\alpha_1 - \alpha_0) \\
&= f(t,\alpha_1).
\end{aligned}$$

Here we have used the mean value theorem and the fact $\phi_x(t,x)$ is increasing in x.

Similarly, we can prove that $\beta_1' \geq f(t,\beta_1)$ and therefore by Theorem 1.1.1, it follows that $\alpha_1(t) \leq \beta_1(t)$ on J which shows that (1.4.39) is valid.

Assume now that for some $k > 1$,

$$\alpha_k' \leq f(t,\alpha_k),\ \beta_k' \geq f(t,\beta_k) \text{ and } \alpha_k(t) \leq \beta_k(t) \text{ on } J.$$

We shall show that

$$\alpha_k(t) \leq \alpha_{k+1}(t) \leq \beta_{k+1}(t) \leq \beta_k(t) \text{ on } J, \tag{1.4.40}$$

where α_{k+1} and β_{k+1} are the solutions of linear IVPs

$$\alpha_{k+1}' = f(t,\alpha_k) + [F_x(t,\alpha_k) - \phi_x(t,\beta_k)]\,(\alpha_{k+1} - \alpha_k), \quad \alpha_{k+1}(0) = x_0, \tag{1.4.41}$$

and

$$\beta_{k+1}' = f(t,\beta_k) + [F_x(t,\alpha_k) - \phi_x(t,\beta_k)]\,(\beta_{k+1} - \beta_k), \quad \beta_{k+1}(0) = x_0. \tag{1.4.42}$$

As before, we set $p = \alpha_k - \alpha_{k+1}$ so that

$$p' \leq [F_x(t,\alpha_k) - \phi_x(t,\beta_k)]\,p, \text{ and } p(0) = 0.$$

Thus we have $\alpha_k(t) \leq \alpha_{k+1}(t)$ on J. On the other hand, letting $p = \alpha_{k+1} - \beta_k$ yields

$$p' \leq f(t,\alpha_k) + [F_x(t,\alpha_k) - \phi_x(t,\beta_k)]\,(\alpha_{k+1} - \alpha_k) - f(t,\beta_k).$$

Since $\beta_k \geq \alpha_k$, (1.4.36) and (A_2), yield after some computation,

$$f(t,\beta_k) \geq f(t,\alpha_k) + [F_x(t,\alpha_k) - \phi_x(t,\beta_k)]\,(\beta_k - \alpha_k),$$

which shows that

$$\begin{aligned} p' &\leq f(t,\alpha_k) + [F_x(t,\alpha_k) - \phi_x(t,\beta_k)](\alpha_{k+1} - \alpha_k) \\ &\quad - [f(t,\alpha_k) + (F_x(t,\alpha_k) - \phi_x(t,\beta_k))(\beta_k - \alpha_k)] \\ &= [F_x(t,\alpha_k) - \phi_x(t,\beta_k)]\, p. \end{aligned}$$

This proves that $p(t) \leq 0$ since $p(0) = 0$ and therefore we have $\alpha_k \leq \alpha_{k+1} \leq \beta_k$ on J.

In a similar manner, we can prove that $\alpha_k \leq \beta_{k+1} \leq \beta_k$ on J. Now to show that $\alpha'_{k+1} \leq f(t,\alpha_{k+1})$, we proceed as before. Utilizing (1.4.36), (1.4.41) and the fact $\alpha_{k+1} \geq \alpha_k$, we get

$$\begin{aligned} \alpha'_{k+1} &\leq f(t,\alpha_{k+1}) + \phi(t,\alpha_{k+1}) - \phi(t,\alpha_k) - \phi_x(t,\beta_k)(\alpha_{k+1} - \alpha_k) \\ &\leq f(t,\alpha_{k+1}) + \phi_x(t,\beta_k)(\alpha_{k+1} - \alpha_k) - \phi_x(t,\beta_k)(\alpha_{k+1} - \alpha_k) \\ &= f(t,\alpha_{k+1}). \end{aligned}$$

Similar arguments yield $\beta'_{k+1} \geq f(t,\beta_{k+1})$ and hence Theorem 1.1.1 shows that $\alpha_{k+1}(t) \leq \beta_{k+1}(t)$on J which proves (1.4.40) is true. Hence by induction we have for all n,

$$\alpha_0 \leq \alpha_1 \leq \alpha_2 \leq \cdots \leq \alpha_n \leq \beta_n \leq \cdots \leq \beta_2 \leq \beta_1 \leq \beta_0 \text{ on } J,$$

and therefore employing standard techniques, it can be shown that the sequences $\{\alpha_n(t)\}$, $\{\beta_n(t)\}$ converge uniformly and monotonically to the unique solution $x(t)$ of (1.4.1) on J.

Next we shall prove that the convergence is quadratic. For this purpose, consider

$$p_{n+1}(t) = x(t) - \alpha_{n+1}(t) \geq 0, \quad q_{n+1}(t) = \beta_{n+1}(t) - x(t) \geq 0.$$

Note that $p_{n+1}(0) = 0$ and $q_{n+1}(0) = 0$. We have

$$\begin{aligned} p'_{n+1} &= x' - \alpha'_{n+1} \\ &= f(t,x) - [f(t,\alpha_n) + \{F_x(t,\alpha_n) - \phi_x(t,\beta_n)\}(\alpha_{n+1} - \alpha_n)] \\ &= F(t,x) - F(t,\alpha_n) - [F_x(t,\alpha_n) - \phi_x(t,\beta_n)](\alpha_{n+1} - \alpha_n) \\ &\quad - [\phi(t,x) - \phi(t,\alpha_n)] \\ &\leq [F_x(t,x) - F_x(t,\alpha_n)]\, p_n + [F_x(t,\alpha_n) - \phi_x(t,\beta_n)]\, p_{n+1} \\ &\quad + [\phi_x(t,\beta_n) - \phi_x(t,\alpha_n)]\, p_n \\ &= F_{xx}(t,\sigma)p_n^2 + [f_x(t,\alpha_n) + \phi_x(t,\alpha_n) - \phi_x(t,\beta_{n+1})]\, p_{n+1} \\ &\quad + \phi_{xx}(t,\gamma)[q_n + p_n]p_n \end{aligned}$$

where $\alpha_n < \sigma < x$ and $\alpha_n < \gamma < \beta_n$. Hence we obtain, since $\phi_x(t, \alpha_n) \leq \phi_x(t, \beta_n)$

$$\begin{aligned} p'_{n+1} &\leq (N + L)p_n^2 + Mp_{n+1} + L(p_n q_n + p_n^2) \\ &\leq (N + \frac{5}{2}L)p_n^2 + Mp_{n+1} + \frac{1}{2}Lq_n^2, \end{aligned}$$

where $|f_x(t,x)| \leq M$, $|f_{xx}(t,x)| \leq N$ and $|\phi_{xx}(t,x)| \leq L$ on Ω. Consequently, Gronwall's inequality yields the estimate

$$\begin{aligned} &\max_J |x(t) - \alpha_{n+1}(t)| \\ \leq \ & \frac{e^{MT}}{M} \left[(N + \frac{5}{2}L) \max_J |x(t) - \alpha_n(t)|^2 + \frac{1}{2}L \max_J |\beta_n(t) - x(t)|^2 \right]. \end{aligned}$$

Similarly, using suitable computation, we arrive at

$$\begin{aligned} &\max_J |\beta_{n+1}(t) - x(t)| \\ \leq \ & \frac{e^{MT}}{M} \left[(\frac{3}{2}N + \frac{5}{2}L) \max_J |\beta_n(t) - x(t)|^2 + \frac{1}{2}(N + L) \max_J |x(t) - \alpha_n(t)|^2 \right]. \end{aligned}$$

The proof is complete.

In the next result, we show that it is possible to develop monotone sequences that converge to the solution quadratically when f admits a decomposition into a sum of two functions F and G with $F + \psi$ concave and $G + \phi$ convex for some concave function ψ and for some convex function ϕ.

Theorem 1.4.3 extends Theorem 1.3.1 in the set up of Theorem 1.4.1. However, here we follow the direct approach of Theorem 1.4.2 rather than the complicated multistage algorithmic method of Theorem 1.4.1. In order to avoid monotonicity we do not consider various other coupled upper and lower solutions, that are discussed in 1.3.

Remark 1.4.1 In Theorem 1.4.1 and Theorem 1.4.2, condition (A_2) is not really an assumption although it is used as a convenient general mechanism. This is because, there always exists a convex function $\phi(t, x)$ such that $f + \phi$ is convex. For example, if f is not convex, noting that the set Ω is compact, we see that $\min_\Omega f_{xx}(t, x) = -M < 0$, for some $M > 0$. Then it is enough to choose $\phi(t, x) = Mx^2$ to satisfy condition (A_2). It is important to remember this observation even though we follow in the text the same procedure as in Theorem 1.4.1 and 1.4.2.

Theorem 1.4.3 *Assume that*

(A_1) *α_0, $\beta_0 \in C^1[J,R]$ such that $\alpha_0' \leq f(t,\alpha_0)$, $\beta_0' \geq f(t,\beta_0)$ and $\alpha_0 \leq \beta_0$ on J;*

(A_2) *$f \in C[\Omega,R]$, f admits a decomposition $f = F+G$ where F_x, G_x, F_{xx}, G_{xx} exist and are continuous satisfying $F_{xx}(t,x)+\psi_{xx}(t,x) \leq 0$ and $G_{xx}(t,x)+\phi_{xx}(t,x) \geq 0$ on Ω, where ψ, $\phi \in C[\Omega,R]$, $\phi_x(t,x)$, $\psi_x(t,x)$, $\phi_{xx}(t,x)$, $\psi_{xx}(t,x)$ exist, are continuous and $\psi_{xx}(t,x) < 0$, $\phi_{xx}(t,x) > 0$ on Ω.*

Then there exist monotone sequences $\{\alpha_n(t)\}$, $\{\beta_n(t)\}$ which converge uniformly to the unique solution of (1.4.1) *and the convergence is quadratic.*

Proof In view of (A_2), we see that

$$F(t,y) \leq H(t,x) - H_x(t,x)(x-y) - \psi(t,y) \tag{1.4.43}$$

and

$$G(t,x) \geq M(t,y) + M_x(t,y)(x-y) - \phi(t,x) \tag{1.4.44}$$

for $x \geq y$, x, $y \in \Omega$, where $H(t,x) = F(t,x) + \psi(t,x)$ and $M(t,x) = G(t,x) + \phi(t,x)$. Also, it is clear that whenever $\alpha_0(t) \leq x_2 \leq x_1 \leq \beta_0(t)$,

$$f(t,x_1) - f(t,x_2) \leq L(x_1 - x_2), \text{ for some } L > 0. \tag{1.4.45}$$

Let α_1, β_1 be solutions of linear IVPs

$$\begin{aligned} \alpha_1' &= f(t,\alpha_0) + [M_x(t,\alpha_0) + H_x(t,\beta_0) - \psi_x(t,\alpha_0) - \phi_x(t,\beta_0)](\alpha_1 - \alpha_0), \\ \alpha_1(0) &= x_0, \end{aligned} \tag{1.4.46}$$

and

$$\begin{aligned} \beta_1' &= f(t,\beta_0) + [M_x(t,\alpha_0) + H_x(t,\beta_0) - \psi_x(t,\alpha_0) - \phi_x(t,\beta_0)](\beta_1 - \beta_0), \\ \beta_1(0) &= x_0, \end{aligned} \tag{1.4.47}$$

where $\alpha_0(0) \leq x_0 \leq \beta_0(0)$.

We shall prove that $\alpha_0 \leq \alpha_1$ on J. To do this, let $p = \alpha_0 - \alpha_1$ so that $p(0) \leq 0$. Then

$$\begin{aligned} p' &= \alpha_0' - \alpha_1' \\ &\leq f(t,\alpha_0) - [f(t,\alpha_0) + \{M_x(t,\alpha_0) + H_x(t,\beta_0) - \psi_x(t,\alpha_0) \\ &\quad - \phi_x(t,\beta_0)\}(\alpha_1 - \alpha_0)] \\ &= [M_x(t,\alpha_0) + H_x(t,\beta_0) - \psi_x(t,\alpha_0) - \phi_x(t,\beta_0)]\,p. \end{aligned}$$

Since $p(0) \le 0$, we get $\alpha_0(t) \le \alpha_1(t)$ on J. Next we let $p = \alpha_1 - \beta_0$ and note that $p(0) \le 0$. Now

$$\begin{aligned} p' &= \alpha_1' - \beta_0' \\ &\le f(t,\alpha_0) + [M_x(t,\alpha_0) + H_x(t,\beta_0) - \psi_x(t,\alpha_0) - \phi_x(t,\beta_0)] \\ &\quad \times(\alpha_1 - \alpha_0) - f(t,\beta_0). \end{aligned}$$

But since $\beta_0 \ge \alpha_0$, using (1.4.43) and (A_2), we have

$$F(t,\alpha_0) \le H(t,\beta_0) - H_x(t,\beta_0)(\beta_0 - \alpha_0) - \psi(t,\alpha_0)$$

and

$$G(t,\beta_0) \ge M(t,\alpha_0) + M_x(t,\alpha_0)(\beta_0 - \alpha_0) - \phi(t,\beta_0),$$

or

$$F(t,\alpha_0) \le F(t,\beta_0) - H_x(t,\beta_0)(\beta_0 - \alpha_0) + [\psi(t,\beta_0) - \psi(t,\alpha_0)]$$

and

$$G(t,\beta_0) \ge G(t,\alpha_0) + M_x(t,\alpha_0)(\beta_0 - \alpha_0) - [\phi(t,\beta_0) - \phi(t,\alpha_0)].$$

Now, by using the mean value theorem,

$$\psi(t,\beta_0) - \psi(t,\alpha_0) = \psi_x(t,\xi)(\beta_0 - \alpha_0)$$

and

$$\phi(t,\beta_0) - \phi(t,\alpha_0) = \phi_x(t,\delta)(\beta_0 - \alpha_0),$$

where $\alpha_0 < \xi,\ \delta < \beta_0$. Because $\psi_x(t,x)$ is decreasing in $x$and $\phi_x(t,x)$ is increasing in x, it follows that

$$\psi(t,\beta_0) - \psi(t,\alpha_0) \le \psi_x(t,\alpha_0)(\beta_0 - \alpha_0)$$

and

$$\phi(t,\beta_0) - \phi(t,\alpha_0) \le \phi_x(t,\beta_0)(\beta_0 - \alpha_0).$$

Thus we get

$$F(t,\alpha_0) \le F(t,\beta_0) - [H_x(t,\beta_0) - \psi_x(t,\alpha_0)](\beta_0 - \alpha_0)$$

and

$$G(t,\beta_0) \ge G(t,\alpha_0) + [M_x(t,\alpha_0) - \phi_x(t,\beta_0)](\beta_0 - \alpha_0)$$

which in turn yields

$$p' \le [M_x(t,\alpha_0) + H_x(t,\beta_0) - \psi_x(t,\alpha_0) - \phi_x(t,\beta_0)]\, p.$$

Consequently, $p(t) \leq 0$ on J proving $\alpha_1 \leq \beta_0$ on J.

As a result, we have

$$\alpha_0(t) \leq \alpha_1(t) \leq \beta_0(t) \text{ on } J.$$

Similarly, we can show that $\alpha_0(t) \leq \beta_1(t) \leq \beta_0(t)$ on J. We need to prove that $\alpha_1(t) \leq \beta_1(t)$ on J so that it follows

$$\alpha_0(t) \leq \alpha_1(t) \leq \beta_1(t) \leq \beta_0(t) \text{ on } J. \tag{1.4.48}$$

Since $\alpha_0 \leq \alpha_1 \leq \beta_0$, using (1.4.43) and (1.4.46), we see that

$$\begin{aligned}
\alpha_1' &= f(t,\alpha_0) + [M_x(t,\alpha_0) + H_x(t,\beta_0) - \psi_x(t,\alpha_0) - \phi_x(t,\beta_0)](\alpha_1 - \alpha_0) \\
&= [F(t,\alpha_0) + H_x(t,\beta_0)(\alpha_1 - \alpha_0)] + [G(t,\alpha_0) + M_x(t,\alpha_0)(\alpha_1 - \alpha_0)] \\
&\quad -\psi_x(t,\alpha_0)(\alpha_1 - \alpha_0) - \phi_x(t,\beta_0)(\alpha_1 - \alpha_0) \\
&\leq [F(t,\alpha_0) + H_x(t,\alpha_1)(\alpha_1 - \alpha_0)] + [G(t,\alpha_0) + M_x(t,\alpha_0)(\alpha_1 - \alpha_0)] \\
&\quad -\psi_x(t,\alpha_0)(\alpha_1 - \alpha_0) - \phi_x(t,\beta_0)(\alpha_1 - \alpha_0) \\
&\leq [F(t,\alpha_1) + \psi(t,\alpha_1) - \psi(t,\alpha_0)] + [G(t,\alpha_1) + \phi(t,\alpha_1) - \phi(t,\alpha_0)] \\
&\quad -\psi_x(t,\alpha_0)(\alpha_1 - \alpha_0) - \phi_x(t,\beta_0)(\alpha_1 - \alpha_0) \\
&\leq f(t,\alpha_1) + \psi_x(t,\alpha_0)(\alpha_1 - \alpha_0)(\alpha_1 - \alpha_0) + \phi_x(t,\beta_0)(\alpha_1 - \alpha_0) \\
&\quad -\psi_x(t,\alpha_0)(\alpha_1 - \alpha_0) - \phi_x(t,\beta_0)(\alpha_1 - \alpha_0) \\
&= f(t,\alpha_1).
\end{aligned}$$

Here we have used the mean value theorem and the facts that $H_x(t,x)$, $\psi_x(t,x)$ are decreasing in x and $\phi_x(t,x)$ is increasing in x.

Similarly, we can prove that $\beta_1' \geq f(t,\beta_1)$ and therefore by Theorem 1.1.1, it follows that $\alpha_1(t) \leq \beta_1(t)$ on J which shows that (1.4.48) is valid.

Assume now that for some $k > 1$, $\alpha_k' \leq f(t,\alpha_k)$, $\beta_k' \geq f(t,\beta_k)$ and $\alpha_k(t) \leq \beta_k(t)$ on J, we shall show that

$$\alpha_k(t) \leq \alpha_{k+1}(t) \leq \beta_{k+1}(t) \leq \beta_k(t) \text{ on } J, \tag{1.4.49}$$

where α_{k+1} and β_{k+1} are the solutions of linear IVP's

$$\begin{aligned}
\alpha_{k+1}' &= f(t,\alpha_k) + [M_x(t,\alpha_k) + H_x(t,\beta_k) - \psi_x(t,\alpha_k) - \phi_x(t,\beta_k)] \\
&\quad \times(\alpha_{k+1} - \alpha_k), \\
\alpha_{k+1}(0) &= x_0,
\end{aligned} \tag{1.4.50}$$

and

$$\begin{aligned}
\beta_{k+1}' &= f(t,\beta_k) + [M_x(t,\alpha_k) + H_x(t,\beta_k) - \psi_x(t,\alpha_k) - \phi_x(t,\beta_k)] \\
&\quad \times(\beta_{k+1} - \beta_k), \\
\beta_{k+1}(0) &= x_0.
\end{aligned} \tag{1.4.51}$$

Hence setting $p = \alpha_k - \alpha_{k+1}$, it follows as before that

$$p' \leq [M_x(t,\alpha_k) + H_x(t,\beta_k) - \psi_x(t,\alpha_k) - \phi_x(t,\beta_k)]\, p \text{ on } J$$

and $p(0) = 0$ which again implies $p(t) \leq 0$ on J. On the other hand, letting $p = \alpha_{k+1} - \beta_k$ yields as before

$$p' \leq [M_x(t,\alpha_k) + H_x(t,\beta_k) - \psi_x(t,\alpha_k) - \phi_x(t,\beta_k)]\, p.$$

This proves that $p(t) \leq 0$, since $p(0) = 0$ and therefore we have $\alpha_k \leq \alpha_{k+1} \leq \beta_k$ on J. In a similar manner, we can prove that

$$\alpha_k \leq \beta_{k+1} \leq \beta_k \text{ on } J.$$

Now using (1.4.43) and (1.4.50) and the fact $\beta_k \geq \alpha_{k+1} \geq \alpha_k$, we get

$$\begin{aligned}
\alpha'_{k+1} &= f(t,\alpha_k) + [M_x(t,\alpha_k) + H_x(t,\beta_k) - \psi_x(t,\alpha_k) - \phi_x(t,\beta_k)]\,(\alpha_{k+1} - \alpha_k) \\
&= [F(t,\alpha_k) + H_x(t,\beta_k)(\alpha_{k+1} - \alpha_k)] \\
&\quad + [G(t,\alpha_k) + M_x(t,\alpha_k)(\alpha_{k+1} - \alpha_k)] \\
&\quad - \psi_x(t,\alpha_k)(\alpha_{k+1} - \alpha_k) - \phi_x(t,\beta_k)(\alpha_{k+1} - \alpha_k) \\
&\leq [F(t,\alpha_k) + H_x(t,\alpha_{k+1})(\alpha_{k+1} - \alpha_k)] \\
&\quad + [G(t,\alpha_k) + M_x(t,\alpha_k)(\alpha_{k+1} - \alpha_k)] \\
&\quad - \psi_x(t,\alpha_k)(\alpha_{k+1} - \alpha_k) - \phi_x(t,\beta_k)(\alpha_{k+1} - \alpha_k) \\
&\leq [F(t,\alpha_{k+1}) + \psi(t,\alpha_{k+1}) - \psi(t,\alpha_k)] \\
&\quad + [G(t,\alpha_{k+1}) + \phi(t,\alpha_{k+1}) - \phi(t,\alpha_k)] \\
&\quad - \psi_x(t,\alpha_k)(\alpha_{k+1} - \alpha_k) - \phi(t,\beta_k)(\alpha_{k+1} - \alpha_k) \\
&\leq f(t,\alpha_{k+1}).
\end{aligned}$$

Here we have used the mean value theorem and the facts that $H_x(t,x)$, $\psi_x(t,x)$ are decreasing in $x$and $\phi_x(t,x)$ is increasing in x.

Similar arguments yield $\beta'_{k+1} \geq f(t,\beta_{k+1})$ and hence Theorem 1.1.1 shows that $\alpha_{k+1}(t) \leq \beta_{k+1}(t)$ on J which proves (1.4.49).

Hence we obtain by induction

$$\alpha_0 \leq \alpha_1 \leq \alpha_2 \leq \cdots \leq \alpha_n \leq \beta_n \leq \cdots \leq \beta_2 \leq \beta_1 \leq \beta_0 \text{ on } J.$$

Now using standard arguments, it is easy to show that the sequences $\{\alpha_n(t)\}$, $\{\beta_n(t)\}$ converge uniformly and monotonically to the unique solution $x(t)$ of (1.4.1) on J. Finally, to prove quadratic convergence, we consider

$$p_{n+1}(t) = x(t) - \alpha_{n+1}(t) \geq 0, \quad q_{n+1}(t) = \beta_{n+1}(t) - x(t) \geq 0.$$

Note that $p_{n+1}(0) = 0 = q_{n+1}(0)$. We then have

$$\begin{aligned}
p'_{n+1} &= x' - \alpha'_{n+1} \\
&= f(t,x) - f(t,\alpha_n) \\
&\quad - \{M_x(t,\alpha_n) + H_x(t,\beta_n) - \psi_x(t,\alpha_n) - \phi_x(t,\beta_n)\}(\alpha_{n+1} - \alpha_n) \\
&= \{H(t,x) - H(t,\alpha_n)\} + \{M(t,x) - M(t,\alpha_n)\} \\
&\quad - \{\psi(t,x) - \psi(t,\alpha_n)\} - \{\phi(t,x) - \phi(t,\alpha_n)\} \\
&\quad - \{M_x(t,\alpha_n) + H_x(t,\beta_n) - \psi_x(t,\alpha_n) - \phi_x(t,\beta_n)\}(\alpha_{n+1} - \alpha_n) \\
&\leq H_x(t,\alpha_n)(x-\alpha_n) + M_x(t,x)(x-\alpha_n) \\
&\quad -\psi_x(t,x)(x-\alpha_n) - \phi_x(t,\alpha_n)(x-\alpha_n) \\
&\quad - \{M_x(t,\alpha_n) + H_x(t,\beta_n) - \psi_x(t,\alpha_n) - \phi_x(t,\beta_n)\} \\
&\quad \times(\alpha_{n+1} - x + x - \alpha_n) \\
&= H_{xx}(t,\xi)(\alpha_n - \beta_n)(x-\alpha_n) + M_{xx}(t,\delta)(x-\alpha_n)^2 \\
&\quad -\psi_{xx}(t,\gamma)(x-\alpha_n)^2 + \phi_{xx}(t,\theta)(\beta_n - \alpha_n)(x-\alpha_n) \\
&\quad + \{M_x(t,\alpha_n) + H_x(t,\beta_n) - \psi_x(t,\alpha_n) - \phi_x(t,\beta_n)\} p_{n+1} \\
&= \{\phi_{xx}(t,\theta) - H_{xx}(t,\xi)\} p_n(p_n + q_n) + \{M_{xx}(t,\delta) - \phi_{xx}(t,\gamma)\} p_n^2 \\
&\quad + \{M_x(t,\alpha_n) + H_x(t,\beta_n) - \psi_x(t,\alpha_n) - \phi_x(t,\beta_n)\} p_{n+1} \\
&\leq \{\phi_{xx}(t,\theta) - H_{xx}(t,\xi)\} p_n(p_n + q_n) + \{M_{xx}(t,\delta) - \psi_{xx}(t,\gamma)\} p_n^2 \\
&\quad + \{G_x(t,\alpha_n) + F_x(t,\beta_n)\} p_{n+1},
\end{aligned}$$

where $\alpha_n < \xi$, $\theta < \beta_n$ and $\alpha_n < \delta$, $\gamma < x$.

Hence we obtain

$$\begin{aligned}
p'_{n+1} &\leq (A+B+C)p_n(p_n+q_n) + (A+D+C)p_n^2 + (E+K)p_{n+1} \\
&\leq (A+B+C)(2p_n^2 + q_n^2) + (A+C+D)p_n^2 + (E+K)p_{n+1} \\
&= Qp_n^2 + Rq_n^2 + Sp_{n+1},
\end{aligned}$$

where on Ω,

$$\begin{aligned}
|\phi_{xx}(t,x)| &< A,\ |F_{xx}(t,x)| < B,\ |\psi_{xx}(t,x)| < C,\ |G_{xx}(t,x)| < D, \\
|G_x(t,x)| &< E,\ |F_x(t,x)| < K,\ Q = 3A + 2B + 3C + D,\ R = A + B + C
\end{aligned}$$

and

$$S = E + K. \tag{1.4.52}$$

Now Gronwall's inequality implies

$$0 \leq p_{n+1}(t) \leq \int_0^t e^{S(t-s)} \left[Qp_n^2(s) + Rq_n^2(s)\right] ds \text{ on } J$$

which yield the desired result

$$\max_J |x(t) - \alpha_{n+1}(t)| \le \frac{e^{ST}}{S}\left[Q\max_J |x(t) - \alpha_n(t)|^2 + R\max_J |\beta_n(t) - x(t)|^2\right].$$

Similarly,

$$\begin{aligned}
q'_{n+1} &= \beta'_{n+1} - x' \\
&= f(t,\beta_n) + \{M_x(t,\alpha_n) \\
&\quad + H_x(t,\beta_n) - \psi_x(t,\alpha_n) - \phi_x(t,\beta_n)\}(\beta_{n+1} - \beta_n) - f(t,x) \\
&= \{H(t,\beta_n) - H(t,x)\} + \{M(t,\beta_n) - M(t,x)\} \\
&\quad + \{\psi(t,x) - \psi(t,\beta_n)\} + \{\phi(t,x) - \phi(t,\beta_n)\} \\
&\quad + \{M_x(t,\alpha_n) + H_x(t,\beta_n) - \psi_x(t,\alpha_n) - \phi_x(t,\beta_n)\}(\beta_{n+1} - \beta_n) \\
&\le H_x(t,x)(\beta_n - x) + M_x(t,\beta_n)(\beta_n - x) \\
&\quad - \psi_x(t,\beta_n)(\beta_n - x) - \phi_x(t,x)(\beta_n - x) \\
&\quad + \{M_x(t,\alpha_n) + H_x(t,\beta_n) - \psi_x(t,\alpha_n) - \phi_x(t,\beta_n)\} \\
&\quad \times(\beta_{n+1} - x + x - \beta_n) \\
&= \{\phi_{xx}(t,\theta) - H_{xx}(t,\xi)\}(\beta_n - x)^2 \\
&\quad + \{M_{xx}(t,\delta) - \psi_{xx}(t,\gamma)\}(\beta_n - \alpha_n)(\beta_n - x) \\
&\quad + \{G_x(t,\alpha_n) + F_x(t,\beta_n)\}(\beta_{n+1} - x) \\
&= \{\phi_{xx}(t,\theta) - H_{xx}(t,\xi)\}q_n^2 + \{M_{xx}(t,\delta) - \psi_{xx}(t,\gamma)\}(q_n + p_n)q_n \\
&\quad + \{G_x(t,\alpha_n) + F_x(t,\beta_n)\}q_{n+1},
\end{aligned}$$

where $\alpha_n < \xi$, $\theta < \beta_n$ and $\alpha_n < \delta$, $\gamma < x$.

Hence we obtain

$$\begin{aligned}
q'_{n+1} &\le (A+B+C)q_n^2 + (A+D+C)q_n(p_n + q_n) + (E+K)q_{n+1} \\
&\le (A+B+C)q_n^2 + (A+C+D)(2q_n^2 + p_n^2) + (E+K)q_{n+1} \\
&= Q^*q_n^2 + R^*p_n^2 + Sp_{n+1},
\end{aligned}$$

where the constants A, B, C, D, E, K and S are as in (1.4.52) and $Q^* = 3A + B + 3C + 2D$, $R^* = A + C + D$.

An application of Gronwall's inequality yields

$$0 \le q_{n+1}(t) \le \int_0^t e^{S(t-s)}\left[Q^*q_n^2(s) + R^*p_n^2(s)\right] ds \text{ on } J$$

and hence

$$\max_J |\beta_{n+1}(t) - x(t)| \le \frac{e^{ST}}{S}\left[Q^*\max_J |\beta_n(t) - x(t)|^2 + R^*\max_J |\alpha_n(t) - x(t)|^2\right].$$

This completes the proof of the theorem.

1.5 REFINEMENTS

In Section 1.3, we investigated the case when the function $f(t,x)$ in (1.2.1) can be decomposed into a difference of two convex and concave functions and developed several results in that frame work. In Section 1.4, we considered the situation, when $f(t,x)$ is not convex, but $f(t,x)+\phi(t,x)$ is convex for some convex function $\phi(t,x)$. In the present section we first consider the situation, when $N(t,x)$ can be split as the sum of three functions $f(t,x)$, $g(t,x)$, $h(t,x)$. Here $f(t,x)+\phi(t,x)$ is convex for some convex function $\phi(t,x)$, $g(t,x)+\psi(t,x)$ is concave for some concave function $\psi(t,x)$ and $h(t,x)$ is only Lipschitzian. We develop two results in this set up. The first result provides nonlinear iterates due to the Lipschitzian term and gives quadratic convergence. In the second result, we have totally linear iterates and obtain semi-quadratic convergence. Some of the results of Section 1.4 will be special cases of this result. Finally we study the situation when $f(t,x)$ can be split as a convex function and a continuous function. In this case we get weak quadratic convergence only.

For convenience of notation, we consider the IVP,

$$x' = N(t,x), \quad x(0) = x_0 \text{ on } J = [0,T]. \tag{1.5.1}$$

where $N \in C[J \times R \times R]$.

Theorem 1.5.1 *Assume that*

(A_1) $\alpha_0, \beta_0 \in C^1[J,R]$ *such that* $\alpha_0' \leq N(t,\alpha_0)$, $\beta_0' \geq N(t,\beta_0)$ *and* $\alpha_0 \leq \beta_0$ *on* J*;*

(A_2) $N \in C[\Omega, R]$ *where* $N(t,x) = f(t,x) + g(t,x) + h(t,x)$, $f_x(t,x)$, $g_x(t,x)$, $f_{xx}(t,x)$, $g_{xx}(t,x)$ *exist and are continuous satisfying* $F_{xx}(t,x) = f_{xx}(t,x) + \phi_{xx}(t,x) \geq 0$, $G_{xx}(t,x) = g_{xx}(t,x) + \psi_{xx}(t,x) \leq 0$ *on* Ω, *where* $\phi, \psi \in C[\Omega, R]$, *are functions such that* ϕ_x, ψ_x, ϕ_{xx}, ψ_{xx} *exist and are continuous, such that* $\phi_{xx} \geq 0, \psi_{xx} \leq 0$ *on* Ω, *and* $h(t,x)$ *satisfies* $h(t,x_1) - h(t,x_2) \leq k(x_1 - x_2)$ *whenever* $x_1 \geq x_2$, $x_1, x_2 \in \Omega$, *where* $k > 0$ *is the Lipschitz constant.*

Then there exist monotone sequences $\{\alpha_n(t)\}$ *and* $\{\beta_n(t)\}$ *which converge uniformly and monotonically to the unique solution of* (1.5.1) *and the convergence is quadratic.*

Proof In view of (A_2), we see that

$$\begin{aligned} f(t,x) &\geq f(t,y) + F_x(t,y)(x-y) + \phi(t,y) - \phi(t,x), \\ g(t,x) &\leq g(t,y) + G_x(t,y)(x-y) + \psi(t,y) - \psi(t,x) \qquad (1.5.2) \\ \text{for } x &\geq y \text{ and } x,y \in \Omega. \end{aligned}$$

Also from (A_2), it follows that

$$N(t,x) - N(t,y) \leq L(x-y) \qquad (1.5.3)$$

whenever $\alpha_0(t) \leq y \leq x \leq \beta_0(t)$. Consider the IVP,

$$\begin{aligned} u' &= f(t,\alpha_0) + g(t,\alpha_0) + h(t,u) \\ &\quad +[F_x(t,\alpha_0) + G_x(t,\beta_0) - \phi_x(t,\beta_0) - \psi_x(t,\alpha_0)](u-\alpha_0) \\ &\equiv M(t,u,\alpha_0,\beta_0), \qquad (1.5.4) \\ u(0) &= x_0. \end{aligned}$$

We shall now show that α_0 is a lower solution of (1.5.4) and β_0 is an upper solution of (1.5.4). For that, consider

$$\begin{aligned} \alpha_0' &\leq N(t,\alpha_0) \\ &= f(t,\alpha_0) + g(t,\alpha_0) + h(t,\alpha_0) \equiv M(t,\alpha_0,\alpha_0,\beta_0) \\ \alpha_0(0) &\leq x_0, \end{aligned}$$

then $\alpha_0(t)$ is a lower solution of (1.5.4) on J. Similarly, it follows that

$$\begin{aligned} \beta_0'(t) &\geq N(t,\beta_0) \\ &= f(t,\beta_0) + g(t,\beta_0) + h(t,\beta_0) \\ &\geq f(t,\alpha_0) + F_x(t,\alpha_0)(\beta_0-\alpha_0) + \phi(t,\alpha_0) - \phi(t,\beta_0) \\ &\quad +g(t,\alpha_0) + G_x(t,\beta_0)(\beta_0-\alpha_0) + \psi(t,\alpha_0) - \psi(t,\beta_0) + h(t,\beta_0) \\ &\geq f(t,\alpha_0) + g(t,\alpha_0) + h(t,\beta_0) + [F_x(t,\alpha_0) + G_x(t,\beta_0) - \phi_x(t,\beta_0) \\ &\quad -\psi_x(t,\alpha_0)](\beta_0-\alpha_0) \\ &\equiv M(t,\beta_0,\alpha_0,\beta_0), \end{aligned}$$

using (1.5.2) and the fact that ϕ_x is increasing and ψ_x is decreasing on Ω. Therefore, $\beta_0(t)$ is an upper solution of (1.5.4) on J and consequently, by Theorem 1.1.3, there exists a solution, $\alpha_1(t)$, of (1.5.4) such that $\alpha_0(t) \leq \alpha_1(t) \leq \beta_0(t)$ on J. Since M satisfies a Lipschitz condition, it follows that $\alpha_1(t)$ is unique on J.

Consider the IVP,

$$\begin{aligned} v' &= f(t,\beta_0) + g(t,\beta_0) + h(t,v) \\ &\quad +[F_x(t,\alpha_0) + G_x(t,\beta_0) - \phi_x(t,\beta_0) - \psi_x(t,\alpha_0)](v-\beta_0) \\ &\equiv K(t,v,\alpha_0,\beta_0), \qquad\qquad (1.5.5) \\ v(0) &= x_0. \end{aligned}$$

We shall show that α_0 and β_0 are lower and upper solutions of (1.5.5) respectively. That is

$$\beta_0' \geq N(t,\beta_0) = f(t,\beta_0) + g(t,\beta_0) + h(t,\beta_0) \equiv K(t,\beta_0,\alpha_0,\beta_0).$$

Also,

$$\begin{aligned} \alpha_0' &\leq N(t,\alpha_0) \\ &= f(t,\alpha_0) + g(t,\alpha_0) + h(t,\alpha_0) \\ &\leq f(t,\beta_0) - F_x(t,\alpha_0)(\beta_0-\alpha_0) - \phi(t,\alpha_0) + \phi(t,\beta_0) \\ &\quad +g(t,\beta_0) + G_x(t,\beta_0)(\alpha_0-\beta_0) + \psi(t,\beta_0) - \psi(t,\alpha_0) + h(t,\alpha_0) \\ &\leq f(t,\beta_0) + g(t,\beta_0) + h(t,\alpha_0) + [F_x(t,\alpha_0) + G_x(t,\beta_0) - \phi_x(t,\beta_0) \\ &\quad -\psi_x(t,\alpha_0)](\alpha_0-\beta_0) \\ &\equiv K(t,\alpha_0,\alpha_0,\beta_0), \end{aligned}$$

from (1.5.2) and the fact that ϕ_x is increasing and ψ_x is decreasing on Ω. Therefore, $\alpha_0(t)$ is a lower solution of (1.5.5), on J. Thus from Theorem 1.1.3, we see that there exists a solution, $\beta_1(t)$, of (1.5.5) such that $\alpha_0(t) \leq \beta_1(t) \leq \beta_0(t)$ on J. As before, since K satisfies the Lipschitz condition, $\beta_1(t)$ is unique on J.

Next, we must show that $\alpha_1(t) \leq \beta_1(t)$ on J. We do this by showing that $\alpha_1(t)$ and $\beta_1(t)$ are lower and upper solutions of (1.5.1), respectively. Since $\alpha_0(t) \leq \alpha_1(t) \leq \beta_0(t)$, we get

$$\begin{aligned} \alpha_1'(t) &= f(t,\alpha_0) + g(t,\alpha_0) + h(t,\alpha_1) \\ &\quad +[F_x(t,\alpha_0) + G_x(t,\beta_0) - \phi_x(t,\beta_0) - \psi_x(t,\alpha_0)](\alpha_1-\alpha_0) \\ &\leq h(t,\alpha_1) + f(t,\alpha_1) - F_x(t,\alpha_0)(\alpha_1-\alpha_0) - \phi(t,\alpha_0) + \phi(t,\alpha_1) \\ &\quad +g(t,\alpha_1) - G_x(t,\alpha_1)(\alpha_1-\alpha_0) + \psi(t,\alpha_1) - \psi(t,\alpha_0) \\ &\quad +[F_x(t,\alpha_0) + G_x(t,\beta_0) - \phi_x(t,\beta_0) - \psi_x(t,\alpha_0)](\alpha_1-\alpha_0) \\ &\leq f(t,\alpha_1) + \phi_x(t,\beta_0)(\alpha_1-\alpha_0) - G_x(t,\beta_0)(\alpha_1-\alpha_0) + g(t,\alpha_1) \\ &\quad +h(t,\alpha_1) + \psi_x(t,\alpha_0)(\alpha_1-\alpha_0) + [G_x(t,\beta_0) - \phi_x(t,\beta_0) \\ &\quad -\psi_x(t,\alpha_0)](\alpha_1-\alpha_0) \\ &\leq N(t,\alpha_1) \end{aligned}$$

using the fact that G_x and ψ_x are decreasing, and ϕ_x is increasing on Ω and by (1.5.2). Similarly, it follows that

$$\begin{aligned}
\beta_1'(t) &= f(t,\beta_0) + g(t,\beta_0) + h(t,\beta_1) \\
&\quad +[F_x(t,\alpha_0) + G_x(t,\beta_0) - \phi_x(t,\beta_0) - \psi_x(t,\alpha_0)](\beta_1 - \beta_0) \\
&\geq f(t,\beta_1) - F_x(t,\beta_1)(\beta_1 - \beta_0) + \phi(t,\beta_1) - \phi(t,\beta_0) \\
&\quad +g(t,\beta_1) - G_x(t,\beta_0)(\beta_1 - \beta_0) - \psi(t,\beta_0) + \psi(t,\beta_1) \\
&\quad +h(t,\beta_1) + [F_x(t,\alpha_0) + G_x(t,\beta_0) - \phi_x(t,\beta_0) - \psi_x(t,\alpha_0)](\beta_1 - \beta_0) \\
&\geq h(t,\beta_1) + f(t,\beta_1) + g(t,\beta_1) - F_x(t,\beta_1)(\beta_1 - \beta_0) - \phi_x(t,\beta_1)(\beta_0 - \beta_1) \\
&\quad -\psi_x(t,\beta_1)(\beta_0 - \beta_1) + [F_x(t,\alpha_0) - \phi_x(t,\beta_1)) - \psi_x(t,\beta_1)](\beta_1 - \beta_0) \\
&= N(t,\beta_1)
\end{aligned}$$

using (1.5.2), the fact that F_x is increasing on Ω and $\alpha_0(t) \leq \beta_1(t) \leq \beta_0(t)$ on J. Hence by Theorem 1.1.1, we get $\alpha_1(t) \leq \beta_1(t)$ on J. Consequently, we have that $\alpha_0(t) \leq \alpha_1(t) \leq \beta_1(t) \leq \beta_0(t)$ on J.

We now prove by induction that $\alpha_0(t) \leq \alpha_1(t) \leq \cdots \leq \alpha_n(t) \leq \alpha_{n+1}(t) \leq \beta_{n+1}(t) \leq \beta_n(t) \leq \cdots \leq \beta_1(t) \leq \beta_0(t)$ on J. Consider the IVP,

$$\begin{aligned}
u'(t) &= f(t,\alpha_n) + g(t,\alpha_n) + h(t,u) \\
&\quad +[F_x(t,\alpha_n) + G_x(t,\beta_n) - \phi_x(t,\beta_n) - \psi_x(t,\alpha_n)](u - \alpha_n) \\
&\equiv M(t,u,\alpha_n,\beta_n) \qquad (1.5.6) \\
u(0) &= x_0.
\end{aligned}$$

Assume that for some $n > 1$, $\alpha_0(t) \leq \alpha_n(t) \leq \beta_n(t) \leq \beta_0(t)$ on J. We will show that $\alpha_0(t) \leq \alpha_{n+1}(t) \leq \beta_{n+1}(t) \leq \beta_0(t)$ on J. For that purpose, consider

$$\begin{aligned}
\alpha_n'(t) &= f(t,\alpha_{n-1}) + g(t,\alpha_{n-1}) + h(t,\alpha_n) \\
&\quad +[F_x(t,\alpha_{n-1}) + G_x(t,\beta_{n-1}) - \phi_x(t,\beta_{n-1}) - \psi_x(t,\alpha_{n-1})] \\
&\quad \times(\alpha_n - \alpha_{n-1}) \\
&\leq f(t,\alpha_n) - F_x(t,\alpha_{n-1})(\alpha_n - \alpha_{n-1}) - \phi(t,\alpha_{n-1}) + \phi(t,\alpha_n) \\
&\quad +g(t,\alpha_n) - G_x(t,\alpha_n)(\alpha_n - \alpha_{n-1}) + \psi(t,\alpha_n) - \psi(t,\alpha_{n-1}) \\
&\quad +h(t,\alpha_n) \\
&\quad +[F_x(t,\alpha_{n-1}) + G_x(t,\beta_{n-1}) - \phi_x(t,\beta_{n-1}) - \psi_x(t,\alpha_{n-1})] \\
&\quad \times(\alpha_n - \alpha_{n-1}) \\
&\leq f(t,\alpha_n) + g(t,\alpha_n) + h(t,\alpha_n) - G_x(t,\alpha_n)(\alpha_n - \alpha_{n-1}) \\
&\quad +G_x(t,\alpha_n)(\alpha_n - \alpha_{n-1}) + \phi_x(t,\beta_{n-1})(\alpha_n - \alpha_{n-1}) \\
&\quad +\psi_x(t,\alpha_{n-1})(\alpha_n - \alpha_{n-1}) - \phi_x(t,\beta_{n-1})(\alpha_n - \alpha_{n-1})
\end{aligned}$$

$$\begin{aligned}
&-\psi_x(t,\alpha_{n-1})(\alpha_n-\alpha_{n-1})\\
=\;& f(t,\alpha_n)+g(t,\alpha_n)+h(t,\alpha_n)\\
\equiv\;& M(t,\alpha_n,\alpha_n,\beta_n),
\end{aligned}$$

where we have used (1.5.1) and the fact that ϕ_x is increasing on Ω, and ψ_x and G_x are decreasing on Ω. Therefore, $\alpha_n(t)$ is a lower solution of (1.5.6). Similarly we show that $\beta_n(t)$ is an upper solution of (1.5.6) on J. Using (1.5.2), it also follows that

$$\begin{aligned}
\beta_n'(t) &= f(t,\beta_{n-1})+g(t,\beta_{n-1})+h(t,\beta_n)\\
&\quad+[F_x(t,\alpha_{n-1})+G_x(t,\beta_{n-1})-\phi_x(t,\beta_{n-1})-\psi_x(t,\alpha_{n-1})]\\
&\quad\times(\beta_n-\beta_{n-1})\\
&\geq f(t,\alpha_n)+F_x(t,\alpha_n)(\beta_{n-1}-\alpha_n)+\phi(t,\alpha_n)-\phi(t,\beta_{n-1})+g(t,\alpha_n)\\
&\quad+G_x(t,\beta_{n-1})(\beta_{n-1}-\alpha_n)-\psi(t,\beta_{n-1})+\psi(t,\alpha_n)+h(t,\beta_n)\\
&\quad+[F_x(t,\alpha_{n-1})+G_x(t,\beta_{n-1})-\phi_x(t,\beta_{n-1})-\psi_x(t,\alpha_{n-1})]\\
&\quad\times(\beta_n-\beta_{n-1})\\
&\geq f(t,\alpha_n)+g(t,\alpha_n)+h(t,\beta_n)+F_x(t,\alpha_n)(\beta_{n-1}-\alpha_n)\\
&\quad+F_x(t,\alpha_{n-1})(\beta_n-\beta_{n-1})-\phi_x(t,\beta_{n-1})(\beta_{n-1}-\alpha_n)\\
&\quad+G_x(t,\beta_{n-1})(\beta_{n-1}-\alpha_n)+G_x(t,\beta_{n-1})(\beta_n-\beta_{n-1})\\
&\quad-\psi_x(t,\alpha_{n-1})(\beta_{n-1}-\alpha_n)-\psi_x(t,\alpha_{n-1})(\beta_n-\beta_{n-1})\\
&\quad-\phi_x(t,\beta_{n-1})(\beta_n-\beta_{n-1})\\
&= f(t,\alpha_n)+g(t,\alpha_n)+h(t,\beta_n)-F_x(t,\alpha_n)(\alpha_n-\beta_n)\\
&\quad+\phi_x(t,\beta_{n-1})(\alpha_n-\beta_n)-G_x(t,\beta_{n-1})(\alpha_n-\beta_n)\\
&\quad+\psi_x(t,\alpha_{n-1})(\alpha_n-\beta_n)\\
&\geq f(t,\alpha_n)+g(t,\alpha_n)+h(t,\beta_n)+[F_x(t,\alpha_n)+G_x(t,\beta_n)-\phi_x(t,\beta_n)\\
&\quad-\psi_x(t,\alpha_n)](\beta_n-\alpha_n)\\
&\equiv M(t,\beta_n,\alpha_n,\beta_n),
\end{aligned}$$

since G_x and ψ_x are decreasing , while ϕ_x and F_x are increasing on Ω. Using Theorem 1.1.3, the solution, $\alpha_{n+1}(t)$, of (1.5.6) exists such that $\alpha_n(t)\leq \alpha_{n+1}(t)\leq\beta_n(t)$ on J. Consequently, since M satisfies the Lipschitz condition on Ω; $\alpha_{n+1}(t)$ is unique. In the same manner we can show that $\alpha_n(t)\leq \beta_{n+1}(t)\leq\beta_n(t)$ on J. It remains to be shown that $\alpha_{n+1}(t)\leq\beta_{n+1}(t)$ on J. In order to prove this, we see that

$$\begin{aligned}
\alpha_{n+1}'(t) &= f(t,\alpha_n)+g(t,\alpha_n)+h(t,\alpha_{n+1})\\
&\quad+[F_x(t,\alpha_n)+G_x(t,\beta_n)-\phi_x(t,\beta_n)-\psi_x(t,\alpha_n)](\alpha_{n+1}-\alpha_n)
\end{aligned}$$

$$\begin{aligned}
&\le f(t,\alpha_{n+1}) - F_x(t,\alpha_n)(\alpha_{n+1}-\alpha_n) - \phi(t,\alpha_n) + \phi(t,\alpha_{n+1})\\
&\quad +g(t,\alpha_{n+1}) + G_x(t,\alpha_{n+1})(\alpha_n-\alpha_{n+1}) - \psi(t,\alpha_n) + \psi(t,\alpha_{n+1})\\
&\quad +h(t,\alpha_{n+1}) + [F_x(t,\alpha_n) + G_x(t,\beta_n) - \phi_x(t,\beta_n) - \psi_x(t,\alpha_n)]\\
&\quad \times(\alpha_{n+1}-\alpha_n)\\
&\le N(t,\alpha_{n+1}) + \phi_x(t,\beta_n)(\alpha_{n+1}-\alpha_n) + \psi_x(t,\alpha_n)(\alpha_{n+1}-\alpha_n)\\
&\quad -G_x(t,\beta_n)(\alpha_{n+1}-\alpha_n) + G_x(t,\beta_n)(\alpha_{n+1}-\alpha_n) - \phi_x(t,\beta_n)\\
&\quad \times(\alpha_{n+1}-\alpha_n)\\
&\quad -\psi_x(t,\alpha_n)(\alpha_{n+1}-\alpha_n)\\
&= N(t,\alpha_{n+1}).
\end{aligned}$$

This implies that $\alpha_{n+1}(t)$ is a lower solution of (1.5.1) on J. Similarly, we can show that $\beta_{n+1}(t)$ is an upper solution of (1.5.1) on J. Using Theorem 1.1.1, it follows that $\alpha_{n+1}(t) \le \beta_{n+1}(t)$ on J. Therefore, by induction, we have that for all n, $\alpha_0(t) \le \alpha_1(t) \le \cdots \le \alpha_{n-1}(t) \le \alpha_n(t) \le \beta_n(t) \le \beta_{n-1}(t) \le \cdots \le \beta_1(t) \le \beta_0(t)$ on J. Using standard arguments, it is easily seen that the sequences $\{\alpha_n(t)\}$and $\{\beta_n(t)\}$ converge uniformly and monotonically to the unique solution, $x(t)$, of (1.5.5) on J.

Finally, we prove quadratic convergence. Let $p_{n+1}(t) = x(t) - \alpha_{n+1}(t)$, and $q_{n+1}(t) = \beta_{n+1}(t) - x(t)$. Then $p_{n+1}(0) = q_{n+1}(0) = 0$.

$$\begin{aligned}
p'_{n+1}(t) &= x(t) - \alpha_{n+1}(t)\\
&= f(t,x) + g(t,x) + h(t,x) - f(t,\alpha_n) - g(t,\alpha_n) - h(t,\alpha_{n+1})\\
&\quad -[F_x(t,\alpha_n) + G_x(t,\beta_n) - \phi_x(t,\beta_n) - \psi_x(t,\alpha_n)](\alpha_{n+1}-\alpha_n)\\
&\le -F_x(t,x)(\alpha_n - x) - \phi(t,x) + \phi(t,\alpha_n) + G_x(t,\alpha_n)(x-\alpha_n)\\
&\quad +\psi(t,\alpha_n)\\
&\quad -\psi(t,x) + k(x-\alpha_{n+1}) - F_x(t,\alpha_n)(\alpha_{n+1}-\alpha_n)\\
&\quad -G_x(t,\beta_n)(\alpha_{n+1}-\alpha_n)\\
&\quad +\phi_x(t,\beta_n)(\alpha_{n+1}-\alpha_n) + \psi_x(t,\alpha_n)(\alpha_{n+1}-\alpha_n)\\
&\le F_{xx}(t,\sigma)p_n^2 - G_{xx}(t,\gamma)p_n(q_n+p_n) + F_x(t,\alpha_n)p_{n+1}\\
&\quad +G_x(t,\beta_n)p_{n+1}\\
&\quad +\phi_{xx}(t,\xi)(p_n)(q_n+p_n) - \psi_{xx}(t,\eta)p_n^2 + kp_{n+1}\\
&\quad +(\phi_x(t,\beta_n) + \psi_x(t,\alpha_n))p_{n+1}
\end{aligned}$$

where $\sigma, \eta \in (\alpha_n, x)$ and $\xi, \gamma \in (\alpha_n, \beta_n)$. Letting $|f_{xx}| < A$, $|\phi_{xx}| < B$, $|g_{xx}| < C$, $|\psi_{xx}| < D$, $|f_x| < E$, $|g_x| < N$, $|\phi_x| < R$, $|\psi_x| < S$, we obtain

$$p'_{n+1}(t) \le (A+D+B)p_n^2 + (C+D+B)p_n(p_n+q_n)$$

$$
\begin{aligned}
& +(k+E+2R+N+2S)p_{n+1} \\
\leq\ & (A+D+B)p_n^2+(C+D+B)\frac{1}{2}(3p_n^2+q_n^2) \\
& +(k+E+2R+N+2S)p_{n+1} \\
=\ & Vp_n^2+Uq_n^2+Wp_{n+1}
\end{aligned}
$$

where $V=\frac{3}{2}C+\frac{5}{2}D+A+\frac{5}{2}B$, $U=\frac{1}{2}(C+D+B)$ and $W=E+2R+N+2S+k$. Using Gronwall's inequality and in view of $p_{n+1}(0)=0$, we obtain

$$p_{n+1}(t)\leq\int_0^t e^{W(t-s)}\left[Vp_n^2(s)+Uq_n^2(s)\right]ds.$$

Hence, we get the estimate

$$\max_{t\in J}(x(t)-\alpha_{n+1}(t))\leq\frac{e^{WT}}{W}\left[V\max_{t\in J}|x(t)-\alpha_n(t)|^2+U\max_{t\in J}|\beta_n(t)-x(t)|^2\right].$$

Similarly, we get an estimate for q_{n+1},

$$q_{n+1}(t)\leq\int_0^t e^{S_1(t-s)}\left[R_1p_n^2(s)+Q_1q_n^2(s)\right]ds.$$

This yields the estimate,

$$\max_{t\in J}(\beta_{n+1}(t)-x(t))\leq\frac{e^{S_1T}}{S_1}\left[R_1\max_{t\in J}|x(t)-\alpha_n(t)|^2+Q_1\max_{t\in J}|\beta_n(t)-x(t)|^2\right],$$

where $Q_1=\frac{3}{2}A+C+\frac{5}{2}D+\frac{5}{2}B$, $R_1=\frac{1}{2}(A+B+D)$, and $S_1=L+2R+2S+K$.

The next theorem demonstrates that if we try linearizing the previous iterates, quadratic convergence is lost.

Theorem 1.5.2 *Assume that (A_1) and (A_2) of Theorem 1.5.1 hold. Assume further that*

(A_3) $h(t,x)-h(t,y)\geq-k(x-y)$, *where* $x\geq y$, $x,y\in\Omega$.

Then there exist monotone sequences $\{\alpha_n(t)\}$ and $\{\beta_n(t)\}$, where $\alpha_n(t)$, $\beta_n(t)$ are solutions of linear differential equations, which converge uniformly and monotonically to the unique solution, $x(t)$, of (1.5.1) *and the convergence is semi-quadratic.*

Proof By assumption (A_2), $N(t,x)$ is Lipschitzian in x on Ω, where the Lipschitz constant is L, and

$$\begin{aligned} f(t,x) &\geq F(t,y)+F_x(t,y)(x-y)-\phi(t,x) \\ g(t,x) &\leq G(t,y)+G_x(t,y)(x-y)-\psi(t,x) \end{aligned} \tag{1.5.7}$$

for $x \geq y$ and $x, y \in \Omega$. Let

$$\begin{aligned} \alpha_1'(t) &= N(t,\alpha_0)+[F_x(t,\alpha_0)+G_x(t,\beta_0)-\phi_x(t,\beta_0)-\psi_x(t,\alpha_0)-k] \\ &\quad \times(\alpha_1-\alpha_0) \\ \alpha_1(0) &= x_0, \\ \beta_1'(t) &= N(t,\beta_0)+[F_x(t,\alpha_0)+G_x(t,\beta_0)-\phi_x(t,\beta_0)-\psi_x(t,\alpha_0)-k] \\ &\quad \times(\beta_1-\beta_0) \\ \beta_1(0) &= x_0. \end{aligned} \tag{1.5.8}$$

We shall show that $\alpha_0(t) \leq \alpha_1(t)$ on J. Let $p=\alpha_0-\alpha_1$, then $p(0) \leq 0$,

$$\begin{aligned} p'(t) &= \alpha_0'(t)-\alpha_1'(t) \\ &\leq N(t,\alpha_0)-N(t,\alpha_0) \\ &\quad -[F_x(t,\alpha_0)+G_x(t,\beta_0)-\phi_x(t,\beta_0)-\psi_x(t,\alpha_0)-k](\alpha_1-\alpha_0) \\ &= [F_x(t,\alpha_0)+G_x(t,\beta_0)-\phi_x(t,\beta_0)-\psi_x(t,\alpha_0)-k]p \end{aligned}$$

which yields $p(t) \leq 0$ on J. This proves, $\alpha_0(t) \leq \alpha_1(t)$ on J. Let $p(t) = \alpha_1(t)-\beta_0(t)$, then $p(0) \leq 0$, and we get

$$\begin{aligned} p'(t) &= N(t,\alpha_0)+[F_x(t,\alpha_0)+G_x(t,\beta_0)-\phi_x(t,\beta_0)-\psi_x(t,\alpha_0)-k] \\ &\quad \times(\alpha_1-\alpha_0) \\ &\quad -N(t,\beta_0) \\ &\leq N(t,\alpha_0)+[F_x(t,\alpha_0)+G_x(t,\beta_0)-\phi_x(t,\beta_0)-\psi_x(t,\alpha_0)-k] \\ &\quad \times(\alpha_1-\alpha_0) \\ &\quad -f(t,\alpha_0)-[F_x(t,\alpha_0)-\phi_x(t,\beta_0)](\beta_0-\alpha_0)-h(t,\alpha_0)+k(\beta_0-\alpha_0) \\ &\quad -g(t,\alpha_0)-[G_x(t,\beta_0)-\psi_x(t,\alpha_0)](\beta_0-\alpha_0) \\ &= [F_x(t,\alpha_0)+G_x(t,\beta_0)-\phi_x(t,\beta_0)-\psi_x(t,\alpha_0)-k]p. \end{aligned}$$

This implies that $p(t) \leq 0$. Thus, $\alpha_1(t) \leq \beta_0(t)$ on J. Similarly, one can prove that $\alpha_0(t) \leq \beta_1(t) \leq \beta_0(t)$ on J. We now prove that $\alpha_1(t) \leq \beta_1(t)$ by showing that $\alpha_1(t)$ and $\beta_1(t)$ are lower and upper solutions of (1.5.1), respectively. Using (1.5.8), we have

$$\begin{aligned}
\alpha_1'(t) &= N(t,\alpha_0) + [F_x(t,\alpha_0) + G_x(t,\beta_0) - \phi_x(t,\beta_0) - \psi_x(t,\alpha_0) - k](\alpha_1 - \alpha_0) \\
&= [f(t,\alpha_0) + F_x(t,\alpha_0)(\alpha_1 - \alpha_0)] + [g(t,\alpha_0) + G_x(t,\beta_0) \\
&\quad \times(\alpha_1 - \alpha_0)] \\
&\quad -k(\alpha_1 - \alpha_0) - \phi_x(t,\beta_0)(\alpha_1 - \alpha_0) - \psi_x(t,\alpha_0)(\alpha_1 - \alpha_0) + h(t,\alpha_0) \\
&\le [f(t,\alpha_0) + F_x(t,\alpha_0)(\alpha_1 - \alpha_0)] + [g(t,\alpha_0) + G_x(t,\alpha_1)(\alpha_1 - \alpha_0)] \\
&\quad -k(\alpha_1 - \alpha_0) - \phi_x(t,\beta_0)(\alpha_1 - \alpha_0) - \psi_x(t,\alpha_0)(\alpha_1 - \alpha_0) \\
&\le [f(t,\alpha_0) + F_x(t,\alpha_0)(\alpha_1 - \alpha_0)] + [g(t,\alpha_1) + \psi(t,\alpha_1) - \psi(t,\alpha_0)] \\
&\quad +h(t,\alpha_1) - \phi_x(t,\beta_0)(\alpha_1 - \alpha_0) - \psi_x(t,\alpha_0)(\alpha_1 - \alpha_0) \\
&= [f(t,\alpha_1) + \phi(t,\alpha_1) - \phi(t,\alpha_0)] + [g(t,\alpha_1) + \psi(t,\alpha_1) - \psi(t,\alpha_0)] \\
&\quad +h(t,\alpha_1) - \phi_x(t,\beta_0)(\alpha_1 - \alpha_0) - \psi_x(t,\alpha_0)(\alpha_1 - \alpha_0) \\
&\le N(t,\alpha_1) + \phi_x(t,\beta_0)(\alpha_1 - \alpha_0) - \phi_x(t,\beta_0)(\alpha_1 - \alpha_0) - \psi_x(t,\alpha_0) \\
&\quad \times(\alpha_1 - \alpha_0) \\
&\quad +\psi_x(t,\alpha_0)(\alpha_1 - \alpha_0) \\
&= N(t,\alpha_1),
\end{aligned}$$

using (1.5.2) and the fact that ϕ_x is increasing on Ω, while ψ_x is decreasing on Ω. Similarly we can show that

$$\begin{aligned}
\beta_1'(t) &\ge N(t,\beta_1) \\
\beta_1(0) &= x_0,
\end{aligned}$$

and as a result of Theorem 1.1.1, we have that $\alpha_1(t) \le \beta_1(t)$ on J. From this it follows that $\alpha_0(t) \le \alpha_1(t) \le \beta_1(t) \le \beta_0(t)$ on J. Following the same pattern as in Theorem 1.5.1, using induction with $n > 1$, we can now show that $\alpha_0(t) \le \alpha_n(t) \le \beta_n(t) \le \beta_0(t)$ on J. Let $p_n(t) = \alpha_n(t) - \alpha_{n+1}(t)$, then $p_n(0) = 0$ and

$$\begin{aligned}
p_n'(t) &= \alpha_n'(t) - \alpha_{n+1}'(t) \\
&= N(t,\alpha_{n-1}) + [F_x(t,\alpha_{n-1}) + G_x(t,\beta_{n-1}) - \phi_x(t,\beta_{n-1}) \\
&\quad -\psi_x(t,\alpha_{n-1})](\alpha_n - \alpha_{n-1}) - [\psi_x(t,\alpha_{n-1}) + k](\alpha_n - \alpha_{n-1}) \\
&\quad -N(t,\alpha_n) - [F_x(t,\alpha_n) + G_x(t,\beta_n)](\alpha_{n+1} - \alpha_n) \\
&\quad -N(t,\alpha_n)[\phi_x(t,\beta_n) + \psi_x(t,\alpha_n) + k](\alpha_{n+1} - \alpha_n) \\
&\le -[F_x(t,\alpha_{n-1}) - \phi_x(t,\alpha_n) + G_x(t,\alpha_n) - \psi_x(t,\alpha_{n-1}) - k] \\
&\quad \times(\alpha_n - \alpha_{n-1})
\end{aligned}$$

$$
\begin{aligned}
& +[F_x(t,\alpha_{n-1}) + G_x(t,\beta_{n-1}) - \phi_x(t,\beta_{n-1}) - \psi_x(t,\alpha_{n-1})] \\
& \times(\alpha_n - \alpha_{n-1}) \\
& -[F_x(t,\alpha_n) + G_x(t,\beta_n) - \phi_x(t,\beta_n) - \psi_x(t,\alpha_n) - k] \\
& \times(\alpha_{n+1} - \alpha_n) \\
\leq\ & -[F_x(t,\alpha_{n-1}) + G_x(t,\alpha_n) - \phi_x(t,\alpha_n) - \psi_x(t,\alpha_{n-1}) - k] \\
& \times(\alpha_n - \alpha_{n-1}) \\
& +[F_x(t,\alpha_{n-1}) - \phi_x(t,\alpha_n) + G_x(t,\alpha_n) - \psi_x(t,\alpha_{n-1}) - k] \\
& \times(\alpha_n - \alpha_{n-1}) \\
& +[F_x(t,\alpha_n) + G_x(t,\beta_n) - \phi_x(t,\beta_n) - \psi_x(t,\alpha_n) - k] \\
& \times(\alpha_n - \alpha_{n+1}) \\
=\ & [F_x(t,\alpha_n) + G_x(t,\beta_n) - \phi_x(t,\beta_n) - \psi_x(t,\alpha_n) - k] \\
& \times(\alpha_n - \alpha_{n+1}),
\end{aligned}
$$

using (1.5.2) and the fact that ϕ_x is increasing on Ω and G_x is decreasing on Ω. This gives $p_n(t) \leq 0$ on J. Consequently, $\alpha_n(t) \leq \alpha_{n+1}(t)$ on J. Next, we show that $\alpha_{n+1}(t) \leq \beta_{n+1}(t)$ on J. Using (1.5.2),

$$
\begin{aligned}
\alpha'_{n+1}(t) &= N(t,\alpha_n) - [F_x(t,\alpha_n) + G_x(t,\beta_n) - \phi_x(t,\beta_n) - \psi_x(t,\alpha_n) - k] \\
&\quad \times(\alpha_{n+1} - \alpha_n) \\
&= [f(t,\alpha_n) + F_x(t,\alpha_n)(\alpha_{n+1} - \alpha_n)] + [g(t,\alpha_n) + G_x(t,\beta_n) \\
&\quad \times(\alpha_{n+1} - \alpha_n)] \\
&\quad -[\phi_x(t,\beta_n) + \psi_x(t,\alpha_n) + k](\alpha_{n+1} - \alpha_n) + h(t,\alpha_n) \\
&\leq [f(t,\alpha_{n+1}) + \phi(t,\alpha_{n+1}) - \phi(t,\alpha_n)] \\
&\quad +[g(t,\alpha_{n+1}) + \psi(t,\alpha_{n+1}) - \psi(t,\alpha_n)] \\
&\quad -\phi_x(t,\beta_n)(\alpha_{n+1} - \alpha_n) - \psi_x(t,\alpha_n)(\alpha_{n+1} - \alpha_n) + h(t,\alpha_{n+1}) \\
&\leq f(t,\alpha_{n+1}) + \phi_x(t,\beta_n)(\alpha_{n+1} - \alpha_n) - \phi_x(t,\beta_n)(\alpha_{n+1} - \alpha_n) \\
&\quad +\psi_x(t,\alpha_n)(\alpha_{n+1} - \alpha_n) - \psi_x(t,\alpha_n)(\alpha_{n+1} - \alpha_n) \\
&= N(t,\alpha_{n+1}) \\
\alpha_{n+1}(0) &= x_0.
\end{aligned}
$$

In a similar fashion, one can show that

$$
\begin{aligned}
\beta'_{n+1}(t) &\geq N(t,\beta_{n+1}) \\
\beta_{n+1}(0) &= x_0.
\end{aligned}
$$

Since $N(t,x)$ satisfies the Lipschitz condition, it follows from Theorem 1.1.1 that $\alpha_{n+1}(t) \leq \beta_{n+1}(t)$ on J. By induction, we have that $\alpha_0(t) \leq \alpha_1(t) \leq \cdots \leq$

$\alpha_n(t) \le \alpha_{n+1}(t) \le \beta_{n+1}(t) \le \beta_n(t) \le \cdots \le \beta_1(t) \le \beta_0(t)$ on J, for all n. Using standard arguments, it is easy to see that the sequences $\{\alpha_n(t)\}$ and $\{\beta_n(t)\}$ converge uniformly and monotonically to the unique solution of (1.5.1) on J. Finally, we prove the convergence is semi-quadratic.

Let $p_{n+1}(t) = x(t) - \alpha_{n+1}(t) \ge 0$ and $q_{n+1}(t) = \beta_{n+1}(t) - x(t)$. Then $p_{n+1}(0) = q_{n+1}(0) = 0$ and by the mean value theorem as well as the increasing and decreasing natures of F_x, ϕ_x,and G_x ψ_x, respectively

$$\begin{aligned}
p'_{n+1} &= x'(t) - \alpha'_{n+1} \\
&= N(t,x) - N(t,\alpha_n) - [F_x(t,\alpha_n) + G_x(t,\beta_n) - \phi_x(t,\beta_n) \\
&\quad -\psi_x(t,\alpha_n)](\alpha_{n+1} - \alpha_n) - k(\alpha_{n+1} - \alpha_n) \\
&= f(t,x) + g(t,x) + h(t,x) - f(t,\alpha_n) - g(t,\alpha_n) \\
&\quad -h(t,\alpha_n) + \phi(t,x) + \psi(t,x) \\
&\quad -\phi(t,\alpha_n) - \psi(t,\alpha_n) - \phi(t,x) - \psi(t,x) + \phi(t,\alpha_n) + \psi(t,\alpha_n) \\
&\quad -[F_x(t,\alpha_n) - k + G_x(t,\beta_n) - \phi_x(t,\beta_n) - \psi_x(t,\alpha_n)](\alpha_{n+1} - \alpha_n) \\
&\le [F(t,x) - F(t,\alpha_n)] + [G(t,x) - G(t,\alpha_n)] + k(x - \alpha_n) \\
&\quad -\phi_x(t,\alpha_n)(x - \alpha_n) - \psi_x(t,x)(x - \alpha_n) - [F_x(t,\alpha_n) + G_x(t,\beta_n) \\
&\quad -\phi_x(t,\beta_n) - \psi_x(t,\alpha_n) - k](\alpha_{n+1} - \alpha_n) \\
&\le F_x(t,x)(x - \alpha_n) + G_x(t,\alpha_n)(x - \alpha_n) + \phi_{xx}(t,\delta)(x - \alpha_n)(\beta_n - \alpha_n) \\
&\quad -\psi_{xx}(t,\mu)(x - \alpha_n)^2 - [F_x(t,\alpha_n) + G_x(t,\beta_n) - \phi_x(t,\beta_n) - \psi_x(t,\alpha_n) \\
&\quad -k](\alpha_{n+1} - x) - [F_x(t,\alpha_n) + G_x(t,\beta_n)](x - \alpha_n) + 2k(x - \alpha_n) \\
&\le F_{xx}(t,\gamma)(x - \alpha_n)^2 - G_{xx}(t,\sigma)(x - \alpha_n)(\beta_n - \alpha_n) \\
&\quad -\psi_{xx}(t,\mu)(x - \alpha_n)^2 \\
&\quad -[F_x(t,\alpha_n) + G_x(t,\beta_n) - \phi_x(t,\beta_n) - \psi_x(t,\alpha_n) - k](\alpha_{n+1} - x) \\
&\quad +2k(x - \alpha_n) + \phi_{xx}(t,\delta)(x - \alpha_n)(\beta_n - \alpha_n) \\
&= [F_{xx}(t,\gamma) - \psi_{xx}(t,\mu)](x - \alpha_n)^2 \\
&\quad +[G_{xx}(t,\sigma) + \phi_{xx}(t,\delta)](x - \alpha_n)(\beta_n - \alpha_n) \\
&\quad +[F_x(t,\alpha_n) + G_x(t,\beta_n) - \phi_x(t,\beta_n) - \psi_x(t,\alpha_n)](x - \alpha_{n+1}) \\
&\quad +2k(x - \alpha_n)
\end{aligned}$$

Letting $|f_{xx}| < A$, $|\phi_{xx}| < B$, $|g_{xx}| < C$, $|\psi_{xx}| < D$, $|f_x| < E$, $|g_x| < T$, $|\phi_x| < R$, $|\psi_x| < S$, we obtain

$$\begin{aligned}
p'_{n+1} &\le (A + B + D)p_n^2 + (B + D + C)p_n(p_n + q_n) + 2Lp_n \\
&\quad +(E + 2R + T + 2S)p_{n+1} \\
&\le (A + B + D)p_n^2 + (B + D + C)\frac{1}{2}(3p_n^2 + q_n^2) + 2Lp_n
\end{aligned}$$

$$
\begin{aligned}
&+(E+2R+T+2S)p_{n+1}\\
&= Np_n^2+Wq_n^2+Mp_{n+1}+2kp_n\\
p_{n+1}(0) &= 0
\end{aligned}
$$

where $N=A+\frac{5}{2}B+\frac{3}{2}C+\frac{5}{2}D\neq 0$, $W=\frac{1}{2}(B+C+D)$, $M=E+2R+T+2S$. Using Gronwall inequality,

$$
\begin{aligned}
p_{n+1} &\leq \int_0^t e^{M(t-s)}(Wq_n^2(s)+2Np_n^2(s)+2kp_n)ds\\
p_{n+1}(0) &= 0.
\end{aligned}
$$

This gives us

$$
\max_{t\in J}|x-\alpha_{n+1}|\leq \frac{e^{MT}}{M}
$$

$$
\times[W\,|\beta_n-x|^2+2N\max_{t\in J}\max|x-\alpha_n|^2+2k\max_{t\in J}\max|x-\alpha_n|].
$$

Similarly, one can show that

$$
\begin{aligned}
q'_{n+1} &\leq Pq_n^2+Rp_n^2+2kq_n+Qq_{n+1}\\
q_{n+1}(0) &= 0,
\end{aligned}
$$

where $P=A+\frac{5}{2}B+\frac{5}{2}D+C\neq 0$, $R=A+B+D$, and $Q=E+T+2S+2R$. As before using Gronwall inequality , we obtain the estimate,

$$
\max_{t\in J}|\beta_{n+1}-x|\leq \frac{e^{QT}}{Q}
$$

$$
[2P\max_{t\in J}\max|\beta_n-x|^2+R\max_{t\in J}|x-\alpha_n|^2+2k\max_{t\in J}|\beta_n-x|].
$$

This proves the semi-quadratic convergence of the sequences.

In Theorem 1.5.2, if $h(t,x)\equiv 0$ then we have the usual quadratic convergence. On the other hand, if $f+g\equiv 0$ then we have the monotone method, and the convergence is linear.

The next result considers the case in which $h(t,x)$ is continuous on Ω. In this situation, we can not show the existence in a closed set since we need to assume one of the inequalities to be strict. Also, we have only weakly quadratic convergence. We need the following result from the Theory of Differential inequalities. See [27].

Theorem 1.5.3 *Assume that*

(i) $m \in C[R_+, R]$, *and* $D^+m(t) \geq g(t, m(t))$, $t \in R_+$, *where* $g \in C[R_+ \times R, R]$.

(ii) $\rho(t)$ *is the minimal solution of* $u' = g(t,u)$, $u(t_0) = u_0$, $t_0 \geq 0$, *existing on* $[t_0, \infty)$.

Then $m(t) \geq \rho(t)$, *on* $[t_0, \infty)$ *provided that* $m(t_0) \geq u_0$.

Theorem 1.5.4 *Assume that*

(A_1) $\alpha_0, \beta_0 \in C^1[J, R]$ *such that* $\alpha_0' < N(t, \alpha_0)$, $\beta_0' \geq N(t, \beta_0)$ *and* $\alpha_0 < x_0$, $\beta_0 \geq x_0$ *on* J;

(A_2) $N \in C[\Omega, R]$ *where* $N(t,x) = f(t,x) + h(t,x)$, $f_x(t,x)$, , $f_{xx}(t,x)$, *exist and are continuous satisfying* $F_{xx}(t,x) = f_{xx}(t,x) + \phi_{xx}(t,x) > 0$, *on* Ω, *where* $\phi \in C[\Omega, R]$, ϕ_x, ϕ_{xx}, *exist and are continuous on* Ω, $\phi_{xx} \geq 0$ *and* $h(t,x)$ *is continuous in* x *on* Ω.

Then there exists a monotone sequence $\{\alpha_n(t)\}$ *which converges uniformly and monotonically to the minimal solution of (1.5.1) and the convergence is weakly quadratic.*

Proof By assumption (A_2), we see that

$$f(t,x) \geq F(t,y) + F_x(t,y)(x-y) - \phi(t,x) \tag{1.5.9}$$

where $x \geq y$, x, $y \in \Omega$. Consider the IVP

$$\begin{aligned} u' &= f(t,\alpha_0) + h(t,u) + [F_x(t,\alpha_0) - \phi_x(t,\beta_0)](u - \alpha_0) \\ &\equiv K(t, u, \alpha_0, \beta_0) \qquad (1.5.10) \\ u(0) &= x_0. \end{aligned}$$

We have that

$$\begin{aligned} \alpha_0' &< f(t,\alpha_0) + h(t,\alpha_0) \\ &\equiv K(t, \alpha_0, \alpha_0, \beta_0) \qquad (1.5.11) \\ \alpha_0(0) &< x_0, \end{aligned}$$

since α_0 is a lower solution of (1.5.1). We also have that

$$\begin{aligned} \beta_0' &\geq f(t,\beta_0) + h(t,\beta_0) \\ &\geq f(t,\alpha_0) + h(t,\beta_0) + [F_x(t,\alpha_0) - \phi_x(t,\beta_0)](\beta_0 - \alpha_0) \qquad (1.5.12) \\ &\equiv K(t, \beta_0, \alpha_0, \beta_0) \\ \beta(0) &= x_0. \end{aligned}$$

Consequently, by Theorem 1.5.3, $\alpha_0(t) < \alpha_1(t) \leq \beta_0(t)$, where $\alpha_1(t)$ is the minimal solution of (1.5.10). Suppose that $\alpha_{n-1}(t) < \alpha_n(t) \leq \beta_0(t)$. We shall show that $\alpha_n(t) < \alpha_{n+1}(t) \leq \beta_0(t)$. Consider the IVP

$$\begin{aligned} u' &= f(t,\alpha_n) + h(t,u) + [F_x(t,\alpha_n) - \phi_x(t,\beta_0)](u-\alpha_n) \\ &\equiv K(t,u,\alpha_n,\beta_0) \qquad (1.5.13) \\ u(0) &= x_0. \end{aligned}$$

Then

$$\begin{aligned} \alpha_n' &= f(t,\alpha_{n-1}) + h(t,\alpha_n) + [F_x(t,\alpha_{n-1}) - \phi_x(t,\beta_0)](\alpha_n - \alpha_{n-1}) \\ &< f(t,\alpha_n) + h(t,\alpha_n) + [F_x(t,\alpha_{n-1}) - \phi_x(t,\beta_0)](\alpha_n - \alpha_{n-1}) \\ &\quad - F_x(t,\alpha_{n-1})(\alpha_n - \alpha_{n-1}) + \phi_x(t,\beta_0)(\alpha_n - \alpha_{n-1}) \qquad (1.5.14) \\ &= f(t,\alpha_n) + h(t,\alpha_n) \\ &\equiv K(t,\alpha_n,\alpha_n,\beta_0) \\ \alpha_n(0) &= x_0 \end{aligned}$$

and

$$\begin{aligned} \beta_0'(t) &\geq f(t,\beta_0) + h(t,\beta_0) \\ &\geq f(t,\alpha_n) + h(t,\beta_0) + [F_x(t,\alpha_n) - \phi_x(t,\beta_0)](\beta_0 - \alpha_n) \\ &\equiv K(t,\beta_0,\alpha_n,\beta_0) \qquad (1.5.15) \\ \beta_0(0) &= x_0. \end{aligned}$$

From (1.5.14) and Vol 1, [27], we see that $\alpha_n(t)$ is less than the minimal solution of (1.5.1). As a result of (1.5.14) and (1.5.15), $\alpha_0(t) < \alpha_1(t) < \cdots < \alpha_{n-1}(t) < \alpha_n(t) \leq \beta_0(t)$ on J. Using standard arguments, one can show that $\{\alpha_n(t)\}$ converges uniformly to $\alpha^*(t)$ on J, where $\alpha^*(t)$ is a solution of (1.5.1). We claim that α^* is the minimal solution of (1.5.1).

Suppose $x(t)$ is any solution of (1.5.1). Then from (1.5.1) and (1.5.13) we have that

$$\begin{aligned} \alpha_n'(t) &< N(t,\alpha_n) \\ \alpha_n(0) &= x_0, \qquad (1.5.16) \end{aligned}$$

and

$$\begin{aligned} x'(t) &= N(t,x) \\ x(0) &= x_0. \qquad (1.5.17) \end{aligned}$$

Using Theorems 1.22 and 1.23 of [27], one can see that $\alpha_n(t) < x(t)$ on J. As $n \to \infty$, we obtain $\alpha^*(t) \leq x(t)$ on J. By the definition of a minimal solution, the conclusion holds.

Next, we show that the convergence of the sequence to the minimal solution is weakly quadratic. Let $p_{n+1}(t) = \alpha^*(t) - \alpha_{n+1}(t)$, $q_{n+1}(t) = \beta_0(t) - \alpha_{n+1}(t)$ on J. Then $p_{n+1}(0) = 0$ and we obtain

$$\begin{aligned}
p'_{n+1}(t) &= f(t,\alpha^*) + h(t,\alpha^*) - f(t,\alpha_n) - h(t,\alpha_{n+1}) \\
&\quad -[F_x(t,\alpha_n) - \phi_x(t,\beta_0)](\alpha_{n+1} - \alpha_n) \\
&\leq F(t,\alpha^*) - F(t,\alpha_n) - [F_x(t,\alpha_n) - \phi_x(t,\beta_0)](\alpha_{n+1} - \alpha^*) + 2M \\
&\quad -\phi_x(t,\alpha_n)(\alpha^* - \alpha_n) - [F_x(t,\alpha_n) - \phi_x(t,\beta_0)](\alpha^* - \alpha_n) \\
&\leq F_x(t,\alpha^*)(\alpha^* - \alpha_n) - F_x(t,\alpha_n)(\alpha^* - \alpha_n) \\
&\quad +[F_x(t,\alpha_n) - \phi_x(t,\beta_0)]p_{n+1} \\
&\quad +2M - \phi_x(t,\alpha_n)(\alpha^* - \alpha_n) + \phi_x(t,\beta_0)(\alpha^* - \alpha_n) \\
&\leq F_{xx}(t,\eta)(\alpha^* - \alpha_n)^2 + [F_x(t,\alpha_n) - \phi_x(t,\beta_0)]p_{n+1} + 2M \\
&\quad +\phi_{xx}(t,\xi)(\alpha^* - \alpha_n)(\beta_0 - \alpha_n) \\
&\leq Np_n^2 + Bp_{n+1} + \frac{1}{2}V(p_n^2 + q_n^2) \\
&\leq Ap_n^2 + Bp_{n+1} + Cq_n^2 + D,
\end{aligned}$$

where $|f_{xx}| < E$, $|\phi_{xx}| < V$, $|f_x| < R$, $|\phi_x| < H$, and as a result, $N = E + V$, $A = E + \frac{3}{2}V$, $B = R + 2H$, $C = \frac{1}{2}V$ and $D = 2M$. Using Gronwall's inequality once more, we see that

$$\begin{aligned}
p_{n+1}(t) &\leq \int_0^t e^{B(t-s)}(Cq_n^2 + Ap_n^2 + D)ds \\
p_{n+1}(0) &= 0.
\end{aligned}$$

Therefore,

$$\max_{t\in J} |\alpha^* - \alpha_{n+1}| \leq \frac{e^{BT}}{B}[\max_{t\in J} |\alpha^* - \alpha_n|^2 + \max_{t\in J} |\beta_0 - \alpha_n|^2 + 2M].$$

Hence, the convergence is weakly quadratic and the result follows.

1.6 NOTES

Section 1.1, dealing with necessary comparison results and the method of upper, lower solutions is taken from Ladde, Lakshmikantham and Vatsala [21], and from Lakshmikantham, Leela and Sivasundram [32]. The basic idea of quasilinearization due to Bellman [2], and Bellman and Kalaba [3] is described in Section 1.2 in terms of method of upper, lower solutions. All the results of Section 1.3, are taken from Lakshmikantham, Leela and Sivasundram [32]. Theorem 1.4.1 in Section 1.4 deals with the results of Lakshmikantham, Leela and McRae [31]. See also Lakshmikantham and Malek [33] which laid the foundation for the present set up. Theorem 1.4.2 is due to Lakshmikantham [23] and Theorem 1.4.3 is adapted from Lakshmikantham and Shahzad [37]. The work included in Section 1.5 is due to Stutson and Vatsala [66]. See also Lakshmikantham and Köksal [26].

2 FIRST ORDER DIFFERENTIAL EQUATIONS (CONTINUED)

2.0 INTRODUCTION

Having discussed a variety of results relative to IVPs of scalar differential equation so far in chapter 1, we shall now embark to investigate, in Section 2.1, periodic boundary value problems (PBVPs) in the framework of generalized quasilinearization. We prove some typical results of interest to avoid monotony. Anti-periodic BVPs are the contents of Section 2.2, which require new ideas compared to PBVP's, since similar generalization is no longer valid. Section 2.3 deals with interval analysis approach to the method of generalized quasilinearization, where one can compute arbitrary sharp bounds in terms of interval valued functions. This has significance in large computations.

In Section 2.4, we discuss higher order convergence of monotone sequences generated in the method of generalized quasilinearization. We recall that we have been investigating so far quadratic convergence only.

Section 2.5 concentrates on the explicit representation of the solution of the given IVP in the original method of quasilinearization and refines it by a simple transformation so that it can be applied to a wider class of functions. Finally, Section 2.6 is dedicated to the extension of the method of generalized quasilinearization to systems of differential equations which is interesting in itself.

2.1 PERIODIC BOUNDARY VALUE PROBLEMS

The method of lower and upper solutions coupled with monotone iterative technique has been widely employed to discuss periodic boundary value problems (PBVP's) for nonlinear differential equations, see [21] for details. In this section, we shall investigate PBVP for first order scalar differential equation utilizing the method of generalized quasilinearization which offers rapid convergence of approximate solutions to the solution of the PBVP.

Before we proceed, we need some preliminary results which we shall now list. Consider the PBVP

$$x' = f(t, x), \quad x(0) = x(2\pi), \tag{2.1.1}$$

where $f \in C[J \times R, R]$ and $J = [0, 2\pi]$.

A function $\alpha \in C^1[J, R]$ is said to be a lower solution of (2.1.1) if

$$\alpha' \leq f(t, \alpha), \quad \alpha(0) \leq \alpha(2\pi),$$

and an upper solution when the reverse inequalities hold.

The following example shows that, in general, convergence of monotone sequences provided by the monotone iterative technique is not quadratic.

Example 2.1.1 Consider the problem (2.1.1) where f is defined by

$$f(x) = \begin{cases} 1 & \text{if } x < -1, \\ x^2 & \text{if } -1 \leq x \leq 0, \\ 0 & \text{if } x > 0. \end{cases}$$

Then $\alpha = -1/2$, $\beta = 0$ are lower and upper solutions of (2.1.1) and f satisfies one-sided Lipschitz condition required for the monotone technique with $M = 1$, that is

$$f(t, x) - f(t, y) \geq -M(x - y), \quad \alpha(t) \leq y \leq x \leq \beta(t).$$

Also, note that if u is a solution of (2.1.1) then $u \geq 0$ since $f \geq 0$ and $\int_0^{2\pi} f(x(t))dt = 0$. As a consequence $u = 0$ is the minimal solution of (2.1.1).

In this situation, it is easy to see that α_n, defined by

$$\alpha_n' = f(t, \alpha_{n-1}) - K(\alpha_n - \alpha_{n-1}), \quad \alpha_n(0) = \alpha_n(2\pi),$$

is constant and therefore given by

$$\alpha_n = \alpha_{n-1}^2 + \alpha_{n-1}.$$

The sequence α_n is monotone increasing and its limit is $u \equiv 0$, the minimal solution. However, the convergence is not quadratic. Indeed, if $|u - \alpha_n| \leq \lambda|u - \alpha_{n-1}|^2$, for some $\lambda > 0$, where $|u| = \sup_{t\in T} |u_n(t)|$, holds, then for $n \geq 1$ we have

$$|\alpha_n| = -\alpha_{n-1} - \alpha_{n-1}^2 \leq \lambda\alpha_{n-1}^2,$$

and

$$\alpha_{n-1} \leq \frac{-1}{1+\lambda} < 0,$$

which is a contradiction.

We can now prove a result which extends Theorem 1.1.1 relative to PBVP (2.1.1).

Theorem 2.1.1 *Let $f \in C[[0,2\pi] \times R, R]$, v, $w \in C[[0,2\pi], R]$ and for $0 < t \leq 2\pi$,*

$$\begin{aligned} v' &\leq f(t,v), \quad v(0) \leq v(2\pi), \\ w' &\geq f(t,w), \quad w(0) \geq w(2\pi). \end{aligned}$$

Suppose further that either (a) $f(t,u)$is strictly decreasing in u for each t or (b) there exists a function $z \in C^1[[0,2\pi], R]$ such that $z(t) > 0$, $z(0) \geq z(2\pi)$ and for every $\lambda > 0$,

$$\lambda z' > f(t, w + \lambda z) - f(t,w), \quad 0 < t \leq 2\pi.$$

Then $v(t) \leq w(t)$, $t \in [0,2\pi]$.

Proof Assume that (a) holds. If the assertion is not true, then there exists an $\varepsilon > 0$ and a $t_0 \in [0,2\pi]$ such that

$$v(t_0) = w(t_0) + \varepsilon \quad \text{and} \quad v(t) \leq w(t) + \varepsilon, \quad t \in [0,2\pi]. \tag{2.1.2}$$

If $t_0 \in (0,2\pi]$, then since $v(t_0) > w(t_0)$, we have

$$v'(t_0) \geq w'(t_0) \geq f(t_0, w(t_0)) > f(t_0, v(t_0))$$

which is a contradiction. If $t_0 = 0, v(0) = w(0) + \varepsilon$ and hence

$$v(2\pi) \geq v(0) = w(0) + \varepsilon \geq w(2\pi) + \varepsilon.$$

It then follows from (2.1.2) that $v'(2\pi) \geq w'(2\pi)$, which again leads to a contradiction as before.

Suppose next that (b) holds. If $v(t) \leq w(t)$, $t \in [0, 2\pi]$ is not true, then we can find a minimal $\lambda > 0$ and a $t_0 \in [0, 2\pi]$ such that

$$v(t_0) = w(t_0) + \lambda z(t_0) \text{ and } v(t) \leq w(t) + \lambda z(t), \quad t \in [0, 2\pi]. \tag{2.1.3}$$

Hence, if $t_0 \in (0, 2\pi]$, then we obtain

$$\begin{aligned} v'(t_0) &\geq w'(t_0) + \lambda z(t_0) \\ &> f(t_0, w(t_0)) + f(t_0, w(t_0) + \lambda z(t_0)) - f(t_0, w(t_0)) \\ &= f(t_0, v(t_0)) \end{aligned}$$

which is a contradiction. On the other hand, if $t_0 = 0$, we get $v(0) = w(0) + \lambda z(0)$ and hence

$$v(2\pi) \geq v(0) = w(0) + \lambda z(0) \geq w(2\pi) + \lambda z(2\pi).$$

Using (2.1.3), we see that

$$v'(2\pi) \geq w'(2\pi) + \lambda z'(2\pi) \text{ and } v(2\pi) \leq w(2\pi) + \lambda z(2\pi).$$

Thus it follows that $v(2\pi) = w(2\pi) + \lambda z(2\pi)$, which leads to a contradiction as before. The proof is therefore complete.

An existence theorem parallel to Theorem 1.1.3 is also true for PBVP. For this purpose we need the following abstract result which we state below. For details see [6].

Let E be a real Hilbert space. Consider the nonlinear operator equation

$$Lu = Nu \tag{2.1.4}$$

where $L : D(L) \subset E \to E$ is a linear operator and $N : D(N) \subset E \to E$ a nonlinear operator with $D(L) \cap D(N) \neq \emptyset$. Let $E_0 = N(L)$ be such that dim $E_0 < \infty$ and $E = E_0 \oplus E_1$. Suppose that $P : E \to E_0$ is the projection operator and $H : E_1 \to E_1$ the partial inverse of L on E_1. Then it is well known that the problem (2.1.4) is equivalent to the coupled system of operator equations

$$u_1 = H(I - P)N(u_0 + u_1), \tag{2.1.5}$$

$$0 = PN(u_0 + u_1). \tag{2.1.6}$$

Concerning the problem (2.1.4), we have the following existence result:

Theorem 2.1.2 *Suppose that*

(i) $\|Nu\| \le J_0$, $u \in D(N)$;

(ii) There exist r_0, $R_0 > 0$ *such that* $\langle N(u_0 + u_1), u_0 \rangle \ge 0$ *(or* ≤ 0*) whenever* $\|u_0\| = R_0$ *and* $\|u_1\| \le r_0$ *where* $u_0 \in E$ *and* $u_1 \in E_1$.

Then problem (2.1.4) *admits at least one solution.*

We list the following assumptions relative to (2.1.1) for convenience.

(A_1) α, $\beta \in C[[0, 2\pi], R]$, $\alpha(t) \le \beta(t)$ on $[0, 2\pi]$ and $D_-\alpha(t) \le f(t, \alpha(t))$, $D_-\beta(t) \ge f(t, \beta(t))$, $0 < t \le 2\pi$;

(A_2) *(i)* $\alpha(0) \le \alpha(2\pi)$ and $\alpha \in C^1[[0, 2\pi], R]$,
(ii) $\beta(0) \ge \beta(2\pi)$ and $\beta \in C^1[[0, 2\pi], R]$;

(A_3) (i) $D^+\alpha(0) \ge f(0, \alpha(0)) + 1 + |\alpha(0)|$ and $\int_0^{2\pi} f(s, \alpha(s))ds \ge 0$; (ii) $D^+\beta(0) \le f(0, \beta(0)) - (1 + |\beta(0)|)$ and $\int_0^{2\pi} f(s, \beta(s))ds \le 0$;

(A_4) (A_2) *(i)* and (A_3) *(ii)* hold;

(A_5) (A_2) *(ii)* and (A_3) *(i)* hold.

Here D_-, D^+ are Dini derivatives.

Theorem 2.1.3 *Let* v, $w \in C^1[J, R]$ *be lower and upper solution of (2.1.1) such that* $v \le w$ *on* J. *Then there exists a solution* $x(t)$ *of* (2.1.1) *satisfying* $v(t) \le x(t) \le w(t)$ *on* J.

Proof Relative to the couple (α, β), we define a modified function $F(t, x) = f(t, p(t, x)) + r(t, x)$, where

$$p(t, x) = \max[\alpha(t),\ \min\{x, \beta(t))\}],$$

and

$$r(t, x) = \begin{cases} (\beta(t) - x)/(1 + x^2) & \text{if } x > \beta(t), \\ 0 & \text{if } \alpha(t) \le x \le \beta(t), \\ (\alpha(t) - x)/(1 + x^2) & \text{if } x < \alpha(t). \end{cases}$$

Obviously, we have $|F(t, x)| \le J$ on $[0, 2\pi] \times R$, where J depends only on (α, β). Let us now consider the modified PBVP

$$x' = F(t, x), \quad x(0) = x(2\pi). \tag{2.1.7}$$

We let $E = L_2[0, 2\pi]$, $Lx = x'$, $D(L) = \{x \in E : x$ is absolutely continuous, $r \in E$ and $x(0) = x(2\pi)\}$ and N be the nonlinear operator generated by F. Then the PBVP (2.1.7) is equivalent to the operator equation (2.1.4). We note that $E_0 = N(L)$ consists of constant functions and hence E_1, where $E = E_0 \oplus E_1$, is the class of all functions whose average is zero. The projection operator P can be defined by $Pu = \int_0^{2\pi} u(s)ds$. It is easily seen that the partial inverse operator H also exists and the assumptions required in Theorem 2.1.2 are satisfied. Since F is bounded, we can find a positive constant A which depends only on (α, β) such that any solution u_1 of (2.1.5) satisfies the estimate $|u_1(t)| \leq A$ on $[0,2\pi]$. Hence, by Theorem 2.1.2, it is sufficient to choose an $R_0 > 0$ such that

$$\langle N(u_0 + u_1),\ u_0\rangle \geq 0 \quad (\text{or} \leq 0) \tag{2.1.8}$$

for all $|u_0| = R_0$ and $|u_1(t)| \leq A$ on $[0, 2\pi]$. Since $u_0 \in E_0$ and $E_0 = \langle 1 \rangle$, the relation (2.1.8) with ≤ 0 is equivalent to

$$\int_0^{2\pi} F(s, R_0 + u_1(s))ds \leq 0 \text{ and } \int_0^{2\pi} F(s, -R_0 + u_1(s))ds \geq 0. \tag{2.1.9}$$

Let us now choose $R_0 > 0$ large enough so that we have

$$R_0 + u_1(t) > \max_{[0,2\pi]} \beta(t) \text{ and } -R_0 + u_1(t) < \min_{[0,2\pi]} \alpha(t).$$

This choice of R_0, in view of definition of F, reduces the inequalities in (2.1.9) to

$$\int_0^{2\pi} f(s, \beta(s))ds \leq 0 \text{ and } \int_0^{2\pi} f(s, \alpha(s))ds \geq 0. \tag{2.1.10}$$

Since α, β are lower and upper solutions, such that $\alpha(t) \leq \beta(t)$, the PBVP (2.1.7) admits at least one solution $u(t)$.

We shall now show that $\alpha(t) \leq u(t) \leq \beta(t)$ on $[0, 2\pi]$, where $u(t)$ is the solution of PBVP (2.1.7) in which case, $u(t)$ is actually a solution of PBVP (2.1.1). We shall only prove $\alpha(t) \leq u(t)$, while proving $u(t) \leq \beta(t)$ we can use similar arguments. Suppose that $\alpha(t) \leq u(t)$ on $[0, 2\pi]$ is not true. Then there exists a $t_0 \in [0, 2\pi]$ and an $\varepsilon > 0$ such that

$$\alpha(t_0) = u(t_0) + \varepsilon \text{ and } \alpha(t) \leq u(t) + \varepsilon,\ t \in [0, 2\pi]. \tag{2.1.11}$$

If $t_0 \in (0, 2\pi]$, we have $D_-\alpha(t_0) \geq u'(t_0)$. Also, from (2.1.11), we see that $\alpha(t_0) > u(t_0)$ which implies $p(t_0, u(t_0)) = \alpha(t_0)$. Hence

$$f(t_0, \alpha(t_0)) \geq D_-\alpha(t_0) \geq u'(t_0) = f(t_0, \alpha(t_0)) + (\alpha(t_0) - u(t_0))/(1 + u^2(t_0))$$

which leads to a contradiction since $\alpha(t_0) - u(t_0) > 0$. If $t_0 = 0$, we obtain, on the other hand, $D^+\alpha(0) \le u'(0)$ and consequently, using (A_3)(i), one gets

$$\begin{aligned} 1 + |\alpha(0)| + f(0, \alpha(0)) &\le D^+\alpha(0) \le u'(0) \\ &= f(0, \alpha(0)) + (\alpha(0) - u(0))/(1 + u^2(0)). \end{aligned}$$

Since $(\alpha - u)/(1 + u^2) \le 1/2 + |\alpha(0)|$, one is lead to a contradiction. Thus, we have proved $\alpha(t) \le u(t) \le \beta(t)$ on $[0, 2\pi]$ which completes the proof of Theorem in case (A_3) holds.

If (A_2) holds, we get $\int_0^{2\pi} f(s, \beta(s))ds \le \int_0^{2\pi} \beta'(s)ds = \beta(2\pi) - \beta(0) \le 0$ and similarly, $\int_0^{2\pi} f(s, \alpha(s))ds = \alpha(2\pi) - \alpha(0) \ge 0$ so that (2.1.10) is satisfied. Hence, it remains to consider the case $t_0 = 0$ in the proof. Consider (A_2)(i) for example. Then (2.1.11) yields $\alpha(2\pi) \ge \alpha(0) = u(0) + \varepsilon = u(2\pi) + \varepsilon$ and therefore, we obtain $\alpha'(2\pi) \ge u'(2\pi)$. Thus,

$$\begin{aligned} f(2\pi, \alpha(2\pi)) &\ge \alpha'(2\pi) \ge u'(2\pi) \\ &= f(2\pi, \alpha(2\pi)) + (\alpha(2\pi) - u(2\pi))/(1 + u^2(2\pi)) \end{aligned}$$

which is a contradiction as before. It therefore follows that the conclusion of the theorem is also true when (A_2) holds. It is easy to see that the theorem is true when (A_4) or (A_5) holds also. Hence, the proof of the theorem is complete.

Let $\Omega = \{(t, u) \in I \times R : \alpha(t) \le u \le \beta(t)\}$. We introduce the following conditions:

$$f_x(t, x),\ f_{xx}(t, x) \text{ exist and are continuous for } (t, x) \in \Omega, \tag{2.1.12}$$

$$\int_0^{2\pi} f_x(s, \gamma(s))ds < \delta < 0 \text{ for } \gamma \in [\alpha, \beta]. \tag{2.1.13}$$

$$f_{uu}(t, u) \ge -2M,\ M > 0 \text{ for } (t, u) \in \Omega. \tag{2.1.14}$$

If condition (2.1.12) holds, for u, v with $\alpha(t) \le v \le u \le \beta(t)$, define the function

$$g(t, u, v) = f(t, v) + [f_u(t, v) + 2Mv](u - v) - M(u^2 - v^2). \tag{2.1.15}$$

Note that $g(t, u, u) = f(t, u)$, and that if we define $F(t, u) = f(t, u) + Mu^2$, then condition (2.1.14) means that $F_{uu}(t, u) \ge 0$ for $(t, u) \in \Omega$.

We are now in a position to prove the following result :

Lemma 2.1.1 *Suppose that f is continuous on $I \times R$ and satisfies* (2.1.12) *and* (2.1.14). *Then, for any u, v, w such that $\alpha(t) \leq v \leq u \leq \beta(t)$, $\alpha(t) \leq w \leq \beta(t)$ we have that*

$$f(t,u) - f(t,v) \geq [f_u(t,v) + 2Mv](u-v) - M(u^2 - v^2). \tag{2.1.16}$$

Moreover, f verifies

$$f(t,u) - f(t,v) \geq -K(u-v), \; K > 0 \tag{2.1.17}$$

and there exist $L > 0$, $N > 0$ such that

$$f(t,u) - f(t,v) \leq L(u-v) \tag{2.1.18}$$

and

$$g(t,u,w) - g(t,v,w) \leq N(u-v). \tag{2.1.19}$$

Proof Note that

$$f_u(t,u) = f_u(t,v) + \int_v^u f_{uu} du \geq f_u(t,v) - 2M(u-v).$$

Integrating once again, we obtain (2.1.16).

On the other hand, let $L = \sup\{|f_u(t,u)| : (t,u) \in \Omega\}$. Thus, for some $\xi \in (v,u)$ we get that

$$f(t,u) - f(t,v) = f_u(t,\xi)(u-v) \leq L(u-v),$$

and consequently f satisfies (2.1.17) with $K \geq L$.

Now, taking $N = 2M \cdot \sup\{|w - z| : w, z, \in [\alpha(t), \beta(t)]$, for some $t \in I\}$, we have, that there exists $\zeta \in (v,u)$ such that

$$g(t,u,w) - g(t,v,w) = 2Mw(u-v) - M(u^2 - v^2) = 2M(w-\zeta)(u-v) \leq N(u-v).$$

Finally, we note that condition (2.1.16) is equivalent to $f(t,u) \geq g(t,u,v)$.

We recall that for $m \neq 0$ and $h \in C(I)$, the solution of the linear problem

$$u'(t) + mu(t) = h(t), \; t \in I, \; u(0) = u(2\pi) \tag{2.1.20}$$

is given by

$$u(t) = \int_0^{2\pi} G(t,s)h(s)ds \tag{2.1.21}$$

where

$$G(t,s) = \frac{1}{1-e^{-2\pi m}} \begin{cases} e^{-m(t-s)} & : \quad 0 \le s \le t \le 2\pi \\ e^{-m(2\pi+t-s)} & : \quad 0 \le t \le s \le 2\pi. \end{cases} \tag{2.1.22}$$

Note that $G > 0$ for $m > 0$ and that $G < 0$ for $m < 0$.

Theorem 2.1.4 *Consider the nonlinear problem* (2.1.1) *with f satisfying* (2.1.12), (2.1.13) *and* (2.1.14). *Suppose that α, $\beta \in C^1(J)$ are lower and upper solutions of* (2.1.1) *such that $\alpha \le \beta$ on Ω. Then, there exists $w_n \in C^1(J)$, $n = 1, 2, \cdots$ such that the sequence $\{w_n\}$ is monotone and converging uniformly to a solution w of* (2.1.1). *Moreover, the convergence is quadratic.*

Proof Set $w_0 = \alpha$ and consider the PBVP

$$u' = g(t, u, \alpha(t)), \quad u(0) = u(2\pi), \quad t \in I$$

where $g(t,u,v) = f(t,v) + [f_u(t,v) + 2Mv](u - v) - M(u^2 - v^2)$. It is easy to see that $\alpha' \le f(t,\alpha) \equiv g(t,\alpha,\alpha)$, and $\beta' \ge f(t,\beta) \ge g(t,\beta,\alpha)$. Taking into account Theorem 2.1.3, let $w_1 \in [w_0, \beta]$ be the solution of the nonlinear problem $w_1' = g(t, w_1, w_0)$, $w_1(0) = w_1(2\pi)$. Consider the boundary value problem

$$w'(t) = g(t, w(t), w_n(t)),\ t \in J,\ w(0) = w(2\pi). \tag{2.1.23}$$

Now, suppose we have constructed w_n, $n \ge 1$ such that $w_0 \le w_1 \le w_2 \le \cdots \le w_n \le \beta$. Thus, for $t \in J$ we have that

$$w_n'(t) = g(t, w_n(t), w_{n-1}(t)) \le f(t, w_n(t))$$

and

$$\beta'(t) \ge f(t, \beta(t)) \ge g(t, \beta(t), w_n(t)).$$

This means that w_n is a lower solution and β is an upper solution for (2.1.23) respectively. Let w_{n+1} be a solution of (2.1.23) such that $w_{n+1} \in [w_n, \beta]$. Hence, the sequence $\{w_n\}$ is increasing and it has a pointwise limit w.

To show that w is actually a solution, choose any $m \ne 0$ and note that w_{n+1} verifies

$$w_{n+1}'(t) + m w_{n+1}(t) = h_n(t),\ t \in J,\ w_{n+1}(0) = w_{n+1}(2\pi), \tag{2.1.24}$$

where

$$h_n(t) = m w_{n+1}(t) + f(t, w_{n+1}(t)) + F_u(t, w_n(t)) \\ \cdot [w_{n+1}(t) - w_n(t)] - M\left[w_{n+1}^2(t) - w_n^2(t)\right].$$

By (2.1.21), the solution of (2.1.24) is given by

$$w_{n+1}(t) = \int_0^{2\pi} G(t,s)h_n(s)ds. \tag{2.1.25}$$

From this we obtain that $\{w_n\}$ is bounded in $C^1(J)$ and hence this sequence converges uniformly to w. Taking into account that $\{h_n\}$ converges to the function h defined by $h(t) = mw(t) + f(t, w(t))$, and passing to the limit in (2.1.25), we have that

$$w(t) = \int_0^{2\pi} G(t,s)h(s)ds$$

and hence w is a solution of (2.1.1).

Finally, to show that $\{w_n\} \to w$ quadratically, choose a constant $c > 0$ such that

$$0 \le F_{xx} \le c \text{ on } \Omega. \tag{2.1.26}$$

Now, set $W_n = w - w_n$, $n \ge 1$. Thus, for $t \in J$ we have that

$$\begin{aligned} W'_{n+1}(t) &= f(t, w(t)) - g(t, w_{n+1}(t), w_n(t)) \\ &= F(t, w(t)) - F(t, w_n(t)) - F_x(t, w_n(t)) \cdot [w_{n+1}(t) - w_n(t)] \\ &\quad -M\left[w^2(t) - w_{n+1}^2(t)\right]. \end{aligned}$$

Consequently, there exist $\xi = \xi(t)$ and $\zeta = \zeta(t)$ with $w_n(t) \le \zeta \le \xi \le w(t)$ such that

$$\begin{aligned} W'_{n+1}(t) &= F_x(t,\xi) \cdot [w(t) - w_n(t)] - F_x(t, w_n(t)) \cdot [w_{n+1}(t) - w_n(t)] \\ &\quad -M\left[w^2(t) - w_{n+1}^2(t)\right] \\ &= [F_x(t,\xi) - F_x(t, w_n(t))] \cdot [w(t) - w_n(t)] \\ &\quad + [F_x(t, w_n(t)) - M(w(t) + w_{n+1}(t))] \cdot [w(t) - w_{n+1}(t)] \\ &= F_{xx}(t,\zeta) \cdot (\xi - w_n(t)) \cdot (w(t) - w_n(t)) \\ &\quad + [F_x(t, w_n(t)) - M(w(t) + w_{n+1}(t))] \cdot [w(t) - w_{n+1}(t)]. \end{aligned}$$

Therefore, for $n \ge 1$ and $t \in J$ we have that

$$\begin{aligned} W'_{n+1}(t) &= \phi_n(t) \cdot (\xi(t) - w_n(t)) \cdot (w(t) - w_n(t)) + \psi_n(t) \cdot W_{n+1}(t), \\ W_{n+1}(0) &= W_{n+1}(2\pi) \end{aligned}$$

Using (2.1.26) we have

$$W'_{n+1} \le c \cdot W_n^2 + \psi_n \cdot W_{n+1}.$$

As a consequence, for $t \in J$, we can write

$$0 \leq W_{n+1}(t) \leq W_{n+1}(0)e^{\sigma(t)} + ce^{\sigma(t)} \int_0^t e^{-\sigma(s)} W_n^2(s)ds,$$

where $\sigma(t) = \int_0^t \psi_n(r)dr$. From condition (2.1.13), we know that $\sigma(2\pi) < 0$, for n large. Using the periodic boundary conditions we obtain that

$$W_{n+1}(0) \leq \frac{e^{\sigma(2\pi)}}{1 - e^{\sigma(2\pi)}} \int_0^{2\pi} e^{-\sigma(s)} W_n^2(s)ds.$$

Hence, there exists a $\gamma > 0$ such that

$$\|W_{n+1}\| \leq \gamma \|W_n\|^2.$$

This concludes the proof.

Next we consider a very general result which is analogous to Theorem 1.4.3 in this set up.

Theorem 2.1.5 *Assume that*

(A_1) $\alpha_0, \beta_0 \in C^1[J, R]$ *such that for* $0 < t \leq 2\pi$,

$$\begin{aligned} \alpha_0' &\leq f(t, \alpha_0),\ \alpha_0(0) \leq \alpha_0(2\pi), \\ \beta_0' &\geq f(t, \beta_0),\ \beta_0(0) \geq \beta_0(2\pi) \end{aligned}$$

and $\alpha_0 \leq \beta_0$ *in* J;

(A_2) $f \in C[\Omega, I\!R]$, f *admits a decomposition* $f = F + G$, *where* F_x, G_x, F_{xx}, G_{xx} *exist and are continuous satisfying* $F_{xx}(t,x) + \psi_{xx}(t,x) \leq 0$ *and* $G_{xx}(t,x) + \phi_{xx}(t,x) \geq 0$ *on* Ω *where* ϕ, $\psi \in C[\Omega, I\!R]$, ϕ_x, ψ_x, ϕ_{xx}, ψ_{xx} *exist, are continuous and* $\psi_{xx}(t,x) \leq 0$, $\phi_{xx}(t,x) \geq 0$ *on* Ω;

(A_3) $\int_0^{2\pi} [M_x(t, \beta_0) + H_x(t, \alpha_0) - \psi_x(t, \beta_0) - \phi_x(t, \alpha_0)]dt < 0$ *where* $M(t,x) = G(t,x) + \phi(t,x)$ *and* $H(t,x) = F(t,x) + \psi(t,x)$;

(A_4) *there exists a function* $z \in C^1[J, I\!R]$ *such that* $z(t) > 0$, $z(0) \geq z(2\pi)$ *and for every* $\lambda > 0$,

$$\lambda z' > f(t, \omega + \lambda z) - f(t, \omega),\ 0 \leq t \leq 2\pi,$$

holds for every upper solution ω *of* f.

Then there exist monotone sequences $\{\alpha_n(t)\}$, $\{\beta_n(t)\}$ *which converge monotonically and uniformly to the unique solution of* (2.1.1) *and the convergence is quadratic.*

Proof In view of (A_2), we see that

$$\begin{cases} F(t,y) & \leq H(t,x) - H_x(t,x)(x-y) - \psi(t,y), \\ G(t,x) & \geq M(t,y) + M_x(t,y)(x-y) - \phi(t,x), \end{cases} \tag{2.1.27}$$

for $x \geq y$, x, $y \in \Omega$.

Consider the linear PBVPs

$$\begin{aligned} u' &= f(t,\alpha_0) + [M_x(t,\alpha_0) + H_x(t,\beta_0) - \psi_x(t,\alpha_0) - \phi_x(t,\beta_0)]\,(u-\alpha_0) \\ &\equiv g(t,u;\alpha_0,\beta_0), \qquad u(0) = u(2\pi), \end{aligned} \tag{2.1.28}$$

$$\begin{aligned} v' &= f(t,\beta_0) + [M_x(t,\alpha_0) + H_x(t,\beta_0) - \psi_x(t,\alpha_0) - \phi_x(t,\beta_0)]\,(v-\beta_0) \\ &\equiv g^*(t,v;\alpha_0,\beta_0), \qquad v(0) = v(2\pi). \end{aligned} \tag{2.1.29}$$

The assumption (A_1) and inequalities (2.1.27) imply successively, for $0 < t \leq 2\pi$,

$$\begin{aligned} \alpha_0' &\leq f(t,\alpha_0) \equiv g(t,\alpha_0;\alpha_0,\beta_0) \\ \beta_0' &\geq f(t,\beta_0) = F(t,\beta_0) + G(t,\beta_0) \\ &\geq H(t,\alpha_0) + H_x(t,\beta_0)(\beta_0-\alpha_0) - \psi(t,\beta_0) \\ &\quad + M(t,\alpha_0) + M_x(t,\alpha_0)(\beta_0-\alpha_0) - \phi(t,\beta_0) \\ &= f(t,\alpha_0) + [M_x(t,\alpha_0) + H_x(t,\beta_0)](\beta_0-\alpha_0) \\ &\quad + \psi(t,\alpha_0) - \psi(t,\beta_0) + \phi(t,\alpha_0) - \phi(t,\beta_0) \\ &\geq f(t,\alpha_0) + [M_x(t,\alpha_0) + H_x(t,\beta_0) \\ &\quad - \psi_x(t,\beta_0) - \phi_x(t,\alpha_0)](\beta_0-\alpha_0) \\ &\equiv g(t,\beta_0;\alpha_0,\beta_0). \end{aligned}$$

Here we have used the mean value theorem and the facts that $\psi_x(t,x)$ is nonincreasing in x and $\phi_x(t,x)$ is nondecreasing in x. Hence by Theorem 2.1.3, there exists a solution $\alpha_1(t)$ of (2.1.28) such that $\alpha_0(t) \leq \alpha_1(t) \leq \beta_0(t)$ on J.

Similarly, using (2.1.27), (A_1), (A_2) and the mean value theorem, we obtain

$$\begin{aligned} \alpha_0' &\leq f(t,\alpha_0) = F(t,\alpha_0) + G(t,\alpha_0) \\ &\leq H(t,\beta_0) + H_x(t,\beta_0)(\beta_0-\alpha_0) + G(t,\beta_0) + \phi(t,\beta_0) - \phi(t,\alpha_0) \\ &\quad - M_x(t,\alpha_0)(\beta_0-\alpha_0) - \phi(t,\alpha_0) \end{aligned}$$

$$\begin{aligned}
&= f(t,\beta_0) + [H_x(t,\beta_0) + M_x(t,\alpha_0)](\alpha_0 - \beta_0) \\
&\quad +\psi(t,\beta_0) - \psi(t,\alpha_0) + \phi(t,\beta_0) - \phi(t,\alpha_0) \\
&\leq f(t,\beta_0) + [H_x(t,\beta_0) + M_x(t,\alpha_0) - \psi_x(t,\alpha_0) \\
&\quad -\phi_x(t,\beta_0)](\alpha_0 - \beta_0) \\
&= g^*(t,\alpha_0;\alpha_0,\beta_0). \\
\beta_0'(t) &\geq f(t,\beta_0) \equiv g^*(t,\beta_0;\alpha_0,\beta_0), \text{ for } 0 < t \leq 2\pi.
\end{aligned}$$

By Theorem 2.1.3, as before, there exists a solution $\beta_1(t)$ of (2.1.29) such that $\alpha_0(t) \leq \beta_1(t) \leq \beta_0(t)$ on J. Now since $\alpha_1' = g(t,\alpha_1;\alpha_0,\beta_0)$, we obtain using (2.1.27)

$$\begin{aligned}
\alpha_1' &= f(t,\alpha_0) + [M_x(t,\alpha_0) + H_x(t,\beta_0) - \psi_x(t,\alpha_0) - \phi_x(t,\beta_0)](\alpha_1 - \alpha_0) \\
&\leq [F(t,\alpha_0) + H_x(t,\beta_0)(\alpha_1 - \alpha_0)] + [G(t,\alpha_0) + M_x(t,\alpha_0)(\alpha_1 - a_0)] \\
&\quad -\psi_x(t,\alpha_0)(\alpha_1 - \alpha_0) - \phi_x(t,\beta_0)(\alpha_1 - \alpha_0) \\
&\leq [F(t,\alpha_1) + \psi(t,\alpha_1) - \psi(t,\alpha_0)] + [G(t,\alpha_1) + \phi(t,\alpha_1) - \phi(t,\alpha_0)] \\
&\quad -\psi_x(t,\alpha_0)(\alpha_1 - \alpha_0) - \phi_x(t,\beta_0)(\alpha_1 - \alpha_0) \\
&\leq f(t,\alpha_1) + \psi_x(t,\alpha_0)(\alpha_1 - \alpha_0) + \phi_x(t,\beta_0)(\alpha_1 - \alpha_0) \\
&\quad -\psi_x(t,\alpha_0)(\alpha_1 - \alpha_0) - \phi_x(t,\beta_0)(\alpha_1 - \alpha_0) \\
&= f(t,\alpha_1).
\end{aligned}$$

Here we have used the mean value theorem and the facts that $H_x(t,x)$, $\psi_x(t,x)$ is nonincreasing in x and $\phi_x(t,x)$ is nondecreasing in x.

Similarly, we can prove that $\beta_1' \geq f(t,\beta_1)$, and therefore by Theorem 2.1.1, it follows that $\alpha_1(t) \leq \beta_1(t)$ on J and as a result, we have

$$\alpha_0(t) \leq \alpha_1(t) \leq \beta_1(t) \leq \beta_0(t) \quad \text{on } J.$$

Assume now that for some $k > 1$, $\alpha_k' \leq f(t,\alpha_k)$, $\beta_k' \geq f(t,\beta_k)$ for $0 < t \leq 2\pi$ and $\alpha_k(t) \leq \beta_k(t)$ on J.

We shall show that

$$\alpha_k(t) \leq \alpha_{k+1}(t) \leq \beta_{k+1}(t) \leq \beta_k(t) \quad \text{on } J,$$

where α_{k+1}, and β_{k+1} are the solutions of the PBVPs

$$\begin{aligned}
u' &= f(t,\alpha_k) + [M_x(t,\alpha_k) + H_x(t,\beta_k) - \psi_x(t,\alpha_k) \\
&\quad -\phi_x(t,\beta_k)](u - \alpha_k) \\
&\equiv g(t,u;\alpha_k,\beta_k), \qquad u(0) = u(2\pi),
\end{aligned} \tag{2.1.30}$$

and

$$\begin{aligned} v' &= f(t,\beta_k) + [M_x(t,\alpha_k) + H_x(t,\beta_k) - \psi_x(t,\alpha_k) \\ &\quad -\phi_x(t,\beta_k)](v-\beta_k) \\ &\equiv g^*(t,v;\alpha_k,\beta_k), \qquad v(0)=v(2\pi). \end{aligned} \tag{2.1.31}$$

Now using (2.1.27) and the mean value theorem, we get for $0 < t \le 2\pi$,

$$\begin{aligned} \alpha_k' &\le f(t,\alpha_k) \equiv g(t,\alpha_k;\alpha_k,\beta_k), \\ \beta_k' &\ge f(t,\beta_k) \\ &= F(t,\beta_k) + G(t,\beta_k) \\ &\ge H(t,\alpha_k) + H_x(t,\beta_k)(\beta_k-\alpha_k) - \psi(t,\beta_k) + M(t,\alpha_k) \\ &\quad +M_x(t,\alpha_k)(\beta_k-\alpha_k) - \phi(t,\beta_k) \\ &= f(t,\alpha_k) + [M_x(t,\alpha_k) + H_x(t,\beta_k)](\beta_k-\alpha_k) + \psi(t,\alpha_k) \\ &\quad -\psi(t,\beta_k) + \phi(t,\alpha_k) - \phi(t,\beta_k) \\ &\ge f(t,\alpha_k) + [M_x(t,\alpha_k) + H_x(t,\beta_k) - \psi_x(t,\beta_k) - \phi_x(t,\alpha_k)](\beta_k-\alpha_k) \\ &\equiv g(t,\beta_k;\alpha_k,\beta_k). \end{aligned}$$

Thus by Theorem 2.1.3, as before, there exists a solution $\alpha_{k+1}(t)$ of (2.1.30) such that $\alpha_k(t) \le \alpha_{k+1}(t) \le \beta_k(t)$ on J. Similar arguments yield of $\alpha_k' \le g^*(t,\alpha_k;\alpha_k,\beta_k)$ and $\beta_k' \ge g^*(t,\beta_k;\alpha_k,\beta_k)$. Hence in view of Theorem 2.1.3, there exists a solution $\beta_{k+1}(t)$ such that $\alpha_k(t) \le \beta_{k+1}(t) \le \beta_k(t)$ on J.

Now using (2.1.27), (2.1.30) and the fact $\alpha_k \le \alpha_{k+1} \le \beta_k$, we get

$$\begin{aligned} \alpha_{k+1}' &= f(t,\alpha_k) + [M_x(t,\alpha_k) + H_x(t,\beta_k) - \psi_x(t,\alpha_k) \\ &\quad -\phi_x(t,\beta_k)](\alpha_{k+1}-\alpha_k) \\ &= [F(t,\alpha_k) + H_x(t,\beta_k)(\alpha_{k+1}-\alpha_k)] + [G(t,\alpha_k) + M_x(t,\alpha_k)(\alpha_{k+1}-\alpha_k)] \\ &\quad -\psi_x(t,\alpha_k)(\alpha_{k+1}-\alpha_k) - \phi_x(t,\beta_k)(\alpha_{k+1}-\alpha_k) \\ &\le [F(t,\alpha_k) + H_x(t,\alpha_{k+1})(\alpha_{k+1}-\alpha_k)] + [G(t,\alpha_k) + M_x(t,\alpha_k)(\alpha_{k+1}-\alpha_k)] \\ &\quad -\psi_x(t,\alpha_k)(\alpha_{k+1}-\alpha_k) - \phi_x(t,\beta_k)(\alpha_{k+1}-\alpha_k) \\ &\le [F(t,\alpha_{k+1}) + \psi(t,\alpha_{k+1}) - \psi(t,\alpha_k)] + [G(t,\alpha_{k+1}) + \phi(t,\alpha_{k+1}) - \phi(t,\alpha_k)] \\ &\quad -\psi_x(t,\alpha_k)(\alpha_{k+1}-\alpha_k) - \phi_x(t,\beta_k)(\alpha_{k+1}-\alpha_k) \\ &\le f(t,\alpha_{k+1}). \end{aligned}$$

Here we have used the mean value theorem and the facts that $H_x(t,x)$ and $\psi_x(t,x)$ are nonincreasing in x and $\phi_x(t,x)$ is nondecreasing in x.

In a similar manner, we can prove that $\beta_{k+1}' \ge f(t,\beta_{k+1})$. Therefore by Theorem 2.1.1 it follows that $\alpha_{k+1}(t) \le \beta_{k+1}(t)$ on J which shows that

$$\alpha_k(t) \le \alpha_{k+1}(t) \le \beta_{k+1}(t) \le \beta_k(t) \text{ on } J.$$

Hence by induction we have for all n,

$$\alpha_0 \leq \alpha_1 \leq \cdots \leq \alpha_n \leq \beta_n \leq \cdots \leq \beta_1 \leq \beta_0 \text{ on } J.$$

It then follows, employing standard arguments, that the sequences $\{\alpha_n(t)\}$, $\{\beta_n(t)\}$ converge uniformly and monotonically to the unique solution $x(t)$ of (2.1.1).

Finally, to prove quadratic convergence, we consider $p_{n+1}(t) = x(t) - \alpha_{n+1}(t) \geq 0$ and $q_{n+1}(t) = \beta_{n+1}(t) - x(t) \geq 0$. Note that $p_{n+1}(0) = p_{n+1}(2\pi)$ and $q_{n+1}(0) = q_{n+1}(2\pi)$. Then we have, using the mean value theorem together with (A_2)

$$\begin{aligned}
p_n' &= x' - \alpha_n' \\
&= f(t,x) - [f(t,\alpha_n) + \{M_x(t,\alpha_n) + H_x(t,\beta_n) - \phi_x(t,\beta_n) \\
&\quad -\psi_x(t,\alpha_n)\}(\alpha_{n+1} - \alpha_n)] \\
&= \{H(t,x) - H(t,\alpha_n)\} + \{M(t,x) - M(t,\alpha_n)\} - \{\phi(t,x) - \phi(t,\alpha_n)\} \\
&\quad -\{M_x(t,\alpha_n) + H_x(t,\beta_n) - \phi_x(t,\beta_n) - \psi_x(t,\alpha_n)\}(\alpha_{n+1} - \alpha_n) \\
&\leq H_x(t,\alpha_n)(x - \alpha_n) + M_x(t,x)(x - \alpha_n) \\
&\quad -\phi_x(t,x)(x - \alpha_n) - \phi_x(t,\alpha_n)(x - \alpha_n) \\
&\quad -[M_x(t,\alpha_n) + H_x(t,\beta_n) - \phi_x(t,\beta_n) - \psi_x(t,\alpha_n)](\alpha_{n+1} - x + x - \alpha_n) \\
&= H_{xx}(t,\xi)(\alpha_n - \beta_n)(x - \alpha_n) + M_{xx}(t,\delta)(x - \alpha_n)^2 - \psi_{xx}(t,\gamma)(x - \alpha_n)^2 \\
&\quad +\phi_{xx}(t,\theta)(\beta_n - \alpha_n)(x - \alpha_n) + [M_x(t,\alpha_n) + H_x(t,\beta_n) - \psi_x(t,\alpha_n) \\
&\quad -\phi_x(t,\beta_n)]p_{n+1} \\
&\leq [\phi_{xx}(t,\theta) - H_{xx}(t,\xi)](q_n^2 + 2p_n^2) + [M_{xx}(t,\delta) - \psi_{xx}(t,\gamma)]p_n^2 \\
&\quad +[M_x(t,\alpha_n) + H_x(t,\beta_n) - \psi_x(t,\alpha_n) - \phi_x(t,\beta_n)]p_{n+1}^2,
\end{aligned}$$

where $\alpha_n < \xi$, $\theta < \beta_n$, and $\alpha_n < \delta$, $\gamma < x$.

By setting $\lambda(t) = [M_x(t,\alpha_n) + H_x(t,\beta_n) - \phi_x(t,\beta_n) - \psi_x(t,\alpha_n)]$ and $\gamma^*(t) = [\phi_{xx}(t,\theta) - H_{xx}(t,\xi)](q_n^2 + 2p_n^2) + [M_{xx}(t,\delta) - \psi_{xx}(t,\gamma)]p_n^2$ and $v(t) = p_{n+1}(t)$, we get

$$v'(t) \leq \lambda(t)v(t) + \gamma^*(t), \qquad v(0) = v(2\pi),$$

which implies that

$$v(t) \leq e^{\int_0^t \lambda(s)ds}\left[v(0) + \int_0^t \exp\left(-\int_0^s \lambda(\sigma)d\sigma\right)\gamma^*(s)ds\right],$$

where

$$v(0) = v(2\pi) \leq \frac{1}{\exp\left(-\int_0^{2\pi} \lambda(s)ds\right) - 1}\int_0^{2\pi} \exp\left(-\int_0^s \lambda(\sigma)d\sigma\right)\gamma^*(s)ds. \tag{2.1.32}$$

Note that (2.1.32) holds, since in view of assumptions (A_2) and (A_3) we have

$$\int_0^{2\pi} \lambda(s)ds < 0.$$

Now

$$\begin{aligned} v(t) &\leq \exp\left(\int_0^t \lambda(s)ds\right) \\ &\times \left[\frac{1}{\exp\left(-\int_0^{2\pi}\lambda(s)ds\right)-1}\int_0^{2\pi}\exp\left(-\int_0^s \lambda(\sigma)d\sigma\right)\gamma^*(s)ds \right. \\ &\left. + \int_0^t \exp\left(-\int_0^s \lambda(\sigma)d\sigma\right)\gamma^*(s)ds\right] \\ &\leq K\int_0^{2\pi}\exp\left(-\int_0^s \lambda(\sigma)d\sigma\right)\gamma^*(s)ds, \end{aligned}$$

where

$$K = \frac{\exp\left(-\int_0^{2\pi}\lambda(s)ds\right)}{\exp\left(-\int_0^{2\pi}\lambda(s)ds\right)-1}.$$

Hence, we obtain

$$\max_J |p_{n+1}(t)| \leq A^* \max_J |p_n(t)|^2 + B^* \max_J |q_n(t)|^2,$$

where on Ω, $|\phi_{xx}(t,x)| \leq A$, $|F_{xx}(t,x)| \leq B$, $|\psi_{xx}(t,x)| \leq C$, $|G_{xx}(t,.x)| \leq D$, $\left|\exp(-\int_0^s \lambda(\sigma)d\sigma)\right| \leq Q$,

$$A^* = KQ(3A + 2B + 3C + D)$$

and

$$B^* = KQ(A + B + C).$$

Similarly, using suitable computation, we arrive at

$$\max_J |q_{n+1}(t)|^2 = A_1^* \max_J |p_n|^2 + B_1^* \max_J |q_n(t)|^2$$

where $A_1^* = KQ(3A + B + 3C + 2D)$, and $B_1^* = KQ(A + C + D)$.

The next result admits a decomposition which covers a very general situation of interest.

Theorem 2.1.6 *Assume that*

(A_1) $\alpha_0, \beta_0 \in C^1[J, R]$ *such that for* $0 < t \le 2\pi$,

$$\alpha_0' \le f(t, \alpha_0), \quad \alpha_0(0) \le \alpha_0(2\pi),$$
$$\beta_0' \ge f(t, \beta_0), \quad \beta_0(0) \ge \beta_0(2\pi),$$

and $\alpha_0(t) \le \beta_0(t)$ *on* J;

(A_2) $f \in C[\Omega, R]$, *f admits a decomposition* $f = F + G + h$, *where* F_x, G_x, F_{xx}, G_{xx} *exist and are continuous satisfying* $F_{xx}(t,x) + \phi_{xx}(t,x) \ge 0$, $G_{xx}(t,x) + \psi_{xx}(t,x) \le 0$ *on* Ω *where* $\phi, \psi \in C[\Omega, R]$, ϕ_x, $\phi_{xx}, \psi_x, \psi_{xx}$ *exist, are continuous and* $\phi_{xx}(t,x) \ge 0$, $\psi_{xx}(t,x) \le 0$ *on* Ω. *Further we assume* $h(t,x)$ *satisfies a one-sided Lipschitz condition of the form*

$$h(t,x) - h(t,y) \le L(x-y),$$

for $\alpha_0 \le y \le x \le \beta_0$.

(A_3) $\int_0^{2\pi} [M_x(t,\beta_0) + H_x(t,\alpha_0) - \psi_x(t,\beta_0) - \phi_x(t,\alpha_0) + L]dt < 0$ *where* $M(t,x) = F(t,x) + \phi(t,x)$ *and* $H(t,x) = G(t,x) + \psi(t,x)$;

(A_4) *there exists a function* $z \in C^1[J,R]$ *such that* $z(t) > 0$, $z(0) \ge z(2\pi)$ *and for every* $\lambda > 0$

$$\lambda z' > f(t, w + \lambda z) - f(t, w), \quad 0 \le t \le 2\pi$$

holds for every upper solution of (2.1.1).

Then there exist monotone sequences $\{\alpha_n(t)\}$, $\{\beta_n(t)\}$ *which converge monotonically and uniformly to the unique solution of* (2.1.1) *and the convergence is quadratic.*

Proof In view of (A_2), we see that

$$\begin{cases} F(t,x) \ge M(t,y) + M_x(t,y)(x-y) - \phi(t,x) \\ G(t,y) \le H(t,x) + H_x(t,x)(x-y) - \psi(t,y). \end{cases} \tag{2.1.33}$$

for $x \ge y$, $x, y \in \Omega$.

Consider the following two periodic boundary value problems

$$\begin{aligned} u' &= F(t,\alpha_0) + G(t,\alpha_0) + [M_x(t,\alpha_0) + H_x(t,\beta_0) \\ &\quad -\phi_x(t,\beta_0) - \psi_x(t,\alpha_0)](u - \alpha_0) + h(t,u) \\ &\equiv g(t,u;\alpha_0,\beta_0), \qquad\qquad u(0) = u(2\pi) \end{aligned} \tag{2.1.34}$$

$$\begin{aligned} v' &= F(t,\beta_0) + G(t,\beta_0) + [M_x(t,\alpha_0) + H_x(t,\beta_0) \\ &\quad -\phi_x(t,\beta_0) - \psi_x(t,\alpha_0)](v-\beta_0) + h(t,v) \\ &\equiv g^*(t,v;\alpha_0,\beta_0), \qquad\qquad v(0)=v(2\pi). \end{aligned} \tag{2.1.35}$$

By definition of $g(t,u;\alpha_0,\beta_0)$ it follows that

$$\alpha_0' \le f(t,\alpha_0) \equiv g(t,\alpha_0;\alpha_0,\beta_0), \quad \alpha_0(0) \le \alpha_0(2\pi).$$

Now using assumption (A_2) and inequality (2.1.33) it follows that

$$\begin{aligned} \beta_0' &\ge f(t,\beta_0) \\ &= F(t,\beta_0) + G(t,\beta_0) + h(t,\beta_0) \\ &\ge M(t,\alpha_0) + M_x(t,\alpha_0)(\beta_0-\alpha_0) - \phi(t,\beta_0) \\ &\quad +H(t,\alpha_0) + H_x(t,\beta_0)(\beta_0-\alpha_0) - \psi(t,\beta_0) + h(t,\beta_0) \\ &\ge F(t,\alpha_0) + G(t,\alpha_0) + [M_x(t,\alpha_0) + H_x(t,\beta_0)](\beta_0-\alpha_0) \\ &\quad +\phi(t,\alpha_0) - \phi(t,\beta_0) + \psi(t,\alpha_0) - \psi(t,\beta_0) + h(t,\beta_0) \\ &\ge F(t,\alpha_0) + G(t,\alpha_0) + [M_x(t,\alpha_0) + H_x(t,\beta_0) - \phi_x(t,\beta_0) \\ &\quad -\psi_x(t,\alpha_0)](\beta_0-\alpha_0) + h(t,\beta_0) \\ &\equiv g(t,\beta_0;\alpha_0,\beta_0) \qquad \text{for } 0 < t \le 2\pi \text{ and } \beta_0(0) \ge \beta_0(2\pi). \end{aligned}$$

Here we have used the mean value theorem and the facts that $\psi_x(t,x)$ is nonincreasing in x and $\phi_x(t,x)$ is nondecreasing in x. Now using Theorem 2.1.3, there exists a solution $\alpha_1(t)$ of (2.1.34) such that $\alpha_0(t) \le \alpha_1(t) \le \beta_0(t)$ on J.

Similarly, using (2.1.33), (A_1), (A_2) and the mean value theorem, we get

$$\begin{aligned} \alpha_0' &\le f(t,\alpha_0) \\ &\equiv F(t,\alpha_0) + G(t,\alpha_0) + h(t,\alpha_0) \\ &\le F(t,\beta_0) + M_x(t,\alpha_0)(\alpha_0-\beta_0) + \phi(t,\beta_0) - \phi(t,\alpha_0) \\ &\quad +H(t,\beta_0) + H_x(t,\beta_0)(\alpha_0-\beta_0) + \psi(t,\beta_0) - \psi(t,\alpha_0) + h(t,\alpha_0) \\ &\le F(t,\beta_0) + [M_x(t,\alpha_0) + H_x(t,\beta_0) - \phi_x(t,\beta_0) \\ &\quad -\psi_x(t,\alpha_0)](\alpha_0-\beta_0) + G(t,\beta_0) + h(t,\alpha_0) \\ &= g^*(t,\alpha_0;\alpha_0,\beta_0) \qquad \text{for } 0 < t \le 2\pi, \text{ and } \alpha_0(0) \le \alpha_0(2\pi). \end{aligned}$$

Also $\beta_0'(t) \ge f(t,\beta_0) \equiv g^*(t,\beta_0;\alpha_0,\beta_0)$, for $0 < t \le 2\pi$.

Now using Theorem 2.1.3, there exists a solution $\beta_1(t)$ of (2.1.35) such that $\alpha_0(t) \le \beta_1(t) \le \beta_0(t)$ on J.

Next we prove $\alpha_1(t) \le \beta_1(t)$. In order to prove this we show $\alpha_1(t)$ is a lower solution of (2.1.1) and $\beta_1(t)$ is an upper solution of (2.1.1).

Using (2.1.33) and (2.1.34), we get

$$\begin{aligned}
\alpha_1' &= F(t,\alpha_0) + G(t,\alpha_0) + h(t,\alpha_1) + [M_x(t,\alpha_0) + H_x(t,\beta_0) - \psi_x(t,\alpha_0) \\
&\quad -\phi_x(t,\beta_0)](\alpha_1 - \alpha_0) \\
&\le F(t,\alpha_1) + \phi(t,\alpha_1) - \phi(t,\alpha_0) - \phi_x(t,\beta_0)(\alpha_1 - \alpha_0) \\
&\quad +G(t,\alpha_1) + \psi(t,\alpha_1) - \psi(t,\alpha_0) - \psi_x(t,\alpha_0)(\alpha_1 - \alpha_0) + h(t,\alpha_1) \\
&\le F(t,\alpha_1) + \phi_x(t,\alpha_1)(\alpha_1 - \alpha_0) - \phi_x(t,\beta_0)(\alpha_1 - \alpha_0) \\
&\quad +G(t,\alpha_1) + \psi_x(t,\alpha_0)(\alpha_1 - \alpha_0) - \psi_x(t,\alpha_0)(\alpha_1 - \alpha_0) + h(t,\alpha_1) \\
&\le F(t,\alpha_1) + G(t,\alpha_1) + h(t,\alpha_1) \equiv f(t,\alpha_1), \qquad \text{and } \alpha_1(0) = \alpha_1(2\pi).
\end{aligned}$$

Here we have used the mean value theorem and the facts that ϕ_x is nondecreasing in x and ψ_x is nonincreasing in x.

Similarly, we can prove that $\beta_1' \ge f(t,\beta_1)$, $\beta_1(0) = \beta_1(2\pi)$. Therefore from assumption (A_4) and the Comparison Theorem 2.1.1, it follows that $\alpha_1(t) \le \beta_1(t)$ on J. Consequently we have

$$\alpha_0(t) \le \alpha_1(t) \le \beta_1(t) \le \beta_0(t) \quad \text{on } J.$$

Assume now that for some $k > 1$,

$$\alpha_0(t) \le \alpha_k(t) \le \beta_k(t) \le \beta_0(t) \text{ on } J$$

$$\text{and } \alpha_k' \le f(t,\alpha_k),\ \beta_k' \ge f(t,\beta_k) \text{ for } 0 < t \le 2\pi$$

and $\alpha_k(0) = \alpha_k(2\pi)$, $\beta_k(0) = \beta_k(2\pi)$.

We will show that

$$\alpha_k(t) \le \alpha_{k+1}(t) \le \beta_{k+1}(t) \le \beta_k(t) \quad \text{on } J,$$

where α_{k+1}, β_{k+1} are the solutions of the periodic boundary value problems

$$\begin{aligned}
\alpha_{k+1}' &= F(t,\alpha_k) + G(t,\alpha_k) + [M_x(t,\alpha_k) + H_x(t,\beta_k) - \phi_x(t,\beta_k) \\
&\quad -\psi_x(t,\alpha_k)](\alpha_{k+1} - \alpha_k) + h(t,\alpha_{k+1}) \qquad (2.1.36) \\
&\equiv g(t,\alpha_{k+1};\alpha_k,\beta_k), \qquad \alpha_{k+1}(0) = \alpha_{k+1}(2\pi)
\end{aligned}$$

and

$$\begin{aligned}
\beta_{k+1}' &= F(t,\beta_k) + G(t,\beta_k) + [M_x(t,\alpha_k) + H_x(t,\beta_k) - \phi_x(t,\beta_k) \\
&\quad -\psi_x(t,\alpha_k)](\beta_{k+1} - \beta_k) + h(t,\beta_{k+1}) \qquad (2.1.37) \\
&\equiv g^*(t,\beta_{k+1};\alpha_k,\beta_k), \qquad \beta_{k+1}(0) = \beta_{k+1}(2\pi).
\end{aligned}$$

It easily follows by definition of $g(t,u;\alpha_k,\beta_k)$ that

$$\alpha_k' \le f(t,\alpha_k) \equiv g(t,\alpha_k;\alpha_k,\beta_k), \quad \alpha_k(0) = \alpha_k(2\pi)$$

which proves $\alpha_k(t)$ is a lower solution of PBVP (2.1.34).

Since $\alpha_k \le \beta_k$, using (2.1.33), the mean value theorem and the nature of ϕ_x, ψ_x, we get

$$\begin{aligned}
\beta_k' &\ge f(t,\beta_k) \\
&= F(t,\beta_k) + G(t,\beta_k) + h(t,\beta_k) \\
&\ge F(t,\alpha_k) + M_x(t,\alpha_k)(\beta_k - \alpha_k) + \phi(t,\alpha_k) - \phi(t,\beta_k) \\
&\quad +G(t,\alpha_k) + H_x(t,\beta_k)(\beta_k - \alpha_k) + \psi(t,\alpha_k) - \psi(t,\beta_k) \\
&\quad +h(t,\beta_k) \\
&= F(t,\alpha_k) + G(t,\alpha_k) + [M_x(t,\alpha_k) + H_x(t,\beta_k) - \phi_x(t,\beta_k) \\
&\quad -\psi(t,\alpha_k)](\beta_k - \alpha_k) + h(t,\beta_k) \\
&\equiv g(t,\beta_k;\alpha_k,\beta_k), \qquad\qquad \text{and } \beta_k(0) = \beta_k(2\pi).
\end{aligned}$$

This proves $\beta_k(t)$ is an upper solution of (2.1.34). It follows from Theorem 2.1.3, that $\alpha_{k+1}(t)$ the solution of (2.1.34) exists such that $\alpha_k(t) \le \alpha_{k+1}(t) \le \beta_k(t)$ on J.

Similar arguments yield

$$\alpha_k' \le g^*(t,\alpha_k;\alpha_k,\beta_k), \qquad \beta_k' \ge g^*(t,\beta_k;\alpha_k,\beta_k)$$

and $\alpha_k(0) = \alpha_k(2\pi)$, $\beta_k(0) = \beta_k(2\pi)$. Hence in view of Theorem 2.1.3, there exists a solution $\beta_{k+1}(t)$ such that $\alpha_k(t) \le \beta_{k+1}(t) \le \beta_k(t)$ on J. In order to prove that $\alpha_{k+1}(t) \le \beta_{k+1}(t)$ on J, it is enough to prove $\alpha_{k+1}(t)$ is a lower solution of (2.1.34) and $\beta_{k+1}(t)$ is an upper solution of (2.1.34). Using (2.1.34), the mean value theorem, the nonincreasing nature of $H_x(t,x)$ and $\psi_x(t,x)$, and the nondecreasing nature of $\phi_x(t,x)$, we get

$$\begin{aligned}
\alpha_{k+1}' &= F(t,\alpha_k) + G(t,\alpha_k) + [M_x(t,\alpha_k) + H_x(t,\beta_k) - \phi_x(t,\beta_k) \\
&\quad -\psi_x(t,\alpha_k)](\alpha_{k+1} - \alpha_k) + h(t,\alpha_{k+1}) \\
&\le F(t,\alpha_{k+1}) + \phi(t,\alpha_{k+1}) - \phi(t,\alpha_k) - \phi_x(t,\beta_k)(\alpha_{k+1} - \alpha_k) \\
&\quad +G(t,\alpha_{k+1}) + \psi(t,\alpha_{k+1}) - \psi(t,\alpha_k) - \psi_x(t,\alpha_k)(\alpha_{k+1} - \alpha_k) \\
&\quad +h(t,\alpha_{k+1}) \\
&\le F(t,\alpha_{k+1}) + G(t,\alpha_{k+1}) + h(t,\alpha_{k+1}) \\
&\quad +\phi_x(t,\alpha_{k+1})(\alpha_{k+1} - \alpha_k) - \phi_x(t,\beta_k)(\alpha_{k+1} - \alpha_k) \\
&\quad +\psi_x(t,\alpha_k)(\alpha_{k+1} - \alpha_k) - \psi_x(t,\alpha_k)(\alpha_{k+1} - \alpha_k) \\
&\le F(t,\alpha_{k+1}) + G(t,\alpha_{k+1}) + h(t,\alpha_{k+1}) \\
&\equiv f(t,\alpha_{k+1}), \qquad\qquad \text{and } \alpha_{k+1}(0) = \alpha_{k+1}(2\pi).
\end{aligned}$$

In a similar manner, we can prove that $\beta_{k+1}' \ge f(t,\beta_{k+1})$, $\beta_{k+1}(0) = \beta_{k+1}(2\pi)$. Therefore by Theorem 2.1.1, it follows that $\alpha_{k+1}(t) \le \beta_{k+1}(t)$ on J. This proves that

$$\alpha_0(t) \leq \alpha_k(t) \leq \alpha_{k+1}(t) \leq \beta_{k+1}(t) \leq \beta_k(t) \leq \beta_0(t) \text{ on } J.$$

Now using induction argument we can conclude that for all n

$$\alpha_0 \leq \alpha_1 \leq \cdots \leq \alpha_n \leq \beta_n \leq \cdots \leq \beta_1 \leq \beta_0 \text{ on } J.$$

Using standard argument one can show that the sequences $\{\alpha_n(t)\}$, $\{\beta_n(t)\}$ converge uniformly and monotonically to the unique solution $x(t)$ of (2.1.1).

Finally, to prove quadratic convergence set $p_{n+1}(t) = x(t) - \alpha_{n+1}(t) \geq 0$ and $q_{n+1}(t) = \beta_{n+1}(t) - x(t) \geq 0$. Note that $p_{n+1}(0) = p_{n+1}(2\pi)$ and $q_{n+1}(0) = q_{n+1}(2\pi)$.

Using (A_2), (2.1.33) and the mean value theorem and the nature of ϕ_x, M_x as nondecreasing in x and ψ_x, H_x as nonincreasing in x, we get

$$\begin{aligned}
p_n' &= x' - \alpha_n' \\
&= F(t,x) + G(t,x) + h(t,x) - [F(t,\alpha_{n-1}) + G(t,\alpha_{n-1}) + h(t,\alpha_n) \\
&\quad + \{M_x(t,\alpha_{n-1}) + H_x(t,\beta_{n-1}) - \phi_x(t,\beta_{n-1}) - \psi_x(t,\alpha_{n-1})\} \\
&\quad \times(\alpha_n - \alpha_{n-1})] \\
&= M(t,x) + H(t,x) + h(t,x) - [M(t,\alpha_{n-1}) + H(t,\alpha_{n-1}) + h(t,\alpha_n) \\
&\quad + \{M_x(t,\alpha_{n-1}) + H_x(t,\beta_{n-1}) - \phi_x(t,\beta_{n-1}) - \psi_x(t,\alpha_{n-1}) \\
&\quad \times(-p_n + p_{n-1})] + \phi(t,\alpha_{n-1}) - \phi(t,x) + \psi(t,\alpha_{n-1}) - \psi(t,x) \\
&\leq M_{xx}(t,\sigma)p_{n-1}^2 + [H_x(t,\alpha_n) - H_x(t,\beta_{n-1})]p_{n-1} \\
&\quad + [M_x(t,\alpha_{n-1}) + H_x(t,\beta_{n-1}) - \phi_x(t,\beta_{n-1}) - \psi_x(t,\alpha_{n-1}) + L]p_n \\
&\quad - \phi_x(t,\alpha_{n-1})p_{n-1} - \psi_x(t,x)p_{n-1} + \phi_x(t,\beta_{n-1})p_{n-1} \\
&\quad + \psi_x(t,\alpha_{n-1})p_{n-1} \\
&\leq [M_x(t,\alpha_{n-1}) + H_x(t,\beta_{n-1}) - \phi_x(t,\beta_{n-1}) - \psi_x(t,\alpha_{n-1}) + L]p_n \\
&\quad + M_{xx}(t,\sigma)p_{n-1}^2 - H_{xx}(t,\eta)p_{n-1}(q_{n-1} + p_{n-1}) \\
&\quad + \phi_{xx}(t,\delta_1)(q_{n-1} + p_{n-1})p_{n-1} - \psi_{xx}(t,\delta_2)p_{n-1}^2 \\
&\leq [M_x(t,\alpha) + H_x(t,\beta_{n-1}) - \phi_x(t,\beta_{n-1}) - \psi_x(t,\alpha_{n-1}) + L]p_n \\
&\quad + [M_{xx}(t,\sigma) - \frac{3}{2}H_{xx}(t,\eta) + \frac{3}{2}\phi_{xx} - \psi_{xx}(t,\delta_2)]p_{n-1}^2 \\
&\quad + \frac{1}{2}[-H_{xx}(t,\eta) + \phi_{xx}(t,\delta_1)]q_{n-1}^2.
\end{aligned}$$

By setting

$$\lambda(t) = [M_x(t,\alpha_n) + H_x(t,\beta_{n-1}) - \phi_x(t,\beta_{n-1}) - \psi_x(t,\alpha_{n-1}) + L]$$

and

$$\begin{aligned}\gamma^*(t) &= [M_{xx}(t,\sigma) - \tfrac{3}{2}H_{xx}(t,\eta) + \tfrac{3}{2}\phi_{xx} - \psi_{xx}(t,\delta_2)]p_{n-1}^2 \\ &+\tfrac{1}{2}[-H_{xx}(t,\eta) + \phi_{xx}(t,\delta_1)]q_{n-1}^2,\end{aligned}$$

we get

$$v'(t) \le \lambda(t)v(t) + \gamma^*(t), \qquad v(0) = v(2\pi).$$

This implies

$$v(t) \le e^{\int_0^t \lambda(s)ds}[v(0) + \int_0^t \exp(-\int_0^s \lambda(\sigma)d\sigma)\gamma^*(s)ds] \tag{2.1.38}$$

where

$$v(0) = v(2\pi) \le \frac{1}{\exp(-\int_0^{2\pi} \lambda(s)ds) - 1} \int_0^{2\pi} \exp(-\int_0^s \lambda(\sigma)d\sigma)\gamma^*(s)ds. \tag{2.1.39}$$

We note that (2.1.38) holds, since by assumption (A_2) and (A_3) we have $\int_0^{2\pi} \lambda(s)ds < 0$. Using (2.1.39) in (2.1.38) and simplifying

$$v(t) \le K \int_0^{2\pi} \exp(-\int_0^s \lambda(\sigma)d\sigma)\gamma^*(s)ds,$$

where

$$K = \frac{\exp(-\int_0^{2\pi} \lambda(s)ds)}{\exp(-\int_0^{2\pi} \lambda(s)ds) - 1}.$$

Consequently, we obtain

$$\max_J |p_n(t)| \le A^* \max_J |p_{n-1}(t)|^2 + B^* \max_J |q_{n-1}(t)|^2$$

where $|F_{xx}| < A$, $|G_{xx}| < B$, $|\phi_{xx}| < C$, $|\psi_{xx}| < D$, $\left|\exp(-\int_0^s \lambda(\sigma)d\sigma)\right| \le Q$ on Ω,

$$A^* = (A + \frac{3}{2}B + \frac{5}{2}C + \frac{5}{2}D)KQ$$

and

$$B^* = \frac{KQ}{2}(B + C + D).$$

Similarly, using an appropriate computation, we arrive at

$$\max_J |q_n(t)|^2 = A_1^* \max_J |p_{n-1}|^2 + B_1^* \max_J |q_{n-1}(t)|^2$$

where $B_1^* = KQ\ (A + \frac{5}{2}B + \frac{5}{2}C + \frac{5}{2}D)$, $A_1^* = KQ\frac{(C+D)}{2}$. This proves the quadratic convergence.

In the next theorem we assume that $h(t,x)$ satisfies a two sided Lipschitz condition. We develop iterates which are solutions of linear periodic boundary value problem. However in this case we do not have quadratic convergence, instead we have semi-quadratic convergence. The advantage of this theorem is that the monotone method comes as a special case of this, besides the computational aspect of the iterates.

Theorem 2.1.7 *Let the assumption (A_1), (A_2) of Theorem 2.1.1 hold. Further, let $h(t,x)$ satisfy the condition*

$$h(t,x) - h(t,y) \geq -L(x-y) \tag{2.1.40}$$

for $\alpha_0 \leq y \leq x \leq \beta_0$ and let (A_3) be replaced by a weaker assumption

(A_1^*) $\int_0^{2\pi}[M_x(t,\beta_0)+H_x(t,\alpha_0)-\psi_x(t,\beta_0)-\phi_x(t,\alpha_0)-L]dt < 0$, *where $M(t,x)$ and $H(t,x)$ be as in (A_3).*

Then there exist sequences $\{\alpha_n(t)\}$, $\{\beta_n(t)\}$ which converge uniformly and monotonically to the unique solution of (2.1.1). Further the convergence is semi-quadratic.

Proof In view of (A_2), we still have (2.1.33) of Theorem 2.1.6. Consider the linear PBVPs.

$$\begin{aligned} u' &= f(t,\alpha_0) + [M_x(t,\alpha_0) + H_x(t,\beta_0) - \phi_x(t,\beta_0) - \psi_x(t,\alpha_0) - L](u-\alpha_0) \\ &\equiv k(t,u;\alpha_0,\beta_0), \qquad u(0) = u(2\pi), \end{aligned} \tag{2.1.41}$$

$$\begin{aligned} v' &= f(t,\beta_0) + [M_x(t,\alpha_0) + H_x(t,\beta_0) - \phi_x(t,\beta_0) - \psi_x(t,\alpha_0) - L](v-\beta_0) \\ &\equiv k^*(t,v;\alpha_0,\beta_0), \qquad v(0) = v(2\pi). \end{aligned} \tag{2.1.42}$$

The assumption (A_1) and inequalities (2.1.33), (2.1.40) imply that

$$\begin{aligned} \alpha_0' &\leq f(t,\alpha_0) \equiv k(t,\alpha_0;\alpha_0,\beta_0), \qquad \alpha_0(0) \leq \alpha_0(2\pi), \\ \beta_0' &\geq f(t,\beta_0) = F(t,\beta_0) + G(t,\beta_0) + h(t,\beta_0) \\ &\geq M(t,\alpha_0) + M_x(t,\alpha_0)(\beta_0-\alpha_0) + H(t,\alpha_0) + H_x(t,\beta_0)(\beta_0-\alpha_0) \\ &\quad -\psi(t,\alpha_0) - \phi(t,\beta_0) + h(t,\alpha_0) - L(\beta_0-\alpha_0) \\ &= F(t,\alpha_0) + G(t,\alpha_0) + h(t,\alpha_0) \\ &\quad +(M_x(t,\alpha_0) + H_x(t,\beta_0) - L)(\beta_0-\alpha_0) - \phi_x(t,\beta_0)(\beta_0-\alpha_0) \\ &\quad -\psi_x(t,\alpha_0)(\beta_0-\alpha_0) \\ &= f(t,\alpha_0) + [M_x(t,\alpha_0) + H_x(t,\beta_0) - \psi_x(t,\alpha_0) - \phi_x(t,\beta_0) - L](\beta_0-\alpha_0) \\ &= k(t,\beta_0;\alpha_0,\beta_0), \qquad \beta_0(0) \geq \beta_0(2\pi). \end{aligned}$$

Here we have used the mean value theorem, and the fact that ϕ_x and ψ_x are nondecreasing and nonincreasing in x respectively. Hence by Theorem 2.1.3, there exists a solution $\alpha_1(t)$ of (2.1.41) such that $\alpha_0(t) \le \alpha_1(t) \le \beta_0(t)$ on J. Similarly, using (2.1.33), (A_1), (A_2) and (2.1.40), the mean value theorem, we obtain

$$\begin{aligned}
\alpha_0' &\le f(t,\alpha_0) = F(t,\alpha_0) + G(t,\alpha_0) + h(t,\alpha_0) \\
&\le M(t,\beta_0) + M_x(t,\beta_0)(\alpha_0 - \beta_0) - \phi(t,\alpha_0) + H(t,\beta_0) \\
&\quad -H_x(t,\beta_0)(\beta_0 - \alpha_0) - \psi(t,\alpha_0) + h(t,\beta_0) + L(\beta_0 - \alpha_0) \\
&\le F(t,\beta_0) + G(t,\beta_0) + h(t,\beta_0) \\
&\quad +[M_x(t,\alpha_0) + H_x(t,\beta_0) - L](\alpha_0 - \beta_0) + \phi_x(t,\beta_0)(\beta_0 - \alpha_0) \\
&\quad +\psi_x(t,\alpha_0)(\beta_0 - \alpha_0) \\
&= k^*(t,\alpha_0;\alpha_0,\beta_0), \qquad \text{for } 0 < t \le 2\pi,\ \alpha_0(0) \le \alpha_0(2\pi).
\end{aligned}$$

and

$$\beta_0' \ge f(t,\beta_0) \equiv k^*(t,\beta_0;\alpha_0,\beta_0), \quad \text{for } 0 < t \le 2\pi \text{ and } \beta_0(0) \ge \beta_0(2\pi).$$

It easily follows from Theorem 2.1.3 that there exists a solution $\beta_1(t)$ of (2.1.42) such that $\alpha_0(t) \le \beta_1(t) \le \beta_0(t)$ on J. In order to prove $\alpha_1(t) \le \beta_1(t)$ on J, we show that $\alpha_1(t)$ is a lower solution of (2.1.33) and $\beta_1(t)$ is an upper solution of (2.1.1).

We have,

$$\begin{aligned}
\alpha_1' &= f(t,\alpha_0) + [M_x(t,\alpha_0) + H_x(t,\beta_0) - \psi_x(t,\alpha_0) - \phi_x(t,\beta_0) - L](\alpha_1 - \alpha_0) \\
&\le M(t,\alpha_0) + M_x(t,\alpha_0)(\alpha_1 - \alpha_0) + H(t,\alpha_0) + H_x(t,\beta_0)(\alpha_1 - \alpha_0) \\
&\quad +h(t,\alpha_1) - \phi(t,\alpha_0) - \psi(t,\alpha_0) + [-\psi_x(t,\alpha_0) - \phi_x(t,\beta_0)](\alpha_1 - \alpha_0) \\
&\le F(t,\alpha_1) + \phi(t,\alpha_1) - \phi(t,\alpha_0) + G(t,\alpha_1) + \psi(t,\alpha_1) - \psi(t,\alpha_0) \\
&\quad +[-\phi_x(t,\beta_0) - \psi_x(t,\alpha_0)](\alpha_1 - \alpha_0) + h(t,\alpha_1) \\
&\le F(t,\alpha_1) + G(t,\alpha_1) + h(t,\alpha_1) + [\phi_x(t,\alpha_1) - \phi_x(t,\beta_0)](\alpha_1 - \alpha_0) \\
&\quad +[\psi_x(t,\alpha_0) - \psi_x(t,\alpha_0)](\alpha_1 - \alpha_0) \\
&\le f(t,\alpha_1), \qquad \text{on } 0 < t \le 2\pi \text{ and } \alpha_1(0) = \alpha_1(2\pi),
\end{aligned}$$

using (2.1.33), the mean value theorem and the nature of ϕ_x, ψ_x.

Similarly one can prove $\beta_1' \ge f(t,\beta_1)$ and we have $\beta_1(0) = \beta_1(2\pi)$. It follows from Theorem 2.1.1 that $\alpha_1(t) \le \beta_1(t)$ on J, and as a result we have

$$\alpha_0(t) \le \alpha_1(t) \le \beta_1(t) \le \beta_0(t) \text{ on } J.$$

Assume now that for some $k > 1$,

$$\alpha_k' \le f(t,\alpha_k),\ \beta_k' \ge f(t,\beta_k) \qquad \text{for } 0 < t \le 2\pi$$

and $\alpha_k(t) \le \beta_k(t)$ on J.

We shall prove that

$$\alpha_k(t) \le \alpha_{k+1}(t) \le \beta_{k+1}(t) \le \beta_k(t) \text{ on } J,$$

where α_{k+1} and β_{k+1} are the solutions of the linear PBVPs

$$\begin{aligned} u' &= f(t,\alpha_k) + [M_x(t,\alpha_k) + H_x(t,\beta_k) - \phi_x(t,\beta_k) \\ &\quad -\psi_x(t,\alpha_k) - L](u-\alpha_k) \\ &\equiv k(t,u;\alpha_k,\beta_k), \qquad u(0) = u(2\pi) \end{aligned} \tag{2.1.43}$$

$$\begin{aligned} v' &= f(t,\beta_k) + [M_x(t,\alpha_k) + H_x(t,\beta_k) - \phi_x(t,\beta_k) \\ &\quad -\psi_x(t,\alpha_k) - L](v-\beta_k) \\ &\equiv k^*(t,v;\alpha_k,\beta_k), \qquad v(0) = v(2\pi). \end{aligned} \tag{2.1.44}$$

Now using (2.1.33), the mean value theorem and the nature of ϕ_x and ψ_x and the assumption (2.1.40), we get for $0 < t \le 2\pi$

$$\alpha_k' \le f(t,\alpha_k) \equiv k(t,\alpha_k;\alpha_k,\beta_k),$$

and

$$\begin{aligned} \beta_k' &\ge f(t,\beta_k) = F(t,\beta_k) + G(t,\beta_k) + h(t,\beta_k) \\ &\ge M(t,\alpha_k) + M_x(t,\alpha_k)(\beta_k - \alpha_k) - \phi(t,\beta_k) \\ &\quad +H(t,\alpha_k) + H_x(t,\beta_k)(\beta_k - \alpha_k) - \psi(t,\beta_k) \\ &\quad +h(t,\alpha_k) - L(\beta_k - \alpha_k) \\ &\ge F(t,\alpha_k) + G(t,\alpha_k) + h(t,\alpha_k) \\ &\quad +[M_x(t,\alpha_k) + H_x(t,\beta_k) - \phi_x(t,\beta_k) - \psi_x(t,\alpha_k) - L](\beta_k - \alpha_k) \\ &= k(t,\beta_k;\alpha_k,\beta_k). \end{aligned}$$

Thus by Theorem 2.1.3, there exists a solution $\alpha_{k+1}(t)$ of (2.1.43) such that $\alpha_k(t) \le \alpha_{k+1}(t) \le \beta_k(t)$ on J. Using a similar argument, we can show that $\beta_{k+1}(t)$ the solution of (2.1.44) exists such that $\alpha_k(t) \le \beta_{k+1}(t) \le \beta_k(t)$ on J.

In order to prove $\alpha_{k+1}(t) \le \beta_{k+1}(t)$ on J, we prove that $\alpha_{k+1}(t)$ is a lower solution of (2.1.33). Using (2.1.33), (2.1.40), (2.1.43), the fact that $\alpha_k \le \alpha_{k+1} \le \beta_k$, the mean value theorem, and the nature of ϕ_k, ψ_k, we get

$$\begin{aligned} \alpha_{k+1}' &= f(t,\alpha_k) + [M_x(t,\alpha_k) + H_x(t,\beta_k) - \phi_x(t,\beta_k) \\ &\quad -\psi_x(t,\alpha_k) - L](\alpha_{k+1} - \alpha_k) \\ &= F(t,\alpha_k) + M_x(t,\alpha_k)(\alpha_{k+1} - \alpha_k) - \phi_x(t,\beta_k)(\alpha_{k+1} - \alpha_k) \\ &\quad +G(t,\alpha_k) + H_x(t,\beta_k)(\alpha_{k+1} - \alpha_k) - \psi_x(t,\alpha_k)(\alpha_{k+1} - \alpha_k) \\ &\quad +h(t,\alpha_k) - L(\alpha_{k+1} - \alpha_k) \\ &\le F(t,\alpha_{k+1}) + \phi(t,\alpha_{k+1}) - \phi(t,\alpha_k) - \phi_x(t,\beta_k)(\alpha_{k+1} - \alpha_k) \\ &\quad +G(t,\alpha_{k+1}) + \psi(t,\alpha_{k+1}) - \psi(t,\alpha_k) - \psi_x(t,\alpha_k)(\alpha_{k+1} - \alpha_k) \\ &\quad +h(t,\alpha_{k+1}) \\ &\le f(t,\alpha_{k+1}) + [\phi_x(t,\alpha_{k+1}) - \phi_x(t,\beta_k)](\alpha_{k+1} - \alpha_k) \\ &\le f(t,\alpha_{k+1}) \qquad \text{on } 0 < t \le 2\pi \text{ and } \alpha_{k+1}(0) = \alpha_{k+1}(2\pi). \end{aligned}$$

Similarly one can prove that

$$\beta'_{k+1} \geq f(t, \beta_{k+1}) \qquad \text{on } 0 < t \leq 2\pi,\ \beta_{k+1}(0) = \beta_{k+1}(2\pi).$$

Using Theorem 2.1.1 it follows that $\alpha_{k+1}(t) \leq \beta_{k+1}(t)$ on J, which shows that

$$\alpha_k(t) \leq \alpha_{k+1}(t) \leq \beta_{k+1}(t) \leq \beta_k(t) \text{ on } J.$$

Hence by induction we have for all n

$$\alpha_0 \leq \alpha_1 \leq \cdots \leq \alpha_n \leq \beta_n \leq \cdots \leq \beta_1 \leq \beta_0 \text{ on } J.$$

It then follows by the standard argument that the sequences $\{\alpha_n(t)\}$, $\{\beta_n(t)\}$ converge uniformly and monotonically to the unique solution $u(t)$ of (2.1.1).

Finally to prove semi-quadratic convergence, we consider

$$p_n(t) = x(t) - \alpha_n(t) \geq 0 \quad \text{and} \quad q_n(t) = \beta_{n+1}(t) - x(t) \geq 0.$$

It is easy to see that $p_n(0) = p_n(2\pi)$, $q_n(0) = q_n(2\pi)$.

Using (2.1.33), (2.1.40), the mean value theorem and the assumptions on ϕ_x, ψ_x etc., we get

$$\begin{aligned}
p'_n &= x' - \alpha'_n \\
&= f(t,x) - [f(t,\alpha_{n-1}) + [M_x(t,\alpha_{n-1}) + H_x(t,\beta_{n-1}) - \phi_x(t,\beta_{n-1}) \\
&\quad -\psi_x(t,\alpha_{n-1}) - L](\alpha_n - \alpha_{n-1})]] \\
&= M(t,x) - [M(t,\alpha_{n-1}) + M_x(t,\alpha_{n-1})p_{n-1} - \phi_x(t,\beta_{n-1})p_{n-1}] \\
&\quad +\phi(t,\alpha_{n-1}) - \phi(t,x) + H(t,x) \\
&\quad -[H(t,\alpha_{n-1}) + H_x(t,\beta_{n-1})p_{n-1} - \psi_x(t,\alpha_{n-1})] \\
&\quad +\psi(t,\alpha_{n-1}) - \psi(t,x) + h(t,x) - h(t,\alpha_{n-1}) + Lp_{n-1} + [M_x(t,\alpha_{n-1}) \\
&\quad +H_x(t,\beta_{n-1}) - \phi_x(t,\beta_{n-1}) - \psi_x(t,\alpha_{n-1}) - L]p_n \\
&\leq M_{xx}(t,\sigma)p_{n-1}^2 + [\phi_x(t,\beta_{n-1}) - \phi_x(t,x)]p_{n-1} + H_x(t,\alpha_{n-1})p_{n-1} \\
&\quad -H_x(t,\beta_{n-1})p_{n-1} + [\psi_x(t,\alpha_{n-1}) - \psi_x(t,x)]p_{n-1} + 2L(x - \alpha_{n-1}) \\
&\quad +[M_x(t,\alpha_{n-1}) + H_x(t,\beta_{n-1}) - \phi_x(t,\beta_{n-1}) - \psi_x(t,\alpha_{n-1}) - L]p_n \\
&\leq M_{xx}(t,\sigma)p_{n-1}^2 + \phi_{xx}(t,\delta_1)q_{n-1}p_{n-1} - H_{xx}(t,\eta)p_{n-1}(q_{n-1} + p_{n-1}) \\
&\quad +2Lp_{n-1} - \psi_{xx}(t,\delta_2)p_{n-1}^2 + 2Lp_{n-1} + \gamma(t)p_n
\end{aligned}$$

where

$$\gamma(t) = M_x(t,\alpha_{n-1}) + H_x(t,\beta_{n-1}) - \phi_x(t,\beta_{n-1}) - \psi_x(t,\alpha_{n-1}) - L.$$

This implies

$$p'_n \leq \gamma(t)p_n(t) + \sigma(t)$$

where $\sigma(t) = M_{xx}p_{n-1}^2 + \phi_{xx}\frac{1}{2}[p_{n-1}^2 + q_{n-1}^2] - H_{xx}[\frac{3}{2}p_{n-1}^2 + \frac{1}{2}q_{n-1}^2] - \psi_{xx}p_{n-1}^2$, and $p_n(0) = p_n(2\pi)$.

This implies

$$p_n(t) \leq \exp \int_0^t \gamma(s)ds[p_n(0) + \int_0^t \exp(-\int_0^s \gamma(\eta)d\eta)\sigma(s)ds]$$

where

$$p_n(0) = p_n(2\pi) \leq \frac{1}{\exp(-\int_0^{2\pi} \gamma(s)ds) - 1}[\int_0^{2\pi} \exp(-\int_0^s \gamma(\sigma)d\sigma)\sigma(s)ds].$$

This is true by assumption (A_3^*).

Hence we obtain

$$\max_J |p_n(t)| \leq A^* \max_J |p_{n-1}(t)|^2 + B^* \max_J |q_{n-1}(t)|^2 + C^* \max_J p_{n-1}(t)$$

where on Ω, $|F_{xx}| \leq A$, $|G_{xx}| \leq B$, $|\phi_{xx}| \leq C$, $|\psi_{xx}| < D$, $\left|\exp(-\int_0^s \gamma(\eta)d\eta)\right| \leq Q$, $K = \frac{\exp(-\int_0^{2\pi} \gamma(s)ds)}{\exp(-\int_0^{2\pi} \gamma(s)ds)-1}$, $A^* = KQ(A+\frac{3}{2}C+\frac{3}{2}B+\frac{5}{2}D)$, $B^* = KQ(\frac{B+C+D}{2})$, $C^* = 2KQL$.

Similarly, we can arrive at

$$\max_J |q_n(t)|^2 = A_1^* \max_J |p_{n-1}(t)|^2 + B_1^* \max_J |q_{n-1}(t)|^2 + C^* \max_J |q_{n-1}|$$

where $C^* = 2QKL$ and A_1^*, B_1^* are suitable constants as before.

This proves the semi-quadratic convergence.

2.2 ANTI-PERIODIC BOUNDARY VALUE PROBLEMS

In this section, we shall discuss anti-periodic boundary value problem and extend the method of generalized quasilinearization. The basic comparison result similar to the periodic boundary value problem is no longer valid for anti-periodic boundary value problems.

For instance, if we assume

$$m'(t) \leq -Mm(t),\ M > 0,\ t \in J, \quad m(0) \leq -m(T),$$

then $m(t) \leq 0$ on $J = [0, T]$ need not be true. A simple example is $m(t) = -2t + 1$ with $M = 1$ and $T = 1$. The natural definition of upper and lower solutions of the anti-periodic boundary value problem, namely,

$$\alpha' \leq f(t, \alpha),\ t \in J,\ \alpha(0) \leq -\alpha(T) \tag{2.2.1}$$

$$\beta' \geq f(t, \beta),\ t \in J,\ \beta(0) \geq -\beta(T) \tag{2.2.2}$$

is not therefore sufficient to obtain $\alpha \leq \beta$ on $[0, T]$. For example, consider

$$u' = \frac{1}{2}u^2 + \frac{1}{2}\sin(\frac{\pi t}{2}), \quad u(0) = -u(2),\, t \in [0, 2].$$

It is easy to check that $\alpha = -t + 1$ and $\beta = t - 1$ satisfy inequalities (2.2.1) and (2.2.2) but neither $\alpha \leq \beta$ nor $\beta \leq \alpha$ on $[0, 2]$ is true. Hence there is a need for a new definition and a new approach to handle anti-periodic boundary value problem (APBVP for short).

Let us first consider first order APBVP,

$$u' = f(t, u), \;\; u(0) = -u(T),\, t \in J = [0, T],\, T > 0. \tag{2.2.3}$$

Definition 2.2.1 *The functions α_0, $\beta_0 \in C^1[J, R]$ are called weakly coupled lower and upper solutions of (2.2.3) if*

$$\begin{aligned} \alpha_0' &\leq f(t, \alpha_0), \quad t \in J, \;\; \alpha_0(0) \leq -\beta_0(T) \\ \beta_0' &\geq f(t, \beta_0), \quad t \in J, \;\; \beta_0(0) \geq -\alpha_0(T). \end{aligned}$$

To obtain merely the existence of unique solution of (2.2.3) using monotone iterative technique, we shall construct the sequences $\{\alpha_n\}$, $\{\beta_n\}$ by choosing α_n and β_n as the solutions of some linear initial value problems. However, in order to take advantage of the method of quasilinearization, we consider α_n and β_n as the solutions of weakly coupled linear systems. Then, we prove that the so-constructed monotone sequences converge quadratically to the unique solution of (2.2.3) under the assumption of convexity on $f(t, u)$ in u. The method of generalized quasilinearization is also employed to relax the convexity assumption on f and to obtain quadratic convergence as well.

We need the following lemmas before we proceed further.

Lemma 2.2.1 *Let $m \in C^1[J, R]$.*

(*i*) *Suppose that $m'(t) \leq \phi(t)m(t)$, $t \in J$, $m(0) \leq m(T)$ and $\int_0^T \phi(t)dt < 0$. Then $m(t) \leq 0$ on J. If the above inequalities are all reversed, the conclusion is still true.*

(*ii*) *Suppose $m'(t) = \phi(t)m(t)$, $t \in J_0 = (0, T)$ and $m(0) = m(T)$. If $\int_0^T \phi(t)dt \neq 0$, then $m(t) = 0$ on J.*

Proof (i) From the differential inequality, we have

$$m(t) \le m(0)\exp\left(\int_0^t \phi(s)ds\right).$$

Letting $t = T$, using $m(0) \le m(T)$, we have,

$$m(0) \le m(T) \le m(0)\exp\left(\int_0^T \phi(s)ds\right)$$

which implies $m(0) \le 0$ since $\int_0^T \phi(s)ds < 0$. Therefore, $m(t) \le 0$ on J. If the inequalities are reversed, we obtain

$$m(t) \le m(T)\exp\left(-\int_t^T \phi(s)ds\right),$$

which implies

$$m(T) \le m(0) \le m(T)\exp\left(-\int_0^T \phi(s)ds\right).$$

Thus, $m(T) \le 0$ since $\int_0^T \phi(s)ds > 0$. We arrive at $m(t) \le 0$ on J.

(ii) Since $m(t) = m(0)\exp\left(\int_0^t \phi(s)ds\right)$, we have on setting $t = T$,

$$m(0) = m(T) = m(0)\exp\left(\int_0^T \phi(s)ds\right)$$

which implies $m(0) = 0$ because $\int_0^T \phi(s)ds \neq 0$. Therefore, $m(t) = 0$ on J.

Lemma 2.2.2 *Assume that u, $v \in C^1[J,R]$ and*

$$u' = \phi_1(t)u + \psi_1(t),\ t \in J,\ u(0) = -v(T), \tag{2.2.4}$$

$$v' = \phi_2(t)v + \psi_2(t),\ t \in J,\ v(0) = -u(T), \tag{2.2.5}$$

where ϕ_i, $\psi_i \in C[J,R]$, $i = 1,2$ and $\sum_{i=1}^{2}\int_0^T \phi_i(s)ds \neq 0$. Then there exists a unique solution (u,v) to linear system (2.2.4) – (2.2.5).

Proof From (2.2.4)-(2.2.5), we get

$$u(t) = \exp\left(\int_0^t \phi_1(s)ds\right)$$

$$\times \left[u(0) + \int_0^t \exp\left(-\int_0^s \phi_1(\xi)d\xi\right) \psi_1(s)ds\right] \tag{2.2.6}$$

$$v(t) = \exp\left(\int_0^t \phi_2(s)ds\right)$$

$$\times \left[v(0) + \int_0^t \exp\left(-\int_0^s \phi_2(\xi)d\xi\right) \psi_2(s)ds\right]. \tag{2.2.7}$$

Let $t = T$. Then

$$u(T) = \exp\left(\int_0^T \phi_1(s)ds\right) \left[u(0) + \int_0^T \exp\left(-\int_0^s \phi_1(\xi)d\xi\right) \psi_1(s)ds\right]$$

$$v(T) = \exp\left(\int_0^T \phi_2(s)ds\right) \left[v(0) + \int_0^T \exp\left(-\int_0^s \phi_2(\xi)d\xi\right) \psi_2(s)ds\right].$$

Thus, we have

$$u(T) + \exp\left(\int_0^T \phi_1(s)ds\right) v(T) = \int_0^T \exp\left(-\int_0^T \phi_1(\xi)d\xi\right) \psi_1(s)ds \tag{2.2.8}$$

$$v(T) + \exp\left(\int_0^T \phi_2(s)ds\right) u(T) = \int_0^T \exp\left(-\int_0^T \phi_2(\xi)d\xi\right) \psi_2(s)ds. \tag{2.2.9}$$

Since $1 - \exp\left(\sum_{i=1}^2 \int_0^T \phi_i(s)ds\right) \neq 0$, there exists unique solution $(u(T), v(T))$ to system (2.2.8)-(2.2.9). Therefore, $(u(t), v(t))$, given in (2.2.6) and (2.2.7), is the unique solution of (2.2.4)-(2.2.5).

Lemma 2.2.3 *Suppose that m_1, $m_2 \in C^1[J, R]$ and ϕ_1, $\phi_2 \in C[J, R]$ are such that*

(i) $m_1'(t) \leq \phi_1(t)m_1(t)$, $t \in J$, $m_1(0) \leq m_2(T)$
$m_2'(t) \leq \phi_2(t)m_2(t)$, $t \in J$, $m_2(0) \leq m_1(T)$;

(ii) $\sum_{i=1}^2 \int_0^T \phi_i(s)ds < 0$.

Then $m_1(t) \leq 0$ and $m_2(t) \leq 0$ on J. If the inequalities in (i) and (ii) are reversed, the conclusion is still true.

Proof From (i), we get

$$m_1(t) \leq m_1(0) \exp\left(\int_0^T \phi_1(s)ds\right) \text{ and } m_2(t) \leq m_2(0) \exp\left(\int_0^T \phi_2(s)ds\right).$$

Let $t = T$. Then,

$$m_1(T) \le m_1(0) \exp\left(\int_0^T \phi_1(s)ds\right) \text{ and } m_2(T) \le m_2(0) \exp\left(\int_0^T \phi_2(s)ds\right).$$

Therefore,

$$\begin{aligned} m_1(0) &\le m_2(T) \\ &\le m_2(0) \exp\left(\int_0^T \phi_2(s)ds\right) \\ &\le m_1(T) \exp\left(\int_0^T \phi_2(s)ds\right) \\ &\le m_1(0) \exp\left(\sum_{i=1}^{2} \int_0^T \phi_i(s)ds\right). \end{aligned}$$

Since $\sum_{i=1}^{2} \int_0^T \phi_i(s)ds \neq 0$, we get $m_1(0) \le 0$. Similarly, $m_2(0) \le 0$. Thus, we have $m_1(t) \le 0$ and $m_2(t) \le 0$ on J. If the inequalities in (i) and (ii) are all reversed, we then obtain

$$m_1(t) \le m_1(T) \exp\left(-\int_0^T \phi_1(s)ds\right) \text{ and } m_2(t) \le m_2(T) \exp\left(-\int_0^T \phi_2(s)ds\right).$$

The rest of the proof follows argument similar to Lemma 1.7.1 and we omit the details.

Lemma 2.2.4 *Let $m_1, m_2 \in C^1[J, R]$ be such that*

$$\begin{aligned} m_1'(t) &\le \phi_1(t)(m_1(t) + \gamma_1) \text{ on } J,\ m_1(0) \ge m_2(T), \\ m_2'(t) &\le \phi_2(t)(m_2(t) + \gamma_2) \text{ on } J,\ m_2(0) \ge m_1(T), \end{aligned}$$

where

$$\begin{aligned} \gamma_1 &= \frac{m_2(0) - m_1(T)}{1 - \exp\left(\int_0^T \phi_1(t)dt\right)}, \\ \gamma_2 &= \frac{m_1(0) - m_2(T)}{1 - \exp\left(\int_0^T \phi_2(t)dt\right)}. \end{aligned}$$

If $\int_0^t \phi_i(s)ds < 0$, $t \in J$, $i = 1, 2$, then $m_1(t) \le 0$ and $m_2(t) \le 0$ on J.

Proof From a straight forward calculation, we get

$$m_1(t) \le m_1(0)\exp\left(\int_0^t \phi_1(s)ds\right) - \gamma_1\left(1-\exp\left(\int_0^t \phi_1(s)ds\right)\right)$$

and

$$m_2(t) \le m_2(0)\exp\left(\int_0^t \phi_2(s)ds\right) - \gamma_2\left(1-\exp\left(\int_0^t \phi_2(s)ds\right)\right).$$

Let $t = T$. Then,

$$m_1(T) \le m_1(0)\exp\left(\int_0^T \phi_1(s)ds\right) - [m_2(0) - m_1(T)]$$

and

$$m_2(T) \le m_2(0)\exp\left(\int_0^T \phi_2(s)ds\right) - [m_2(0) - m_2(T)].$$

Therefore,

$$m_2(0) \le m_1(0)\exp\left(\int_0^T \phi_1(s)ds\right) \text{ and } m_1(0) \le m_2(0)\exp\left(\int_0^T \phi_2(s)ds\right),$$

which implies $m_1(0) \le 0$ and $m_2(0) \le 0$. Since γ_1, $\gamma_2 \ge 0$, we get $m_1(t) \le 0$ and $m_2(t) \le 0$ on J.

Lemma 2.2.5 *Let m_1, $m_2 \in C^1[J, R]$ be such that*

$$\begin{aligned} m_1'(t) &\ge \phi_1(t)\left(m_1(t) + \gamma_1\right) \text{ on } J,\ m_1(0) \le m_2(T), \\ m_2'(t) &\ge \phi_2(t)\left(m_2(t) + \gamma_2\right) \text{ on } J,\ m_2(0) \le m_1(T), \end{aligned}$$

where

$$\begin{aligned} \gamma_1 &= \frac{m_2(T) - m_1(0)}{1 - \exp\left(-\int_0^T \phi_1(t)dt\right)}, \\ \gamma_2 &= \frac{m_1(T) - m_2(0)}{1 - \exp\left(-\int_0^T \phi_2(t)dt\right)}. \end{aligned}$$

If $\int_0^t \phi_i(s)ds < 0$, $t \in J$, $i = 1, 2$, then $m_1(t) \le 0$ and $m_2(t) \le 0$ on J.

Proof The proof is similar to that of Lemma 2.2.4.

It is well known that if one-sided Lipschitz condition

$$f(t,u) - f(t,v) \leq N(u-v),\ N > 0, \tag{2.2.10}$$

is assumed and if v, $w \in C^1[J,R]$ are lower and upper solutions of the initial value problem (IVP for short) $u' = f(t,u)$, $u(0) = u_0$, respectively, that is, $v' \leq f(t,v)$ and $w' \geq f(t,w)$ on J. Then $v(0) \leq w(0)$ implies $v \leq w$ on J. Therefore, (2.2.10) assures the uniqueness of the solution of IVP to the right. A careful study of the proof of this result shows that the following is also true.

Lemma 2.2.6 *Let v, $w \in C^1[J,R]$ be such that $v' \geq f(t,v)$, $w' \leq f(t,w)$ on J. If*

$$f(t,u) - f(t,v) \geq -N(u-v),\ N > 0, \tag{2.2.11}$$

then $v(T) \leq w(T)$ implies $v(t) \leq w(t)$ on J.

It is easy to notice that condition (2.2.11) assures the uniqueness of the solution to the left of T,for the initial value problem $u' = f(t,u)$, $u(T) = u_T$. The following lemma gives a uniqueness result when either (2.2.10) or (2.2.11) holds.

Lemma 2.2.7 *Assume that $f \in C[J \times R, R]$. If either* (2.2.10) *or* (2.2.11) *holds, then* (2.2.3) *possesses at most one solution on J.*

Proof Suppose that (2.2.3) has two solutions u_1 and u_2 and that (2.2.10) holds. Let $A = \{t_k : u_1(t_k) = u_2(t_k),\ t_k \in J\}$. A is closed. We will show that $A = \emptyset$, assuming $A \neq J$, since, otherwise, u_1 and u_2 will coincide. Let $t_0 \in A$. If $t_0 = 0$ or $t_0 = T$, then $u_1(0) = u_2(0)$. By known result for initial value problem, $A = J$. If $0 < t_0 < T$, then $u_1(t_0) = u_2(t_0)$, which implies $[t_0, T] \subset A$. Thus, $u_1(T) = u_2(T)$ and it follows $u_1(0) = u_2(0)$, which implies $A = J$. This is a contradiction again. If $A = \emptyset$, we may assume $u_1 < u_2$ on J. Let $m(t) = u_2(t) - u_1(t)$. Then $m'(t) \leq Lm(t)$, which implies $m(T) \leq m(0)e^{TL}$. Since $m(0) = -m(T)$, we get $m(T) \leq 0$ and this is also a contradiction. Same argument holds if we assume $u_2 < u_1$. Therefore, $A = J$. The case related to (2.2.11) can be proved similarly by using Lemma 2.2.6. Thus, we omit the details.

Let α_0, $\beta_0 \in C^1[J,R]$. We list the following conditions on α_0 and β_0 for later use.

(H_1) $\alpha_0' \leq f(t,\alpha_0)$, $t \in J$, $\alpha_0(0) \leq -\beta_0(T)$,
$\beta_0' \geq f(t,\beta_0)$, $t \in J$, $\beta_0(0) \geq -\alpha_0(T)$.

(H_2) $\alpha_0' \leq f(t, \alpha_0) - N\gamma_1$, $t \in J$, $\alpha_0(0) \geq -\beta_0(T)$,
$\beta_0' \geq f(t, \beta_0) + N\gamma_2$, $t \in J$, $\beta_0(0) \leq -\alpha_0(T)$,
where $N > 0$ is the Lipschitz constant for $f(t, u)$ in u and
$\gamma_1 = [\alpha_0(0) + \beta_0(T)] / \left(1 - e^{-NT}\right)$,
$\gamma_2 = [-\beta_0(0) - \alpha_0(T)] / (1 - e^{-NT})$.

(H_3) $f \in C^{0,1}[J \times R, R]$ and
$\alpha_0' \leq f(t, \alpha_0) + f_u(t, \alpha_0)\gamma_1$, $t \in J$, $\alpha_0(0) \geq -\beta_0(T)$,
$\beta_0' \geq f(t, \beta_0) - f_u(t, \alpha_0)\gamma_2$, $t \in J$, $\beta_0(0) \leq -\alpha_0(T)$,
where $\gamma_1 = [-\beta_0(0) - \alpha_0(T)] / \left(1 - \exp\left(\int_0^T f_u(t, \alpha_0)dt\right)\right)$ and
$\gamma_2 = [\alpha_0(0) + \beta_0(T)] / \left(1 - \exp\left(\int_0^T f_u(t, \alpha_0)dt\right)\right)$.

(H_4) $f \in C^{0,1}[[0, T] \times R, R]$ and
$\alpha_0' \leq f(t, \alpha_0) - f_u(t, \alpha_0)\gamma_1$, $t \in J$, $\alpha_0(0) \geq -\beta_0(T)$,
$\beta_0' \geq f(t, \beta_0) + f_u(t, \alpha_0)\gamma_2$, $t \in J$, $\beta_0(0) \leq -\alpha_0(T)$,
where $\gamma_1 = [\alpha_0(0) + \beta_0(T)] / \left(1 - \exp\left(-\int_0^T f_u(t, \alpha_0)dt\right)\right)$ and
$\gamma_2 = [-\beta_0(0) - \alpha_0(T)] / \left(1 - \exp\left(-\int_0^T f_u(t, \alpha_0)dt\right)\right)$.

Notice that (H_4) reduces to (H_2) if $f_u(t, \alpha_0) \equiv N$.
Let α_0, $\beta_0 \in C^1[J, R]$. We define

$$\Omega = \{(t, u) : \alpha_0 \leq u \leq \beta_0,\ t \in J\},$$

when $\alpha_0 \leq \beta_0$ on J, and we define

$$\Omega_1 = \{(t, u) : \beta_0 \leq u \leq \alpha_0,\ t \in J\}$$

when $\beta_0 \leq \alpha_0$ on J.
We are now in a position to prove the following results.

Theorem 2.2.1 *Assume that*

(i) (H_1) *holds and* $\alpha_0 \leq \beta_0$ *on* J,

(ii) $f \in C^{0,1}[\Omega, R]$ *and* $\int_0^T f_u(t, \theta(y))dt \neq 0$ *for any* $(t, \theta) \in \Omega$.

Then, there exist monotone sequences $\{\alpha_n\}$ *and* $\{\beta_n\}$ *which converge uniformly to the unique solution of* (2.2.3).

Proof Since $f \in C^{0,1}[\Omega, R]$, there exists a constant $N > 0$ such that for $u \geq v$,

$$|f(t,u) - f(t,v)| \leq N\,|u - v|\,. \tag{2.2.12}$$

Consider the following linear initial value problems:

$$\alpha_n' = f(t, \alpha_{n-1}) - N(\alpha_n - \alpha_{n-1}),\ \alpha_n(0) = -\beta_{n-1}(T), \tag{2.2.13}$$
$$\beta_n' = f(t, \beta_{n-1}) - N(\beta_n - \beta_{n-1}),\ \beta_n(0) = -\alpha_{n-1}(T), \tag{2.2.14}$$

on J for $n = 1, 2, \cdots$. The existence and uniqueness of the solution of (2.2.13) or (2.2.14) are guaranteed by known result for initial value problems. We will first show $\alpha_0 \leq \alpha_1 \leq \beta_1 \leq \beta_0$ on J. Letting $p = \alpha_0 - \alpha_1$, we get $p' \leq -Np$ and $p(0) \leq 0$ which implies $m(t) \leq 0$ on J. Similarly, $\beta_1 \leq \beta_0$ on J. Let $p = \alpha_1 - \beta_1$. Then, by condition (2.2.12),

$$p' = f(t, \alpha_0) - f(t, \beta_0) - Np + N(\alpha_0 - \beta_0) \leq -Np \text{ on } J$$

and

$$p(0) = \alpha_0(T) - \beta_0(T) \leq 0,$$

which implies $p(t) \leq 0$ on J.

Now, assume $\alpha_{n-1} \leq \alpha_n \leq \beta_n \leq \beta_{n-1}$ on J, we will show $\alpha_n \leq \alpha_{n+1} \leq \beta_{n+1} \leq \beta_n$ on J. Let $p = \alpha_n - \alpha_{n+1}$. Then

$$\begin{aligned} p' &= f(t, \alpha_n) - f(t, \alpha_{n-1}) - Np + N(\alpha_n - \alpha_{n-1}) \leq -Np \text{ on } J \\ p(0) &= -\beta_{n-1}(T) + \beta_n(T) \leq 0. \end{aligned}$$

Therefore, $p(t) \leq 0$ on J. Similarly, $\beta_{n+1} \leq \beta_n$ on J. Let $p = \alpha_{n+1} - \beta_{n+1}$, we get

$$\begin{aligned} p' &= f(t, \alpha_n) - f(t, \beta_n) - N(\alpha_{n+1} - \alpha_n) + N(\beta_{n+1} - \beta_n) \\ &\leq -N(\alpha_n - \beta_n) - Np + N(\alpha_n - \beta_n) = -Np \text{ on } J \\ p(0) &= -\beta_n(T) + \alpha_n(T) \leq 0. \end{aligned}$$

Then, $p(t) \leq 0$ on J. Therefore, we obtain monotone sequences $\{\alpha_n\}$ and $\{\beta_n\}$ satisfying

$$\alpha_0 \leq \alpha_1 \leq \cdots \leq \alpha_n \leq \beta_n \leq \cdots \leq \beta_1 \leq \beta_0 \text{ on } J.$$

It follows from the standard arguments that $\{\alpha_n\}$ and $\{\beta_n\}$ converge uniformly to their respective limit functions. Let $\rho(t) = \lim_{n\to\infty} \alpha_n(t)$ and $r(t) = \lim_{n\to\infty} \beta_n(t)$. Then $\rho(t)$ and $r(t)$ satisfy

$$\rho' = f(t, \rho),\ t \in J,\ \rho(0) = -r(T), \tag{2.2.15}$$
$$r' = f(t, r),\ t \in J,\ r(0) = -\rho(T), \tag{2.2.16}$$

respectively. We claim that $\rho \equiv r \equiv u$ on J and thus u is the solution of (2.2.3). The uniqueness follows from Lemma 2.2.7. To this end, we observe that $\rho \leq r$ on J and if $\rho(t_0) = r(t_0)$ for some $t_0 \in J$, then $\rho \equiv r$ by the comparison principle for IVP. Therefore we may assume $\rho < r$ on J. Let $m(t) = \rho(t) - r(t)$. By the mean value theorem,

$$\int_0^T f_u(t, \theta(t))dt = \int_0^T \frac{f(t, \rho(t)) - f(t, r(t))}{\rho(t) - r(t)} dt = \int_0^T \frac{m'(t)}{m(t)} dt = 0,$$

which contradicts assumption (ii). This completes the proof.

Theorem 2.2.2 *Assume that*

(i) (H_1) *or* (H_2) *holds and* $\alpha_0 \geq \beta_0$ *on* J*,*

(ii) $f \in C^{0,1}[\Omega_1, R]$ *and* $\int_0^T f_u(t, \theta(t))dt \neq 0$ *for any* $(t, \theta) \in \Omega_1$*.*

Then, there exist monotone sequences $\{\alpha_n\}$ *and* $\{\beta_n\}$ *which converge uniformly to the unique solution of* (2.2.3).

Proof We consider the linear weakly coupled system

$$\begin{aligned} \alpha_n' &= f(t, \alpha_{n-1}) + N(\alpha_n - \alpha_{n-1}), \; \alpha_n(0) = -\beta_n(T), \\ \beta_n' &= f(t, \beta_{n-1}) + N(\beta_n - \beta_{n-1}), \; \beta_n(0) = -\alpha_n(T), \end{aligned}$$

on J for $n = 1, 2, \cdots$. The rest of the proof is essentially contained in that of the Theorem 2.2.1 and we omit the details here.

Example 2.2.1 (1) Let $f(t, u) = -\tan(u) + 1$ and $T = 2\pi$ in (2.2.3). It is easy to check that $f_u < 0$ for $-\pi/2 < u < \pi/2$ and $f(t, u) = 0$ at $u = \pi/4$. Let $\alpha_0 = -\pi/4$ and $\beta_0 = \pi/4$. Then, α_0 and β_0 satisfy (H_1). In addition, $\int_0^{2\pi} f_u(t, \theta(t))dt < 0$ for any $(t, \theta) \in \Omega$, where $\Omega = \{(t, u) : -\pi/4 \leq u \leq \pi/4, t \in J\}$. By Theorem 2.2.1, there exists a unique solution of (2.2.3).

(2) Let $f(t, u) = \sin(u) + \sin(t)$ and $T = 2\pi$ in (2.2.3). Choose $\alpha_0 = \pi/2$ and $\beta_0 = -\pi/2$. Then, α_0 and β_0 satisfy (H_1). Moreover, $\int_0^{2\pi} f_u(t, \theta(t))dt = \int_0^{2\pi} \cos(\theta(t))dt > 0$ since we have $-\pi/2 < \theta < \pi/2$. By Theorem 2.2.2, (2.2.3) has a unique solution.

As a special case, we obtain an existence result for (2.2.3) with a class of functions f satisfying some properties mentioned in the above examples, as stated in the following corollary.

Corollary 2.2.1 *Assume that $f \in C^{0,1}[J \times [-b,b], R]$, $b > 0$ and that there exists a constant M, $0 < M \le b$ such that either $f(t,-M) \ge 0$ and $f(t,M) \le 0$ or, $f(t,-M) \le 0$ and $f(t,M) \ge 0$ for $t \in J$. Assume further $\int_0^T f_u(t,\theta(t))dt \neq 0$ for any $(t,\theta) \in [0,T] \times [-M,M]$. Then, there exists a unique solution of (2.2.3) between $-M$ and M.*

We consider the example

$$u' = 1/(1-u),\ u(0) = -u(T),\ t \in J.$$

It is easy to see that $u(t) = 1 - \sqrt{(T+2)^2 - 8t}/2 < 1$ is the unique solution if $T < 2$. But the solution ceases to exists if $T \ge 2$ since $u(t)$ can not go beyond 1. This example suggests that for some functions f, the solution of (2.2.3) exists if T is small enough and may not exist if T is too large. By critical length, we mean the unique $T^* > 0$ which is the length such that the unique solution of (2.2.3) exists if $T < T^*$ and does not exist if $T > T^*$. Consider problem (2.2.3). We assume that $f \in C^{2,1}[[0,T] \times (-\infty, c), (0,\infty)]$ satisfies $f_u > 0$, $f_t \ge 0$, $f_{tt} \ge 0$. We note that it is possible that $\lim_{u \to c} f(t,u) = +\infty$.

Lemma 2.2.8 (*i*) (2.2.3) *has a unique solution if T is small enough.*

(*ii*) *If* (2.2.3) *has a unique solution for $T = T_1$, then for any $T_2 < T_1$, it also has a unique solution.*

(*iii*) *Let $u_1(t)$, $u_2(t)$ be the solutions of* (2.2.3) *on $[0,T_1]$ and $[0,T_2]$, respectively. It $T_1 > T_2$, then $u_1(T_1) > u_2(T_2)$.*

Proof (i) We only need to find two functions α_0, β_0 which satisfy the assumptions of Theorem 2.2.2. Let $0 < M < c$. Choose $T > 0$ small enough such that

$$f(T,-M/2)T \le M/2 \tag{2.2.17}$$

and

$$f(t,-M/2) + f_t(t,-M/2)T \le f(t,M/2). \tag{2.2.18}$$

Notice that (2.2.17) holds since $H(T) \equiv f(T,-M/2)T$ is monotone increasing in T and $H(0) = 0$ and that (2.2.18) holds since for fixed t, $f(t,-M/2) < f(t,M/2)$ and $f_t(t,-\frac{M}{2})t \le G(T) \equiv f_t(T,-M/2)T$, where $G(T)$ is monotone increasing in T and $G(0) = 0$. Let $\alpha_0 = M - f(T,-M/2) + f(t,-M/2)t$ and $\beta_0(t) = f(t,-M/2)t - M$, then

$$\alpha_0' = f(t,-M/2) + f_t(t,-M/2)t$$

$$\begin{aligned}
&\leq f(t,-M/2)+f_t(t,-M/2)T\\
&\leq f(t,M/2)\\
&\leq f(t,\alpha_0),\\
\beta_0' &= f(t,-M/2)+f_t(t,-M/2)t\\
&\geq f(t,-M/2)\\
&\geq f(t,\beta_0)
\end{aligned}$$

and

$$\alpha_0(0) = M - f(T,-M/2)T = -\beta_0(T),\ \beta_0(0) = -M = -\alpha_0(T).$$

Therefore, (2.2.3) has a unique solution on J by Theorem 2.2.2.

(ii) Suppose that $u_1(t)$ is the unique solution of (2.2.3) on $[0,T_1]$. Let $\beta_0(t) = u_1(t)$ and $\alpha_0(t) = u_1(t) + u_1(T_1) - u_1(T_2)$, $0 \leq t \leq T_2 \leq T_1$. Since $u_1(t)$ is strictly increasing in t, $u_1(T_1) > u_2(T_2)$ and thus, $\alpha_0(t) > \beta_0(t)$. We find that

$$\begin{aligned}
\alpha_0'(t) &= u_1'(t) = f(t,u_1) < f(t,\alpha_0) \text{ and } \alpha_0(0) = -\beta_0(T),\\
\beta_0'(t) &= u_1'(t) = f(t,\beta_0) \text{ and } \beta_0(0) = -\alpha_0(T).
\end{aligned}$$

By Theorem 2.2.2, there exists a unique solution $u_2(t)$ of (2.2.3) on $[0,T_2]$.

(iii) From (ii), $u_1(0) \leq u_2(0) < 0$. If $u_1(0) = u_2(0)$, then $u_1(t) = u_2(t)$ on $[0,T_2]$. Since $u_1(t)$ is strictly increasing in t, we have $u_1(T_1) > u_1(T_2)$, which is a contradiction. Therefore, $u_1(0) < u_2(0)$ and equivalently, $u_1(T_1) > u_2(T_2)$.

Theorem 2.2.3 *There exists a unique critical length T^* for* (2.2.3).

Proof From Lemma 2.2.8 (i), (2.2.3) has a unique solution $u(t)$ if T is small enough and $u < c$. It follows from Lemma 2.2.8 (ii) that if for $T = T_1$, (2.2.3) has a unique solution. Further (2.2.3) has a unique solution for $T = T_2 < T_1$. Now, assume for any $T > 0$, (2.2.3) has a unique solution $u(t)$ and $u < c$. Then, there exist a unique $t_1 \in (0,T)$ such that $u(t_1) = 0$. Integrating the equation in (2.2.3) from 0 to t_1 and from t_1 to T, respectively, we obtain

$$u(T) = \int_{t_1}^{T} f(s,u(s))ds \geq f(0,-c)(T-t_1)$$

and

$$u(0) = -\int_0^{t_1} f(s,u(s))ds \leq -f(0,-c)t_1.$$

Since $u(0) = -u(T)$, we have $u(T) \geq \max\{f(0,-c)t_1,\ f(0,-c)(T-t_1)\}$. If T is large enough, $u(T)$ will go beyond c. This contradiction proves the theorem.

We shall next apply the method of quasilinearization to (2.2.3). The desired quadratic convergence of monotone sequences is obtained when the convexity of $f(t,u)$ in u is assumed. We also extend this result by applying the generalized quasilinearization method and relax the assumption on $f(t,u)$. The main results are contained in the following theorems.

Theorem 2.2.4 *Assume that*

(i) (H_1) *or* (H_3) *holds and* $\alpha_0 \le \beta_0$ *on* J.

(ii) $f \in C^{0,2}[\Omega, R]$ *satisfying* $f_{uu} \ge 0$ *on* Ω *and either* $\int_0^T f_u(s,\beta_0(s))ds < 0$ *if* $(H1)$*holds or for* $t \in J_0$, $\int_0^T f(u(s,\beta_0(s)))ds < 0$ *if* (H_3) *holds.*

Then, there exist monotone sequences $\{\alpha_n\}$ *and* $\{\beta_n\}$ *which converge uniformly to the unique solution of* (2.2.3) *and moreover, the convergence of the sequence is quadratic.*

Proof Since $f_{uu} \ge 0$, we easily get the inequality for any $u \ge v$, where (t,u), $(t,v) \in \Omega$,

$$f(t,u) \ge f(t,v) + f_u(t,v)(u-v). \tag{2.2.19}$$

Consider the following weakly coupled linear system

$$\begin{aligned}\alpha_n' &= f(t,\alpha_{n-1}) + f_u(t,\alpha_{n-1})(\alpha_n - \alpha_{n-1})\\ &\text{on} \quad J,\ \alpha_n(0) = -\beta_n(T)\end{aligned} \tag{2.2.20}$$

$$\begin{aligned}\beta_n' &= f(t,\beta_{n-1}) + f_u(t,\alpha_{n-1})(\beta_n - \beta_{n-1})\\ &\text{on} \quad J,\ \beta_n(0) = -\alpha_n(T)\end{aligned} \tag{2.2.21}$$

for $n = 1, 2, \cdots$. The existence of unique solution of (2.2.20)-(2.2.21) is guaranteed by Lemma 2.2.2 when $n = 1$, since f_u is nondecreasing in u, and $\int_0^T f_u(t,\beta_0(t))dt < 0$. First, we shall prove that $\alpha_0 \le \alpha_1 \le \beta_1 \le \beta_0$ on J if either (H_1) or (H_2) holds. Let $m_1 = \alpha_0 - \alpha_1$ and $m_2 = \beta_1 - \beta_0$. If (H_1) holds, then

$$\begin{aligned}m_1' &\le f_u(t,\alpha_0)m_1 \text{ on } J \text{ and } m_1(0) \le m_2(T)\\ m_2' &\le f_u(t,\alpha_0)m_2 \text{ on } J \text{ and } m_2(0) \le m_1(T).\end{aligned}$$

By Lemma 2.2.3, $m_1(t) \le 0$ and $m_2(t) \le 0$ on J. If (H_2) holds, we have

$$\begin{aligned}m_1' &\le f_u(t,\alpha_0)(m_1 + \gamma_1) \text{ on } J \text{ and } m_1(0) \ge m_2(T)\\ m_2' &\le f_u(t,\alpha_0)(m_2 + \gamma_2) \text{ on } J \text{ and } m_2(0) \ge m_1(T),\end{aligned}$$

where

$$\begin{aligned}\gamma_1 &= [m_2(0) - m_1(T)] \Big/ \left(1 - \exp\left(\int_0^T f_u(t,\alpha_0)dt\right)\right), \text{ and}\\ \gamma_2 &= [m_1(0) - m_2(T)] \Big/ \left(1 - \exp\left(\int_0^T f_u(t,\alpha_0)dt\right)\right).\end{aligned}$$

Since $\int_0^t f_u(t,\alpha_0(t))dt < 0$, $m_1(t) \le 0$ and $m_2(t) \le 0$ on J by Lemma 2.2.4.

To show $\alpha_1 \le \beta_1$ on J, we let $m = \alpha_1 - \beta_1$. It follows that

$$\begin{aligned}m' &= f(t,\alpha_0) - f(t,\beta_0) + f_u(t,\alpha_0)(\alpha_1 - \alpha_0) - f_u(t,\alpha_0)(\beta_1 - \beta_0)\\ &\le f_u(t,\alpha_0)(\alpha_0 - \beta_0) + f_u(t,\alpha_0)(\alpha_1 - \alpha_0) - f_u(t,\alpha_0)(\beta_1 - \beta_0)\\ &= f_u(t,\alpha_0)m \text{ on } J, \text{ and } m(0) = m(T).\end{aligned}$$

By Lemma 2.2.1 (i), we get $m(t) \le 0$ on J.

Assume that $\alpha_{n-1} \le \alpha_n \le \beta_n \le \beta_{n-1}$. We shall show that $\alpha_n \le \alpha_{n+1} \le \beta_{n+1} \le \beta_n$ on J. The existence of unique solutions α_{n+1}, β_{n+1} of (2.2.20)-(2.2.21) is guaranteed by Lemma 2.2.2. Let $m_1 = \alpha_n - \alpha_{n+1}$ and $m_2 = \beta_{n+1} - \beta_n$. Using (2.2.19), we get

$$\begin{aligned}m_1' &= f(t,\alpha_{n-1}) - f(t,\alpha_n) + f_u(t,\alpha_{n-1})(\alpha_n - \alpha_{n-1}) - f_u(t,\alpha_n)(\alpha_{n+1} - \alpha_n)\\ &\le f_u(t,\alpha_{n-1})(\alpha_{n-1} - \alpha_n) + f_u(t,\alpha_{n-1})(\alpha_n - \alpha_{n-1})\\ &\quad - f_u(t,\alpha_n)(\alpha_{n+1} - \alpha_n)\\ &= f_u(t,\alpha_n)m_1 \text{ on } J \text{ and } m_1(0) = m_2(T),\\ m_2' &= f(t,\beta_n) - f(t,\beta_{n-1}) - f_u(t,\alpha_{n-1})(\beta_n - \beta_{n-1}) + f_u(t,\alpha_n)(\beta_{n+1} - \beta_n)\\ &\le f_u(t,\beta_n)(\beta_n - \beta_{n-1}) - f_u(t,\alpha_{n-1})(\beta_n - \beta_{n-1}) + f_u(t,\alpha_n)(\beta_{n+1} - \beta_n)\\ &\le f_u(t,\beta_n)(\beta_n - \beta_{n-1}) - f_u(t,\beta_n)(\beta_n - \beta_{n-1}) + f_u(t,\alpha_n)(\beta_{n+1} - \beta_n)\\ &= f_u(t,\alpha_n)m_2 \text{ on } J, \text{ and } m_2(0) = m_1(T).\end{aligned}$$

Since $\int_0^T f_u(t,\alpha_0(t))dt < 0$, $m_1(t) \le 0$ and $m_2(t) \le 0$ on J by Lemma 2.2.3. Next, letting $m = \alpha_{n+1} - \beta_{n+1}$, we have

$$\begin{aligned}m' &= f(t,\alpha_n) - f(t,\beta_n) + f_u(t,\alpha_n)(\alpha_{n+1} - \alpha_n) - f_u(t,\alpha_n)(\beta_{n+1} - \beta_n)\\ &\le f_u(t,\alpha_n)(\alpha_n - \beta_n) + f_u(t,\alpha_n)(\alpha_{n+1} - \alpha_n) - f_u(t,\alpha_n)(\beta_{n+1} - \beta_n)\\ &= f_u(t,\alpha_n)m \text{ on } J \text{ and } m(0) = m(T).\end{aligned}$$

Since f_u is nondecreasing in u and $\alpha_n \le \beta_0$, $\int_0^T f_u(t,\alpha_n(t))dt < 0$. By Lemma 2.2.1, $m(t) \le 0$ on J. Therefore, we obtain monotone sequences $\{\alpha_n\}$ and

$\{\beta_n\}$ which converge uniformly to their respective limit functions. Let $\rho(t) = \lim_{n\to\infty} \alpha_n(t)$ and $r(t) = \lim_{n\to\infty} \beta_n(t)$. Then the pair $(\rho(t), r(t))$ satisfies

$$\begin{aligned} \rho' &= f(t,\rho),\ t \in J,\ \rho(0) = -r(T) \\ r' &= f(t,r),\ t \in J,\ r(0) = -\rho(T). \end{aligned}$$

Since $\int_0^T f_u(t,\theta(t))dt < 0$ for any $(t,\theta) \in \Omega$, it follows from the same argument as in Theorem 2.2.1 that $\rho \equiv r \equiv u$ is the unique solution of (2.2.3).

The proof will be completed if we show the convergence of $\{\alpha_n\}$ and $\{\beta_n\}$ to u is quadratic. Let $p_n = u - \alpha_n$, $q_n = \beta_n - u$, $n \geq 1$. Note that $p_n \geq 0$ and $q_n \geq 0$ on J for all $n \geq 1$. From (2.2.3), (2.2.20) and (2.2.21), we obtain

$$\begin{aligned} p_n' &= f(t,u) - f(t,\alpha_{n-1}) - f_u(t,\alpha_{n-1})(\alpha_n - \alpha_{n-1}) \\ &= f_u(t,\theta_1)(u - \alpha_{n-1}) - f_u(t,\alpha_{n-1})(u - \alpha_{n-1}) + f_u(t,\alpha_{n-1})(u - \alpha_n) \\ &= f_{uu}(t,\theta_2)(\theta_1 - \alpha_{n-1})(u - \alpha_{n-1}) + f_u(t,\alpha_{n-1})(u - \alpha_n) \\ &\leq f_{uu}(t,\theta_2)(u - \alpha_{n-1})^2 + f_u(t,\alpha_{n-1})(u - \alpha_n) \\ &= f_{uu}(t,\theta_2)p_{n-1}^2 + f_u(t,\alpha_{n-1})p_n \text{ on } J, \end{aligned}$$

where $\alpha_{n-1} \leq \theta_2 \leq \theta_1 \leq u$ on J, and similarly,

$$q_n' \leq f_{uu}(t,\eta_2)[2q_{n-1}^2 + p_{n-1}^2] + f_u(t,\alpha_{n-1})q_n \text{ on } J,$$

where $u \leq \eta_2 \leq \beta_{n-1}$ on J. The boundary conditions are

$$p_n(0) = q_n(T) \text{ and } q_n(0) = p_n(T).$$

Let $R_n = p_n + q_n$, we get

$$\begin{aligned} R_n' &\leq f_u(t,\alpha_{n-1})R_n + \left(f_{uu}(t,\theta_2)p_{n-1}^2 + f_{uu}(t,\eta_2)[2q_{n-1}^2 + p_{n-1}^2]\right) \quad (2.2.22) \\ &\qquad \text{on } J_0, R_n(0) = R_n(T). \end{aligned}$$

Let $\gamma(t) = f_{uu}(t,\theta_2)p_{n-1}^2 + f_{uu}(t,\eta_2)[2q_{n-1}^2 + p_{n-1}^2]$ and $\lambda(t) = \int_0^t f_u(s,\alpha_{n-1}(s))ds$. Then $\gamma(t) \geq 0$ for $0 \leq t \leq T$ and $\lambda(T) < 0$. Solving for $R_n(t)$ from (2.2.22), we get

$$R_n(t) \leq e^{\lambda(t)}\left[R_n(0) + \int_0^t e^{-\lambda(s)}\gamma(s)ds\right].$$

Let $t = T$, we have

$$R_n(0) = R_n(T) \leq e^{\lambda(T)}\left[R_n(0) + \int_0^T e^{-\lambda(s)}\gamma(s)ds\right]$$

which implies,

$$R_n(t) \le \frac{\int_0^T e^{-\lambda(s)}\gamma(s)ds}{e^{-\lambda(T)}-1},$$

since $\lambda(T) < 0$.

Therefore,

$$R_n(t) \le e^{\lambda(t)}\left[\frac{\int_0^T e^{-\lambda(s)}\gamma(s)ds}{e^{-\lambda(T)}-1} + \int_0^t e^{-\lambda(s)}\gamma(s)ds\right]$$
$$\le K_1 \int_0^T e^{-\lambda(s)}\gamma(s)ds,$$

where $K_1 \ge K_2/(1-e^{\lambda(T)})$ and $K_2 \ge e^{\lambda(t)}$ on J. Assume $|f_{uu}(t,u)| \le C_1$ and $\int_0^T e^{-\lambda(s)}ds \le C_2$ on Ω, then

$$\max_J R_n(t) \le K(\max_J p_{n-1}^2(t) + \max_J q_{n-1}^2(t)),$$

where $K = 2K_1C_1C_2$. Since $\max_J p_n(t) \le \max_J R_n(t)$ and $\max_J q_n(t) \le \max_J R_n(t)$, we get the desired quadratic convergence. The proof is therefore complete.

Remark 2.2.1 If we do not require quadratic convergence of the sequences, under exactly the same conditions as in Theorem 2.2.1, we can construct the monotone sequences, say $\{w_n\}$ and $\{v_n\}$, by using the method employed in Theorem 2.2.4, namely, for $n \ge 1$,

$$\begin{aligned} w_n' &= f(t,w_{n-1}) - N(w_n - w_{n-1}) \text{ on } J_0, \quad w_n(0) = -v_n(T), \\ v_n' &= f(t,v_{n-1}) - N(v_n - v_{n-1}) \text{ on } J_0, \quad v_n(0) = -w_n(T), \end{aligned}$$

where $w_0 = \alpha_0$ and $v_0 = \beta_0$, and the sequences converge to the unique solution of (2.2.3). Actually, we can show that $\alpha_n \le w_n \le v_n \le \beta_n$ for $n \ge 1$, where $\{\alpha_n\}$ and $\{\beta_n\}$ are the sequences constructed in Theorem 2.2.1. This means the convergence of $\{w_n\}$ and $\{v_n\}$ to the solution is faster than that of $\{\alpha_n\}$ and $\{\beta_n\}$ of Theorem 2.2.1, respectively. But the situation is much simpler in Theorem 2.2.1 because we deal with only linear initial value problems.

The next result complements convergence as in Theorem 2.2.4.

Theorem 2.2.5 *Assume that*

(i) (H_1) *or* (H_4) *holds and* $\alpha_0 \ge \beta_0$ *on* J.

(*ii*) $f \in C^{0,2}[\Omega_1, R]$ *satisfying* $f_{uu} \geq 0$ *on* Ω_1 *and either* $\int_0^T f_u(s, \beta_0(s))ds > 0$ *if* (H_1)*holds or for* $t \in [0, T)$, $\int_0^T f(u(s, \beta_0(s)))ds > 0$ *if* (H_4) *holds.*

Then, there exist monotone sequences $\{\alpha_n\}$ *and* $\{\beta_n\}$ *which converge uniformly to the unique solution of* (2.2.3) *and moreover, the convergence of the sequence is quadratic.*

Proof Since $f_{uu} \geq 0$, we easily get the inequality for any $u \geq v$, where (t, u), $(t, v) \in \Omega$,

$$f(t, u) \leq f(t, v) + f_u(t, u)(u - v). \tag{2.2.23}$$

Consider the following weakly coupled linear system

$$\alpha_n' = f(t, \alpha_{n-1}) + f_u(t, \alpha_{n-1})(\alpha_n - \alpha_{n-1}),\ \alpha_n(0) = -\beta_n(T) \tag{2.2.24}$$

$$\beta_n' = f(t, \beta_{n-1}) + f_u(t, \alpha_{n-1})(\beta_n - \beta_{n-1}),\ \beta_n(0) = -\alpha_n(T) \tag{2.2.25}$$

for $n = 1, 2, \cdots$. The existence of unique solution of (2.2.24)-(2.2.25) is guaranteed by Lemma 2.2.2 when $n = k + 1$ provided that it is true for $n = k$, $k \leq 1$. We only point out that to establish the monotonicity of $\{\alpha_n\}$ and $\{\beta_n\}$, we need to utilize Lemma 2.2.1, the second part of Lemma 2.2.3, Lemma 2.2.5 and (2.2.23) following a similar argument as in the proof of Theorem 2.2.4. The existence of the unique solution is also obtained consequently.

It remains to show the convergence of $\{\alpha_n\}$ and $\{\beta_n\}$ to u is quadratic. Let $p_n = \alpha_n - u$, $q_n = \beta_n - u$, $n \geq 1$. Note that $p_n \geq 0$ and $q_n \geq 0$ on J. As in Theorem 2.2.4,

$$p_n' \geq -f_{uu}(t, \theta_2)p_{n-1}^2 + f_u(t, \alpha_{n-1})p_n \text{ on } J \text{ and } p_n(0) = p_n(T),$$

where $\alpha_{n-1} \geq \theta_2 \geq \theta_1 \geq u$ on Ω_1, and

$$q_n' \geq -f_{uu}(t, \eta_2)[2q_{n-1}^2 + p_{n-1}^2] + f_u(t, \alpha_{n-1})q_n \text{ on } J \text{ and } q_n(0) = q_n(T)$$

where $u \geq \eta_2 \geq \beta_{n-1}$ on Ω_1. Let $R_n = p_n + q_n$. Then

$$R_n' \geq f_u(t, \alpha_{n-1})R_n - \gamma(t) \text{ and } R_n(0) = R_n(T), \tag{2.2.26}$$

where $\gamma(t) = f_{uu}(t, \theta_2)p_{n-1}^2 + f_{uu}(t, \eta_2)[2q_{n-1}^2 + p_{n-1}^2]$. Let $\lambda(t) = \int_t^T \times f_u(s, \alpha_{n-1}(s))ds$. It follows from a direct computation that

$$\begin{aligned} R_n(t) &\leq e^{-\lambda(t)}\left[\int_0^T e^{\lambda(s)}\gamma(s)ds \big/ \left(e^{\lambda(0)} - 1\right) + \int_t^T e^{\lambda(s)}\gamma(s)ds\right] \\ &\leq K_1 \int_0^T e^{\lambda(s)}\gamma(s)ds, \end{aligned}$$

where $K_1 \geq e^{\lambda(0)-\lambda(t)} / \left(e^{\lambda(0)} - 1\right)$ on J. Assume that $\int_0^T e^{\lambda(s)} ds \leq C_2$ and $|f_{uu}(t,u)| \leq C_1$on Ω. Then,

$$\max_J R_n(t) \leq K \left(\max_J p_{n-1}^2(t) + \max_J q_{n-1}^2(t) \right),$$

where $K = 2K_1C_1C_2$. This completes the proof.

Remark 2.2.2 In the case f admits a decomposition as $f = F + G$ where F, $G \in C^{0,2}[\Lambda, R]$ satisfy $F_{uu} \geq 0$ and $G_{uu} \leq 0$ ($\Lambda = \Omega$ if $\alpha_0 \leq \beta_0$ and $\Lambda = \Omega_1$ if $\alpha_0 \geq \beta_0$), the conclusions of Theorem 2.2.4 and 2.2.5 remain true under suitable assumptions.

Concerning Theorem 2.2.3, we shall give below two problems in which we do not assume f is convex but there exists a convex function $\phi(t,x)$ such that $f + \phi$ is convex. As remarked earlier we can always find such a $\phi(t,x)$ since the set Ω is compact.

Problem 2.2.1 Assume that

(*i*) (H_1) holds and $\alpha_0 \leq (\geq) \beta_0$ on J.

(*ii*) $f \in C^{0,2}[\Omega, R]$ and $\phi \in C^2[R, R]$ satisfy $\phi_{uu} \geq 0$ and $f_{uu} + \phi_{uu} \geq 0$ on Ω.

(*iii*) $\int_0^T [f_u(t,\beta_0) + \phi_u(\beta_0) - \phi_u(\alpha_0)]dt < (>) 0$.

Then, there exist monotone sequences $\{\alpha_n\}$ and $\{\beta_n\}$ which converge uniformly to the unique solution of (2.2.3) and moreover, the convergence of the sequence is quadratic.

Problem 2.2.2 Assume that

(*i*) (H_1) holds and $\alpha_0 \leq (\geq) \beta_0$ on J.

(*ii*) $f \in C^{0,2}[\Omega, R]$ and $f = F + G$ satisfying $F_{uu} + \phi_{uu} \geq 0$ and $\phi_{uu} \geq 0$ on Ω, $G_{uu} + \eta_{uu} \leq 0$and $\eta_{uu} \leq 0$ on Ω where ϕ, $\eta \in C^2[R, R]$ satisfy $\phi_{uu} \geq 0$ and $\eta_{uu} \geq 0$.

(*iii*) $\int_0^T [F_u(t,\beta_0) + \phi_u(t,\beta_0) - \phi_u(t,\alpha_0) + G_u(t,\alpha_0) + \eta_u(t,\alpha_0) - \eta_u(t,\beta_0)]dt < (>) 0$.

Then, there exist monotone sequences $\{\alpha_n\}$ and $\{\beta_n\}$ which converge uniformly to the unique solution of (2.2.3) and moreover, the convergence of the sequence is quadratic.

2.3 INTERVAL ANALYSIS AND QUASILINEARIZATION

Interval analysis is an approach to scientific computing that treats an interval as a kind of a number. It is a language for inequalities and has direct applicability to important problems of significance in large computations. By using interval arithmetic and interval valued functions, one can compute arbitrary sharp upper and lower bounds for exact solutions of nonlinear problems simultaneously. See [52] for details.

Before we can combine the two techniques some properties of interval mathematics need to be listed. The following properties are needed for the development of the results in this section.

1. Let $X_1 = [\underline{X_1}, \overline{X_1}]$ and $X_2 = [\underline{X_2}, \overline{X_2}]$ be intervals then $X_1 \subseteq X_2$ if and only if $\underline{X_2} \leq \underline{X_1} \leq \overline{X_1} \leq \overline{X_2}$. (The bars over and under the endpoints of the interval are standard notation for the upper and lower endpoints of the interval, respectively.)

2. An interval is said to be degenerative if $X = [\underline{X}, \overline{X}]$ and $\underline{X} = \overline{X}$. The width of an interval is given by $w[X] = \overline{X} - \underline{X}$. In a degenerative interval the width is zero.

3. A continuous interval function $Y(t)$ on a closed set is defined by $Y(t) = \left[\underline{Y(t)}, \overline{Y(t)}\right]$ where the endpoints $\underline{Y(t)}, \overline{Y(t)} \in C[J, I\!R]$, and $J = [0, T]$. $y(t) \in Y(t)$ with $y(t) \in C[J, I\!R]$ if and only if $\underline{Y(t)} \leq y(t) \leq \overline{Y(t)}$ for all $t \in J$.

4. A united extension of a real function or operator, $g(x)$, is given by $\overline{g}(X) = \cup_{x \in X} \{g(x)\}$, where $\overline{g}(X)$ is an interval function or operator.

5. A function or operator, F, is called inclusion monotonic if $Y \subseteq X$ implies $F(Y) \subseteq F(X)$. United extensions are always inclusion monotonic.

6. Let $p : C[J, I\!R] \to C[J, I\!R]$ be an integral operator on some interval J. We define an integral operator by

$$P(Y(t)) = [p(y(t)) : y(t) \in Y(t) \text{ for } t \in J]$$

where Y is an interval function. P is the united extension of p by definition and is therefore inclusion monotonic.

In order to combine the two techniques, the method of quasilinearization and the interval method, we need the following theorem.

Theorem 2.3.1 *Suppose that there exists a sequence of interval operators* $\{P_i\}$ *where* $P_i : CI[J] \to CI[J]$ *($CI[J]$ is a continuous interval function defined on J) and an interval function $X_0(t)$ such that the following are true:*

$$P_0(X_0(t)) \subseteq X_0(t), \text{ for} t \in J; \tag{2.3.1}$$

$$P_n\left(\prod_{i=0}^{n-1} P_i\left(X_0(t)\right)\right) \subseteq \prod_{i=0}^{n-1} P_i(X_0(t)), \textit{ for } n \in N,\ t \in J; \textit{ in other words}$$

$$P_{n+1}\left(P_n\left(P_{n-1}\cdots\left(P_0(X_0(t)\right)\right)\right) \subseteq \left(P_n\left(P_{n-1}\cdots\left(P_0\left(X_0(t)\right)\right)\right)\right); \tag{2.3.2}$$

$$P_n(X(t)) \to P(X(t))\text{exists}; \tag{2.3.3}$$

$$w\left[P_n\left(X(t)\right)\right] \le k\left(w\left(X(t)\right)\right)^2 \text{for} n \in N, X(t) \subseteq X_0(t), \text{on} J, \tag{2.3.4}$$

$K \geq 0$ *($w$is the interval width function);*

$$w_0\left[P\left(X(t)\right)\right] \le Lw_0\left[X(t)\right] \text{for} L \tag{2.3.5}$$

such that $0 < L < 1$ *(w_0 is a weighted width function given by* $w_0(X) = \max_J w\left(X(t)\right) e^{-\delta t}$*, where δ is an arbitrary constant to be chosen).*

Then the successive iterates given by $X_n(t) = \prod_{i=0}^{n-1} P_i\left(X_0(t)\right)$ *form a decreasing sequence of interval valued functions, and* $\lim_{n\to\infty} X_n(t) = X(t)$ *exists as an interval function on J. Furthermore, $X(t)$ is a degenerative interval which is the unique solution of* $P\left(X(t)\right) = X(t)$ *on J and the sequence* $\{X_n(t)\}$ *converges quadratically.*

Proof The uniform convergence of the decreasing sequence is obvious. The endpoints of $\{X_n(t)\}$ are bounded monotone sequences, thus convergent. The convergence can be written as

$$X(t) = \bigcap_{k=0}^{\infty} X_k(t).$$

Since the operators converge, it follows that $P_n\left(X_n(t)\right) \to P\left(X(t)\right)$, for $t \in J$, which implies $P\left(X(t)\right) = X(t)$, for $t \in J$. $X(t)$ is the solution of the interval operator equation. We prove that this solution is unique by using the assumption (2.3.5) in the following manner. Let $X(t)$ be the solution of the operator equation, $P(X(t)) = X(t)$, then look at the weighted width of $X(t)$. Using (2.3.5) gives

$$0 \leq w_0[X] = w_0[PX] \leq Lw_0[X].$$

But, this is only true if $w_0[X] = 0$ which means $X(t)$ is a degenerate interval and the solution is unique. Finally, we can show quadratic convergence of the sequence by using assumption (2.3.4) which yields

$$w[X_{n+1}] \leq k(w[X_n])^2 \text{ for } n \in N,\ k \geq 0.$$

The proof is complete.

We start with the IVP

$$x' = f(t,x),\ x(0) = x_0 \tag{2.3.6}$$

where $f \in C[J \times \mathbb{R}, \mathbb{R}]$ and $J = [0,T]$, with the following assumptions:

(A_1) $\alpha_0,\ \beta_0 \in C^1[J,\mathbb{R}]$, $\alpha_0(t) \leq \beta_0(t)$, and $\alpha_0'(t) \leq f(t,\alpha_0(t))$, $\beta_0'(t) \geq f(t,\beta_0(t))$ on J with $\alpha_0(0) \leq x_0 \leq \beta_0(0)$.

(A_2) $f,\ \phi \in C^{0,2}[\Omega,\mathbb{R}]$, $f_{xx}(t,x) + \phi_{xx}(t,x) \geq 0$ on Ω and $\phi_{xx}(t,x) > 0$ on Ω where $\Omega = [(t,x) : \alpha_0(t) \leq x \leq \beta_0(t),\ t \in J]$.

Assumption (A_2) yields the following useful inequalities

$$f_x(t,x) \geq F_x(t,y) - \phi_x(t,x),\ \alpha_0(t) \leq y \leq x \leq \beta_0(t) \text{ on } J, \tag{2.3.7}$$

which can also be written as

$$F_x(t,x) \geq F_x(t,y),\ \alpha_0(t) \leq y \leq x \leq \beta_0(t) \text{ on } J, \tag{2.3.8}$$

and

$$\begin{aligned} f(t,x) &\geq F(t,y) + F_x(t,y)(x-y) - \phi(t,x), \\ \text{for } \alpha_0(t) &\leq y \leq x \leq \beta_0(t) \text{ on } J, \end{aligned} \tag{2.3.9}$$

where $F(t,x) = f(t,x) + \phi(t,x)$. Because of the convexity condition on ϕ we also have

$$\phi_x(t,x) \geq \phi_x(t,y),\ \alpha_0(t) \leq y \leq x \leq \beta_0(t) \text{ on } J. \tag{2.3.10}$$

This leads to the main result.

Theorem 2.3.2 *Consider the initial value problem given by (2.3.6), with assumption (A_1) and (A_2). There exists a decreasing sequence of interval functions $\{X_n(t)\}$, produced from the initial interval function $X_0(t) = [\alpha_0(t), \beta_0(t)]$, such that $\lim_{n\to\infty} X_n(t) = X(t)$ exists with $X(t) = [x(t), x(t)]$, where $x(t)$ is the unique solution of (2.3.6). In addition, the sequence $\{X_n(t)\}$ converges quadratically.*

Proof Suppose that (A_1) and (A_2) hold and consider the following sequence of initial value problems

$$\begin{aligned} x'(t) &= (F_x(t, \alpha_i(t) - \phi_x(t, \beta_i(t)))\, x(t) + f(t, x(t)) - (F_x(t, \alpha_i(t)) - \phi_x(t, \beta_i(t)))\, x(\\ x(0) &= x_0, \text{ for } i = 0, \cdots, k. \end{aligned}$$

This new IVP has two parts, the linear term $(F_x(t, \alpha_i(t) - \phi_x(t, \beta_i(t)))\, x(t)$ and the perturbation term $f(t, x(t)) - (F_x(t, \alpha_i(t)) - \phi_x(t, \beta_i(t)))\, x(t)$. By close inspection we see that this IVP is the same as (2.3.6), however the added term gives us some important properties. First, the differential equation has some linear terms which we will use to help develop an integral operator. Second, and most importantly, the term $F_x(t, \alpha_i(t)) - \phi_x(t, \beta_i(t))$ in the perturbation portion of the equation will aid in making our operator increasing, using the convexity property of the operator.

Using variation of parameters, we solve the new IVP for $x(t)$. Initially we solve the linear term $x'(t) = (F_x(t, \alpha_i(t)) - \phi_x(t, \beta_i(t)))\, x(t)$ which gives

$$x(t) = x_0 e^{\int_0^t G_x(s, \alpha_i(s), \beta_i(s)) ds}$$

where $G_x(t, \alpha_i(t), \beta_i(t)) = F_x(t, \alpha_i(t)) - \phi_x(t, \beta_i(t))$ with $\alpha_i(t) \le \beta_i(t)$. Upon completion of the technique the following sequence of integral operators, $p_i : C[J, \mathbb{R}] \to C[J, \mathbb{R}]$, given by

$$\begin{aligned} (p_i(x, \alpha_i, \beta_i))\,(t) &= x_0 e^{\int_0^t G_x(s, \alpha_i(s), \beta_i(s)) ds} \\ &\quad + \int_0^t H\left(\xi, \alpha_i(\xi), \beta_i(\xi), x(\xi)\right) e^{\int_\xi^t G_x(s, \alpha_i(s), \beta_i(s)) ds} d\xi \end{aligned} \tag{2.3.11}$$

where $H(t, \alpha_i(t), \beta_i(t), x(t)) = f(t, x(t)) - G_x(t, \alpha_i(t), \beta_i(t)) x(t)$ with $\alpha_i(t) \le x(t) \le \beta_i(t)$ on J are obtained. These operators are used to define a sequence of interval integral operators $P_i : CI[J] \to CI[J]$. Given an interval function $X(t) \subseteq [\alpha_0(t), \beta_0(t)]$ we define

$$\begin{aligned} &P_i(X, \alpha_i, \beta_i)(t) \\ &= \left[p_i(x, \alpha_i, \beta_i) \ : \ x(t) \in X(t),\ \alpha_i(t) = \underline{X(t)},\ \beta_i(t) = \overline{X(t)} \text{ for } t \in J\right]. \end{aligned}$$

For notational purposes we now let

$$p_i x(t) = (p_i(x, \alpha_i, \beta_i))(t)$$

and

$$P_i X(t) = (P_i(X, \alpha_i, \beta_i))(t).$$

These new interval operators are inclusion monotonic,because they are united extensions of the original operators. In addition, each p_i is an increasing operator. This follows because $H(t, \alpha_i(t), \beta_i(t), x)$ is increasing in x for each p_i when $\alpha_i \leq x \leq \beta_i(t)$ on J. This is shown by differentiating H with respect to x :

$$\begin{aligned} H_x(t, \alpha_i(t), \beta_i(t), x) &= f_x(t, x) - G_x(t, \alpha_i(t), \beta_i(t)) \\ &= f_x(t, x) - F_x(t, \alpha_i(t)) + \phi_x(t, \beta_i(t)), \end{aligned}$$

which by the increasing nature of ϕ, (2.3.10), becomes

$$H_x(t, \alpha_i(t), \beta_i(t)) \geq F_x(t, x) - F_x(t, \alpha_i(t)).$$

Then using (2.3.8), $H_x(t, \alpha, \beta, x) \geq 0$ is obtained. Since H is increasing in x it follows that each p_i is increasing in x. This provides the following useful procedure

$$P_i\left[\underline{X(t)}, \overline{X(t)}\right] = \left[p_i\underline{X(t)}, p_i\overline{X(t)}\right] \text{ for } i = 0, \cdots, k. \tag{2.3.12}$$

This allows us to apply (2.3.11) to just the endpoints of the interval to determine the result of applying the interval integral operator to the entire interval. With this in mind we now proceed to develop a sequence of interval functions using our operator.

Beginning with our initial interval function $X_0(t) = [\alpha_0(t), \beta_0(t)]$ comprised of the upper and lower solutions of (2.3.6), we must prove that $P_0X_0(t) \subseteq X_0(t)$. Using (2.3.12) permits us to show $\alpha_0(t) \leq p_0\alpha_0(t)$ and $\beta_0(t) \geq p_0\beta_0(t)$ in order to obtain the above inclusion, where p_0 is the operator

$$\begin{aligned} (p_0 x)(t) &= x_0 e^{\int_0^t G_x(s, \alpha_0(s), \beta_0(s))ds} \\ &\quad + \int_0^t H(\xi, \alpha_0(\xi), \beta_0(\xi), x(\xi)) e^{\int_\xi^t G_x(s, \alpha_0(s), \beta_0(s))ds} d\xi. \end{aligned}$$

$\alpha_0(t) \leq p_0\alpha_0(t)$ is shown by solving the differential inequality,

$$\begin{aligned} \alpha_0'(t) \leq\ & (F_x(t, \alpha_0(t) - \phi_x(t, \beta_0(t)))\, \alpha_0(t) \\ & + f(t, \alpha_0(t)) - (F_x(t, \alpha_0(t)) - \phi_x(t, \beta_0(t)))\, \alpha_0(t) \end{aligned}$$

$\alpha_0(0) = \alpha_0$ with $\alpha_0 \leq x_0 \leq \beta_0$. The IVP now has linear terms and can be solved using variation of parameters. The same process is used to obtain operator (2.3.11). The result is, for $t \in J$,

$$\begin{aligned}\alpha_0 &\leq x_0 e^{\int_0^t G_x(s,\alpha_0(s),\beta_0(s))ds} \\ &+ \int_0^t H(\xi, \alpha_0(\xi), \beta_0(\xi), \alpha_0(\xi)) e^{\int_\xi^t G_x(s,\alpha_0(s),\beta_0(s))ds} d\xi = p_0\alpha_0.\end{aligned}$$

A similar technique is used to show $\beta_0(t) \geq p_0\beta_0(t)$ using a different differential inequality.

We then need to show that $p_0\alpha_0(t)$ and $p_0\beta_0(t)$ are still lower and upper solutions of (2.3.6) respectively. We start with

$$\begin{aligned}p_0\alpha_0(t) &= x_0 e^{\int_0^t G_x(s,\alpha_0(s),\beta_0(s))ds} \\ &+ \int_0^t H(\xi, \alpha_0(\xi), \beta_0(\xi), \alpha_0(\xi)) e^{\int_\xi^t G_x(s,\alpha_0(s),\beta_0(s))ds} d\xi\end{aligned}$$

and then differentiate to get

$$p_0\alpha_0'(t) = G_x(t, \alpha_0(t), \beta_0(t))p_0\alpha_0(t) + H(t, \alpha_0(t), \beta_0(t), \alpha_0(t)).$$

Using the definitions of G and H we arrive at

$$\begin{aligned}p_0\alpha_0'(t) &= f(t, \alpha_0(t)) - f_x(t, \alpha_0(t))(\alpha_0(t) - p_0\alpha_0(t)) \\ &- (\phi_x(t, \beta_0(t)) - \phi_x(t, \alpha_0(t)))\,(p_0\alpha_0(t) - \alpha_0(t))\,.\end{aligned}$$

By using (2.3.9) with $p_0\alpha_0(t) \geq \alpha_0(t)$ we obtain,

$$\begin{aligned}p_0\alpha_0'(t) &\leq f(t, p_0\alpha_0(t)) + \phi(t, p_0\alpha_0(t)) \\ &-\phi(t, \alpha_0(t)) - \phi_x(t, \beta_0(t))(p_0\alpha_0(t) - \alpha_0(t)).\end{aligned}$$

An application of the mean value theorem with equation (2.3.10), gives the desired inequality,

$$p_0\alpha_0'(t) \leq f(t, p_0\alpha_0(t)).$$

Following the same procedure we obtain the other needed inequality, namely

$$p_0\beta_0'(t) \geq f(t, p_0\beta_0(t)).$$

For notation purposes, we allow $[\alpha_1(t), \beta_1(t)] = [p_0\alpha_0(t), p_0\beta_0(t)]$ as we continue the process.

To complete the convergence, it must be shown that (2.3.2) holds. Again for notation, we let $X_n(t) = [\alpha_n(t), \beta_n(t)] = \prod_{i=0}^{n-1} P_i X_0(t)$. The next operator is obtained by substituting $\alpha_n(t)$ and $\beta_n(t)$ in the appropriate places in (2.3.11), i.e.

$$\begin{aligned}(p_n(x,\alpha_n,\beta_n))(t) &= x_0 e^{\int_0^t G_x(s,\alpha_n(s),\beta_n(s))ds} \\ &\quad + \int_0^t H(\xi,\alpha_n(\xi),\beta_n(\xi),x(\xi)) e^{\int_\xi^t G_x(s,\alpha_n(s),\beta_n(s))ds} d\xi.\end{aligned}$$

Assuming the upper and lower solutions $\alpha_n'(t) \leq f(t,\alpha_n(t))$ and $\beta_n'(t) \geq f(t,\beta_n(t))$ on J using the operator p_n, we can show that $\alpha_n(t) \leq p_n\alpha_n(t)$ and $\beta_n(t) \geq p_n\beta_n(t)$.

Also, $\alpha_n(t) \leq p_n\alpha_n(t)$ is shown as above. Solve the differential inequality

$$\begin{aligned}\alpha_n'(t) &\leq (F_x(t,\alpha_n(t)) - \phi_x(t,\beta_n(t)))\,\alpha_n(t) \\ &\quad + f(t,\alpha_n(t)) - (F_x(t,\alpha_n(t)) - \phi_x(t,\beta_n(t)))\,\alpha_n(t)\end{aligned}$$

with $\alpha_n(0) = \alpha_n$ and $\alpha_n \leq x_0$ to obtain for $t \in J$,

$$\begin{aligned}\alpha_n &\leq x_0 e^{\int_0^t G_x(s,\alpha_n(s),\beta_n(s))ds} \\ &\quad + \int_0^t H(\xi,\alpha_n(\xi),\beta_n(\xi),x(\xi)) e^{\int_\xi^t G_x(s,\alpha_n(s),\beta_n(s))ds} d\xi = p_n\alpha_n.\end{aligned}$$

Similarly for $\beta_n(t) \geq p_n\beta_n(t)$ except for the starting inequality. This proves (2.3.2) and we have $P_nX_n(t) \subseteq X_n(t)$.

To show that $p_n\alpha_n(t)$ and $p_n\beta_n(t)$ are still lower and upper solutions of (2.3.6), we start with

$$\begin{aligned}p_n\alpha_n(t) &= x_0 e^{\int_0^t G_x(s,\alpha_n(s),\beta_n(s))ds} \\ &\quad + \int_0^t H(\xi,\alpha_n(\xi),\beta_n(\xi),\alpha_n(\xi)) e^{\int_\xi^t G_x(s,\alpha_n(s),\beta_n(s))ds} d\xi.\end{aligned}$$

Differentiate and simplify to get

$$p_n\alpha_n'(t) = G_x(t,\alpha_n(t),\beta_n(t))p_n\alpha_n(t) + H(t,\alpha_n(t),\beta_n(t),\alpha_n(t)).$$

Apply the definitions of G and H to arrive at

$$\begin{aligned}p_n\alpha_n'(t) &= f(t,\alpha_n(t)) - f_x(t,\alpha_n(t))(\alpha_n(t) - p_n\alpha_n(t)) \\ &\quad - (\phi_x(t,\beta_n(t)) - \phi_x(t,\alpha_n(t)))\,(p_n\alpha_n(t) - \alpha_n(t)).\end{aligned}$$

Using (2.3.9) with $p_n\alpha_n(t) \geq \alpha_n(t)$, we obtain,

$$p_n\alpha_n'(t) \leq f(t,p_n\alpha_n(t)) + \phi(t,p_n\alpha_n(t)) - \phi(t,\alpha_n(t))$$

$$-\phi_x(t,\beta_n(t))(p_n\alpha_n(t)-\alpha_n(t)).$$

Now if the mean value theorem is used in conjunction with (2.3.10), the desired inequality

$$p_n\alpha_n'(t)\le f(t,p_n\alpha_n(t)),$$

results.

We now assign

$$X_{n+1}(t)=[\alpha_{n+1}(t),\beta_{n+1}(t)]=[p_n\alpha_n(t),p_n\beta_n(t)].$$

By induction we now have an embedded sequence of intervals, such that

$$\underline{X_0(t)}\ \le\ \underline{X_1(t)}\le\cdots\le\underline{X_n(t)}\le\overline{X_n(t)}\le\cdots\le\overline{X_1(t)}\le\overline{X_0(t)}$$

$$\text{or } X_n(t)\ \subseteq\ \cdots\subseteq X_1(t)\subseteq X_0(t) \text{ for } t\in J.$$

The convergence of the sequence $\{X_n(t)\}$ is established.

Next we prove uniqueness of the solution by showing that (2.3.5) is satisfied for each P_i. Begin with the width of an operator interval and apply (2.3.12) and the definition of the operator p_i,

$$\begin{aligned} w\left[P_n\left[X(t)\right]\right] &= w\left[p_n\underline{X(t)},p_n\overline{X(t)}\right] \\ &= \int_0^t\left[H(\xi,\alpha_n(\xi),\beta_n(\xi),\overline{X(\xi)})-H(\xi,\alpha_n(\xi),\beta_n(\xi),\underline{X(\xi)}\right] \\ &\quad\times e^{\int_0^t G_x(s,\alpha_n(s),\beta_n(s))ds}d\xi. \end{aligned}$$

Apply the mean value theorem and the definition of H and G to get

$$\begin{aligned} w\left[P_n\left[X(t)\right]\right] &\le \int_0^t\left[f_x(\xi,\sigma(\xi))-(F_x(\xi,\alpha_n(\xi))-\phi_x(\xi,\beta_n(\xi)))\right] \\ &\quad\times\left(\overline{X(\xi)}-\underline{X(\xi)}\right)e^{\int_\xi^t G_x(s,\alpha_n(s),\beta_n(s))ds}d\xi, \end{aligned}$$

where $\underline{X(t)}<\sigma(t)<\overline{X(t)}$ on J. We introduce the new weighted width defined by $w_0(X)=\max_J w(X(t))e^{-\delta t}$, where δ is an arbitrary constant to be chosen. Insert the weighted term and take the maximum of $w\left[X(t)\right]$ over J, yielding

$$\begin{aligned} w\left[P_n(X(t))\right] &\le w_0[X]\int_0^t\left[f_x(\xi,\sigma(\xi))-(F_x(\xi,\alpha_n(\xi))-\phi_x(\xi,\beta_n(\xi)))\right] \\ &\quad\times(e^{\delta\xi})e^{\int_\xi^t G_x(s,\alpha_n(s),\beta_n(s))ds}d\xi. \end{aligned}$$

Using equation (2.3.9) with $\sigma(t) < \overline{X(t)} = \beta_n(t)$ we obtain

$$\begin{aligned} w\left[P_n(X(t))\right] &\le w_0[X]\int_0^t \left[F_x(\xi,\beta_n(\xi)) - F_x(\xi,\alpha_n(\xi)) + \phi_x(\xi,\beta_n(\xi))\right. \\ &\quad \left. -\phi_x(\xi,\alpha_n(\xi))\right]\cdot(e^{\delta\xi})e^{\int_\xi^t G_x(s,\alpha_n(s),\beta_n(s))ds}d\xi. \end{aligned}$$

Then, on the left hand side insert the weighted width and take the maximum of $w\left[X(t)\right]$ over J giving

$$w_0\left[P_n\left[X\right]\right](e^{\delta t}) \le (M+N)\, w_0[X]\int_0^t (e^{\delta\xi})e^{Q(t-\xi)}d\xi,$$

where $F_x(t,x(t)) - F_x(t,y(t)) \le M$ and $\phi_x(t,x(t)) - \phi_x(t,y(t)) \le N$ for $\alpha_0(t) \le y(t) \le x(t) \le \beta_0(t)$ on J and $|G_x(t,x(t))| \le Q$ for $\alpha_0(t) \le x(t) \le \beta_0(t)$ on J. Simplifying we arrive at

$$w_0\left[P_n\left[X\right]\right] \le \frac{(M+N)e^{QT}}{\delta} w_0\left[X\right].$$

Since δ is arbitrary, set $\delta = 2(M+N)e^{QT}$, which gives the desired contraction (2.3.5).

Having shown the uniqueness of the solution, we proceed to prove (2.3.3). To verify that the interval operators converge uniformly, all we need show is the convergence of the p_i's to an operator that solves (2.3.6). Start with

$$\begin{aligned} \lim_{i\to\infty} \quad & \left[(p_i x)(t) = x_0 e^{\int_0^t G_x(s,\alpha_i(s),\beta_i(s))ds}\right. \\ & \left. + \int_0^t H(\xi,\alpha_i(\xi),\beta_i(\xi),x(\xi))e^{\int_\xi^t G_x(s,\alpha_i(s),\beta_i(s))ds}d\xi\right]. \end{aligned}$$

Because of the uniqueness of solution, proven above, we are allowed to make the following substitutions for the endpoints of the current intervals, $\alpha_i(t) \to x(t)$ and $\beta_i(t) \to x(t)$. It follows that

$$\begin{aligned} (p(x))(t) &= x_0 e^{\int_0^t G_x(s,x(s),x(s))ds} \\ &\quad + \int_0^t H(\xi,x(\xi),x(\xi),x(\xi))e^{\int_\xi^t G_x(s,x(s),x(s))ds}d\xi. \end{aligned}$$

Differentiate with respect to t to obtain

$$x'(t) = (p(x))'(t) = x(t)G_x(t,x(t),x(t)) + H(t,x(t),x(t),x(t)).$$

Apply the definition of H in terms of G to produce the desired differential equation,

$$x'(t) = (p(x))'(t) = f(t, x(t)).$$

Upon reintegration the desired limit of the operators is obtained,

$$(p(x))(t) = x_0 + \int_0^t f(s, x(s))ds.$$

The uniform convergence of the p_i's guarantees the uniform convergence of the P_i's. The p_i's converge to the desired interval operator, the solution of (2.3.6). Now apply Theorem 2.3.1 which states that $\{X_n(t)\}$ converges to the unique solution of (2.3.6).

Finally, we show that the convergence of the sequence of interval functions is quadratic. Consider the width operator, w, which gives the width of the enclosed interval, and consider also any interval, $X_n(t)$, such that $X_n(t) \subseteq X_0(t)$. Start with

$$w\left[P_n\left(X_n(t)\right)\right] = p_n\overline{X_n(t)} - p_n\underline{X_n(t)}.$$

We then apply the definition of our operator p_n, but instead of using $\alpha_n(t)$ and $\beta_n(t)$, we make the substitution $\underline{X_n(t)} = \alpha_n(t)$ and $\overline{X_n(t)} = \beta_n(t)$ to represent the endpoints of the current interval we are operating on. This yields

$$\begin{aligned} w\left[P_n\left(X_n(t)\right)\right] & \\ = & \int_0^t \left[H\left(\xi, \underline{X_n(\xi)}, \overline{X_n(\xi)}, \overline{X_n(\xi)}\right) - H\left(\xi, \underline{X_n(\xi)}, \overline{X_n(\xi)}, \underline{X_n(\xi)}\right)\right] \\ & \times e^{\int_\xi^t G_x(s, \underline{X_n(s)}, \overline{X_n(s)})ds} d\xi. \end{aligned}$$

Apply the definition of H and the mean value theorem giving

$$\begin{aligned} w\left[P_n\left(X_n(t)\right)\right] \leq & \int_0^t \left[f_x\left(\xi, \sigma(\xi)\right) - G_x\left(\xi, \underline{X_n(\xi)}, \overline{X_n(\xi)}\right)\right] \\ & \times \left(\overline{X_n(\xi)} - \underline{X_n(\xi)}\right) e^{\int_\xi^t G_x(s, \underline{X_n(s)}, \overline{X_n(s)})ds} d\xi \end{aligned}$$

where $\underline{X_n(t)} < \delta(t) < \overline{X_n(t)}$, on J. Using the definition of G, we get

$$\begin{aligned} w\left[P_n\left(X_n(t)\right)\right] \leq & \int_0^t \left[f_x(\xi, \sigma(\xi)) - F_x(\xi, \underline{X_n(\xi)}) + \phi_x(t, \overline{X_n(\xi)})\right] \\ & \times \left(\overline{X_n(\xi)} - \underline{X_n(\xi)}\right) e^{\int_\xi^t G_x(s, \underline{X_n(s)}, \overline{X_n(s)})ds} d\xi. \end{aligned}$$

Using inequality (2.3.9) and the increasing nature of ϕ, gives

$$\begin{aligned} w\left[P_n\left(X_n(t)\right)\right] & \\ &\leq \int_0^t \Big[F_x\left(\xi,\overline{X_n(\xi)}\right) - F_x\left(\xi,\underline{X_n(\xi)}\right) + \phi_x\left(t,\overline{X_n(\xi)}\right) \\ &\quad -\phi_x\left(t,\underline{X_n(\xi)}\right)\Big]\left(\overline{X_n(\xi)} - \underline{X_n(\xi)}\right) \\ &\quad \times e^{\int_\xi^t G_x(s,\underline{X_n(s)},\overline{X_n(s)})ds}d\xi. \end{aligned}$$

Apply the mean value theorem to F and ϕ to get

$$\begin{aligned} w\left[P_n\left(X_n(t)\right)\right] \leq & \int_0^t \left[F_{xx}\left(\xi,\mu(\xi)\right) + \phi_{xx}\left(t,\delta(\xi)\right)\right] \\ & \times\left(\overline{X_n(\xi)} - \underline{X_n(\xi)}\right) e^{\int_\xi^t G_x(s,\underline{X_n(s)},\overline{X_n(s)})ds}d\xi. \end{aligned}$$

Then,

$$w\left[P_n\left(X_n(t)\right)\right] \leq (M+N)\int_0^t w\left[X_n(t)\right]^2 e^{Q(t-\xi)}d\xi$$

where $\phi_{xx}(t,x) \leq N$, $F_{xx}(t,x) \leq M$ and $|G_x(t,y,x)| \leq Q$ on J when $\alpha_0(t) \leq y \leq x \leq \beta_0(t)$. Take the maximum and simplify, yielding

$$\max_J\left[P_n\left(X_n(t)\right)\right] \leq \frac{(M+N)\,e^{QT}}{Q}\max_j w\left[X_n(t)\right]^2.$$

This is the desired result. It indicates that the operator associated with each P_i produces inequality (2.3.4) in Theorem 2.3.1, thus the sequence converges quadratically. The proof of Theorem 2.3.2 is complete.

In the next result we consider the differential equation when the forcing function is the sum of a convex and a concave function.

We begin with the IVP

$$x' = f(t,x) + g(t,x),\ x(0) = x_0 \tag{2.3.13}$$

where $f,\ g \in C\left[J\times I\!R, I\!R\right]$ and $J = [0,T]$; with the following assumptions:

(A_3) $\alpha_0,\ \beta_0 \in C^1\left[J, I\!R\right]$, $\alpha_0(t) \leq \beta_0(t)$, with $\alpha_0'(t) \leq f(t,\alpha_0(t)) + g(t,\alpha_0(t))$, and $\beta_0'(t) \geq f(t,\beta_0(t)) + g(t,\beta_0(t))$, on J with $\alpha_0(0) \leq x_0 \leq \beta_0(0)$.

(A_4) $f,\ g \in C^{0.,2}\left[\Omega, I\!R\right]$, $f_{xx}(t,x) \geq 0$, $g_{xx}(t,x) \leq 0$ on Ω and where $\Omega = \left[(t,x) : \alpha_0(t) \leq x \leq \beta_0(t),\ t \in J\right]$.

Assumption (A_4) provides the following inequalities for f

$$f_x(t,x) \geq f_x(t,y),\ \alpha_0(t) \leq y \leq x \leq \beta_0(t) \text{ on } J, \tag{2.3.14}$$

$$f(t,x) - f(t,y) \geq f_x(t,y)(x-y),\ \alpha_0(t) \leq y \leq x \leq \beta_0(t) \text{ on } J, \tag{2.3.15}$$

as well as for g

$$g_x(t,x) \leq g_x(t,y),\ \alpha_0(t) \leq y \leq x \leq \beta_0(t) \text{ on } J, \tag{2.3.16}$$

$$g(t,x) - g(t,y) \geq g_x(t,y)(x-y),\ \alpha_0(t) \leq y \leq x \leq \beta_0(t) \text{ on } J, \tag{2.3.17}$$

We now proceed to the main result.

Theorem 2.3.3 *Consider the initial value problem given by* (2.3.13), *with assumption* (A_3) *and* (A_4). *There exists a decreasing sequence of interval functions* $\{X_n(t)\}$, *produced from the initial interval function* $X_0(t) = [\alpha_0(t), \beta_0(t)]$, *such that* $\lim_{n\to\infty} X_n(t) = X(t)$ *exists with* $X(t) = [x(t), x(t)]$, *where* $x(t)$ *is the unique solution of* (2.3.13). *In addition, the sequence* $[X_n(t)]$ *converges quadratically.*

Proof With (A_3) and (A_4) holding, consider the sequence of initial value problems

$$\begin{aligned} x'(t) &= (f_x(t,\alpha_i(t) + g_x(t,\beta_i(t)))\,x(t) + f(t,x(t)) - (f_x(t,\alpha_i(t)) + g_x(t,\beta_i(t)))\,x(t) \\ x(0) &= x_0 \end{aligned}$$

for $i = 0, \cdots, k$. This new IVP has a linear term $x'(t) = (f_x(t,\alpha_i(t) + g_x(t,\beta_i(t)))\,x(t)$ and the perturbation term $f(t,x(t)) - (f_x(t,\alpha_i(t)) + g_x(t,\beta_i(t)))\,x(t)$. The IVP is the same as (2.3.13). However we have created some important properties. First, the differential equation has a linear term so we can find an integral operator that is a solution. More importantly, the perturbation term $f_x(t,\alpha_i(t)) + g_x(t,\beta_i(t))$ will make our operator increasing, and it will allow the operator to contain both convexity and concavity property supplied by our assumptions (2.3.14)-(2.3.17).

Using variation of parameters, we solve the new IVP for $x(t)$. Initially we solve the linear term $x'(t) = (f_x(t,\alpha_i(t)) + g_x(t,\beta_i(t)))\,x(t)$ which gives

$$x(t) = x_0 e^{\int_0^t H_x(s,\alpha_i(s),\beta_i(s))ds}$$

where $H_x(t,\alpha_i(t),\beta_i(t)) = f_x(t,\alpha_i(t)) + g_x(t,\beta_i(t))$ with $\alpha_i(t) \leq \beta_i(t)$. Upon completion of the technique, the following sequence $p_i : C[J,\mathbb{R}] \to C[J,\mathbb{R}]$ of integral operators is produced

$$\begin{aligned}(p_i(x,\alpha_i,\beta_i))(t) &= x_0 e^{\int_0^t H_x(s,\alpha_i(s),\beta_i(s))ds} \\ &\quad + \int_0^t D(\xi,\alpha_i(\xi),\beta_i(\xi),x(\xi)) \qquad (2.3.18) \\ &\quad \times e^{\int_\xi^t H_x(s,\alpha_i(s),\beta_i(s))ds} d\xi \qquad (2.3.19)\end{aligned}$$

where $D(t,\alpha_i(t),\beta_i(t),x) = f(t,x) + g(t,x) - H_x(t,\alpha_i(t),\beta_i(t))x$ with $\alpha_i(t) \leq x \leq \beta_i(t)$. A sequence of integral interval operators, $P_i : CI[J] \to CI[J]$, is defined using the p_is. Using an interval function defined by $X(t) \subseteq [\alpha_i(t),\beta_i(t)]$ we define

$$\begin{aligned}P_i(X,\alpha_i,\beta_i)(t) &= \\ [p_i(x,\alpha_i,\beta_i) &: x(t) \in X(t),\ \alpha_i(t) = \underline{X(t)},\ \beta_i(t) = \overline{X(t)} \text{ for } t \in J\Big].\end{aligned}$$

For notational purposes we let

$$p_i x(t) = (p_i(x,\alpha_i,\beta_i))(t)$$

and

$$P_i X(t) = (P_i(X,\alpha_i,\beta_i))(t).$$

These interval integral operators are united extensions of the original integral operators which makes them inclusion monotonic. We can show that the p_i's are increasing operators by showing that $G(t,\alpha_i(t),\beta_i(t),x)$ is increasing in x for each p_i when $\alpha_0(t) \leq x \leq \beta_0(t)$ on J. Differentiating D with respect to x gives

$$D_x(t,\alpha_i(t),\beta_i(t),x) = f_x(t,x) + g_x(t,x) - H_x(t,\alpha_i(t),\beta_i(t)).$$

Using this definition of H we get

$$D_x(t,\alpha_i(t),\beta_i(t),x) \geq f_x(t,x) - f_x(t,\alpha_i(t)) - (g_x(t,\beta_i(t)) - g_x(t,x)).$$

It is easy to show that $H_x(t,\alpha,\beta,x) \geq 0$ using (2.3.14) and (2.3.16). Since D is increasing in x it follows that each p_i is also increasing in x. This allows us to apply our new operators in the following manner

$$P_i\left[\underline{X(t)},\overline{X(t)}\right] = \left[p_i\underline{X(t)},p_i\overline{X(t)}\right] \text{ for } i = 0,\cdots,k. \qquad (2.3.20)$$

We start building our sequence of interval functions from the function $X_0(t) = [\alpha_0(t), \beta_0(t)]$. We show that $P_0 X_0(t) \subseteq X_0(t)$ in order to establish the decreasing nature of the sequence. All that is necessary is to apply (2.3.20) and show that $\alpha_0(t) \leq p_0\alpha_0(t)$ and $\beta_0(t) \geq p_0\beta_0(t)$, where p_0 is the operator

$$\begin{aligned}(p_0 x)(t) &= x_0 e^{\int_0^t h_x(s,\alpha_0(s),\beta_0(s))ds} \\ &\quad + \int_0^t D(\xi, \alpha_0(\xi), \beta_0(\xi), x(\xi)) \\ &\quad e^{\int_\xi^t h_x(s,\alpha_0(s),\beta_0(s))ds} d\xi.\end{aligned}$$

If we solve the differential inequality,

$$\begin{aligned}\alpha_0'(t) &= (f_x(t, \alpha_0(t) + g_x(t, \beta_0(t)))\,\alpha_0(t) \\ &\quad + f(t, \alpha_0(t)) - (f_x(t, \alpha_0(t)) + g_x(t, \beta_0(t)))\,\alpha_0(t)\end{aligned}$$

where $\alpha_0(0) = \alpha_0$, by using variation of parameters on the linear terms with $\alpha_0 \leq x_0 \leq \beta_0$, it follows that

$$\begin{aligned}\alpha_0 &\leq x_0 e^{\int_0^t H_x(s,\alpha_0(s),\beta_0(s))ds} \\ &\quad + \int_0^t D(\xi, \alpha_0(\xi), \beta_0(\xi), \alpha_0(\xi)) \\ &\quad \times e^{\int_\xi^t H_x(s,\alpha_0(s),\beta_0(s))ds} d\xi = p_0\alpha_0.\end{aligned}$$

Therefore $\alpha_0(t) \leq p_0\alpha_0(t)$ for all $t \in J$, as necessary. We show $\beta_0(t) \geq p_0\beta_0(t)$ in a similar manner but the differential inequality is slightly varied.

Next, we must show that $p_0\alpha_0(t)$ and $p_0\beta_0(t)$ are lower and upper solutions of (2.3.13) respectively. Starting with

$$\begin{aligned}p_0\alpha_0(t) &= x_0 e^{\int_0^t H_x(s,\alpha_0(s),\beta_0(s))ds} \\ &\quad + \int_0^t D(\xi, \alpha_0(\xi), \beta_0(\xi), \alpha_0(\xi)) \\ &\quad \times e^{\int_\xi^t H_x(s,\alpha_0(s),\beta_0(s))ds} d\xi\end{aligned}$$

we differentiate producing

$$p_0\alpha_0'(t) = H_x(t, \alpha_0(t), \beta_0(t)) p_0\alpha_0(t) + D(t, \alpha_0(t), \beta_0(t), \alpha_0(t)).$$

By using the definitions of D and H we arrive at

$$\begin{aligned}p_0\alpha_0'(t) &= f(t, \alpha_0(t)) + g(t, \alpha_0(t)) - f_x(t, \alpha_0(t))(\alpha_0(t) - p_0\alpha_0(t)) \\ &\quad - g_x(t, \beta_0(t))\,(p_0\alpha_0(t) - \alpha_0(t)).\end{aligned}$$

We apply (2.3.15) and (2.3.17) with $p_0\alpha_0(t) \geq \alpha_0(t)$ and $p_0\beta_0(t) \leq \beta_0(t)$ to obtain,

$$\begin{aligned} p_0\alpha_0'(t) &\leq f(t, p_0\alpha_0(t)) + g(t, p_0\alpha_0(t)) - g_x(t, p_0\alpha_0(t))(p_0\alpha_0(t) - \alpha_0(t)) \\ &\quad + g_x(t, \beta_0(t))(p_0\alpha_0(t) - \alpha_0(t)). \end{aligned}$$

An application of equation (2.3.16) with $p_0\alpha_0(t) \leq \beta_0(t)$, yields the desired inequality,

$$p_0\alpha_0'(t) \leq f(t, p_0\alpha_0(t)) + g(t, p_0\alpha_0(t)).$$

We follow the same procedure to show that $p_0\beta_0(t)$ is an upper solution by giving the inequality $p_0\beta_0'(t) \geq f(t, p_0\beta_0(t)) + g(t, p_0\beta_0(t))$. To simplify the notation we allow $[\alpha_1(t), \beta_1(t)] = [p_0\alpha_0(t), p_0\beta_0(t)]$.

In order to complete the convergence of the sequence we must show that (2.3.2) holds. For simplicity, we let $X_n(t) = [\alpha_n(t), \beta_n(t)] = \prod_{i=0}^{n-1} P_i X_0(t)$. We produce the next operator by inserting $\alpha_n(t)$ and $\beta_n(t)$ in the appropriate places in (2.3.19), i.e.

$$\begin{aligned} (p_n(x, \alpha_n, \beta_n))(t) &= x_0 e^{\int_0^t G_x(s, \alpha_n(s), \beta_n(s))ds} + \int_0^t H(\xi, \alpha_n(\xi), \beta_n(\xi), x(\xi)) \\ &\quad \times e^{\int_\xi^t G_x(s, \alpha_n(s), \beta_n(s))ds} d\xi. \end{aligned}$$

We assume that $\alpha_n(t)$ and $\beta_0(t)$ are the upper and lower solutions $\alpha_n'(t) \leq f(t, \alpha_n(t))$ and $\beta_n'(t) \geq f(t, \beta_n(t))$ on J. We use the result in [52] in order to show that $\alpha_n(t) \leq p_n\alpha_n(t)$ and $\beta_n(t) \geq p_n\beta_n(t)$, which gives us $P_n X_n(t) \subseteq X_n(t)$.

We show $\alpha_n(t) \leq p_n\alpha_n(t)$ by solving the differential inequality

$$\begin{aligned} \alpha_n'(t) &\leq (f_x(t, \alpha_n(t)) + g_x(t, \beta_n(t)))\, \alpha_n(t) \\ &\quad + f(t, \alpha_n(t)) - (f_x(t, \alpha_n(t)) + g_x(t, \beta_n(t)))\, \alpha_n(t) \end{aligned}$$

with $\alpha_n(0) = \alpha_0$. Since $\alpha_n \leq x_0$ we get

$$\begin{aligned} \alpha_n(t) &\leq x_0 e^{\int_0^t H_x(s, \alpha_n(s), \beta_n(s))ds} \\ &\quad + \int_0^t D(\xi, \alpha_n(\xi), \beta_n(\xi), x(\xi)) e^{\int_\xi^t H_x(s, \alpha_n(s), \beta_n(s))ds} d\xi \\ &= p_n\alpha_n. \end{aligned}$$

We obtain $\beta_n(t) \geq p_n\beta_n(t)$ in a similar manner except we use a related starting inequality. This proves (2.3.2) and we have $P_n X_n(t) \subseteq X_n(t)$.

Now we need to show that $p_n\alpha_n(t)$ and $p_n\beta_n(t)$ are also lower and upper solution (2.3.13). We start with

$$\begin{aligned} p_n\alpha_n(t) &= x_0 e^{\int_0^t H_x(s,\alpha_n(s),\beta_n(s))ds} \\ &\quad + \int_0^t D(\xi,\alpha_n(\xi),\beta_n(\xi),\alpha_n(\xi)) \\ &\quad \times e^{\int_\xi^t H_x(s,\alpha_n(s),\beta_n(s))ds} d\xi. \end{aligned}$$

Differentiating the equation with respect to t gives,

$$p_n\alpha_n'(t) = H_x(t,\alpha_n(t),\beta_n(t))p_n\alpha_n(t) + D(t,\alpha_n(t),\beta_n(t),\alpha_n(t)).$$

Applying the definitions of H and D to get

$$\begin{aligned} p_n\alpha_n'(t) &= f(t,\alpha_n(t)) + g(t,\alpha_n(t)) - f_x(t,\alpha_n(t))(\alpha_n(t) - p_n\alpha_n(t)) \\ &\quad - g_x(t,\beta_n(t))\left(p_n\alpha_n(t) - \alpha_n(t)\right). \end{aligned}$$

Using (2.3.15) and (2.3.17) with $p_n\alpha_n(t) \geq \alpha_n(t)$ and $p_n\beta_n(t) \leq \beta_n(t)$, respectively, we obtain,

$$\begin{aligned} p_n\alpha_n'(t) &\leq f(t,p_n\alpha_n(t)) + g(t,p_n\alpha_n(t)) - g_x(t,\alpha_n(t))(p_n\alpha_n(t) - \alpha_n(t)) \\ &\quad + g_x(t,\beta_n(t))(p_n\alpha_n(t) - \alpha_n(t)). \end{aligned}$$

Now utilizing (2.3.16) with $p_n\alpha_n(t) \leq \beta_n(t)$, we have the desired inequality,

$$p_n\alpha_n'(t) \leq f(t,p_n\alpha_n(t)) + g(t,p_n\alpha_n(t)).$$

We now assign

$$X_{n+1}(t) = [\alpha_{n+1}(t),\beta_{n+1}(t)] = [p_n\alpha_n(t),p_n\beta_n(t)]$$

to simplify the notation. We have produced an embedded sequence of intervals,

$$\begin{aligned} \underline{X_0(t)} &\leq \underline{X_1(t)} \leq \cdots \leq \underline{X_n(t)} \leq \overline{X_n(t)} \leq \cdots \leq \overline{X_1(t)} \leq \overline{X_0(t)} \\ \text{or } X_n(t) &\subseteq \cdots \subseteq X_1(t) \subseteq X_0(t) \text{ for } t \in J, \end{aligned}$$

which yields the convergence of the sequence $\{X_n(t)\}$ as desired.

Next we prove uniqueness of the solution. This is obtained by showing that (2.3.5) is satisfied for each P_n. We start by looking at the width an operator interval, and then use (2.3.20) and the definition of the operator p_n, to obtain

$$\begin{aligned} w\left[P_n\left[X(t)\right]\right] &= w\left[p_n\underline{X(t)}, p_n\overline{X(t)}\right] \\ &= \int_0^t \left[D(\xi,\alpha_n(\xi),\beta_n(\xi),\overline{X(\xi)}) - D(\xi,\alpha_n(\xi),\beta_n(\xi),\underline{X(\xi)}\right] \\ &\quad \times e^{\int_0^t D_x(s,\alpha_n(s),\beta_n(s))ds} d\xi. \end{aligned}$$

We utilize the definition of D and then apply the mean value theorem to functions f and g to get

$$\begin{aligned} w\left[P_n\left[X(t)\right]\right] &\le \int_0^t \left[f_x(\xi,\sigma(\xi)) + g_x(\xi,\delta(\xi)) - H_x(\xi,\alpha_n(\xi),\beta_n(\xi))\right] \\ &\quad \times \left(\overline{X(\xi)} - \underline{X(\xi)}\right) e^{\int_\xi^t H_x(s,\alpha_n(s),\beta_n(s))ds} d\xi, \end{aligned}$$

where $\underline{X(t)} < \sigma(t) < \overline{X(t)}$ and $\underline{X(t)} < \delta(t) < \overline{X(t)}$ on J. Using a new weighted width defined by $w_0(X) = \max_J w(X(t))e^{-\gamma t}$, where γ is an arbitrary constant to be chosen, we insert the weight and take the maximum, yielding

$$\begin{aligned} w\left[P_n(X(t))\right] &\le w_0[X] \int_0^t \left[f_x(\xi,\sigma(\xi)) + g_x(\xi,\delta(\xi)) - H_x(\xi,\alpha_n(\xi),\beta_n(\xi))\right] \\ &\quad \times (e^{\gamma\xi}) e^{\int_\xi^t H_x(s,\alpha_n(s),\beta_n(s))ds} d\xi. \end{aligned}$$

Using the definition of H to obtain

$$\begin{aligned} w\left[P_n(X(t))\right] &\le w_0[X] \int_0^t \left[f_x(\xi,\sigma(\xi)) - f_x(\xi,\alpha_n(\xi)) + g_x(\xi,\delta(\xi))\right. \\ &\quad \left. - g_x(\xi,\alpha_n(\xi))\right] (e^{\gamma\xi}) e^{\int_\xi^t H_x(s,\alpha_n(s),\beta_n(s))ds} d\xi. \end{aligned}$$

Then, on the left hand side insert the weighted width and take the maximum of $w\left[X(t)\right]$ over J giving

$$w_0\left[P_n\left[X\right]\right](e^{\gamma t}) \le (M+N)\, w_0[X] \int_0^t (e^{\gamma\xi}) e^{Q(t-\xi)} d\xi,$$

where $f_x(t,x(t)) - f_x(t,y(t)) \le M$ and $g_x(t,y(t)) - g_x(t,x(t)) \le N$ for $\alpha_0(t) \le y(t) \le x(t) \le \beta_0(t)$ on J and $|H_x(t,x(t))| \le Q$ for $\alpha_0(t) \le x(t) \le \beta_0(t)$ on J. Simplifying gives

$$w_0\left[P_n\left[X\right]\right] \le \frac{(M+N)e^{QT}}{\gamma} w_0\left[X\right].$$

Since γ is arbitrary, we can set $\gamma = 2(M+N)e^{QT}$, which gives a contraction mapping as needed in (2.3.5), proving uniqueness.

We proceed to prove (2.3.3), that the interval operators, P_i's, converge uniformly. All we need show is the uniform convergence of the p_i's, since they produce the P_i's. We begin with

$$\lim_{i\to\infty} \Bigg[(p_i x)(t) = x_0 e^{\int_0^t H_x(s,\alpha_i(s),\beta_i(s))ds} + \int_0^t D(\xi,\alpha_i(\xi),\beta_i(\xi),x(\xi)) e^{\int_\xi^t H_x(s,\alpha_i(s),\beta_i(s))ds} d\xi\Bigg].$$

Because of the uniqueness of the solution, proven above, we are allowed to make the following substitutions for the endpoints of the current intervals, $\alpha_i(t) \to x(t)$ and $\beta_i(t) \to x(t)$. Therefore

$$(p(x))(t) = x_0 e^{\int_0^t H_x(s,x(s),x(s))ds} + \int_0^t D(\xi,x(\xi),x(\xi),x(\xi)) e^{\int_\xi^t H_x(s,x(s),x(s))ds} d\xi.$$

Differentiating with respect to t, we obtain

$$x'(t) = (p(x))'(t) = x(t)H_x(t,x(t),x(t)) + D(t,x(t),x(t),x(t)).$$

Apply the definition of D in terms of H to yield the desired original differential equation (2.3.13),

$$x'(t) = (p(x))'(t) = f(t,x(t)).$$

we then integrate to obtain the standard integral representation of our solution,

$$(p(x))(t) = x_0 + \int_0^t f(s,x(s))ds.$$

Furthermore, the uniform convergence of the p_i's guarantees the uniform convergence of the P_i's. We may now apply Theorem 2.3.1 which uses this result to show that $\{X_n(t)\}$ converges to the unique solution of (2.3.13).

Finally, we show that the convergence is quadratic. Start by considering the usual width operator, w. Then we examine any interval, $X_n(t)$, where $X_n(t) \subseteq X_0(t)$. Therefore,

$$w\left[P_n\left(X_n(t)\right)\right] = p_n\overline{X_n(t)} - p_n\underline{X_n(t)}.$$

We then apply the definition of our operator p_n, but instead of using $\alpha_n(t)$ and $\beta_n(t)$, we make the substitution $\underline{X_n(t)} = \alpha_n(t)$ and $\overline{X_n(t)} = \beta_n(t)$ to represent the endpoints of the current interval we are operating on. This gives

$$\begin{aligned} w\left[P_n\left(X_n(t)\right)\right] &= \int_0^t \Big[D\left(\xi, \underline{X_n(\xi)}, \overline{X_n(\xi)}, \overline{X_n(\xi)}\right) \\ &\quad -D\left(\xi, \underline{X_n(\xi)}, \overline{X_n(\xi)}, \underline{X_n(\xi)}\right)\Big] \\ &\quad \times e^{\int_\xi^t H_x(s,\underline{X_n(s)},\overline{X_n(s)})ds} d\xi. \end{aligned}$$

We apply the definition of D and use the mean value theorem on f and g to get

$$\begin{aligned} w\left[P_n\left(X_n(t)\right)\right] &\leq \int_0^t \left[f_x\left(\xi, \sigma(\xi)\right) + g_x(\xi, \delta(\xi))\right. \\ &\quad \left.-f_x(\xi, \underline{X_n(\xi)}) - g_x(\xi, \overline{X_n(\xi)})\right] \\ &\quad \times \left(\overline{X_n(\xi)} - \underline{X_n(\xi)}\right) e^{\int_\xi^t H_x(s,\underline{X_n(s)},\overline{X_n(s)})ds} d\xi \end{aligned}$$

where $\underline{X_n(t)} < \delta(t), \sigma(t) < \overline{X_n(t)}$, on J. Using equation (2.3.14) and (2.3.16) with $\delta(t) < \overline{X_n(t)}$ and $\underline{X_n(t)} < \sigma(t)$, respectively, we obtain

$$\begin{aligned} w\left[P_n\left(X_n(t)\right)\right] &\leq \int_0^t \Big[f_x(\xi, \overline{X_n(\xi)}) - f_x(\xi, \underline{X_n(\xi)}) \\ &\quad +g_x(t, \overline{X_n(\xi)}) - g_x(\xi, \overline{X_n(\xi)}\Big] \\ &\quad \times \left(\overline{X_n(\xi)} - \underline{X_n(\xi)}\right) e^{\int_\xi^t H_x(s,\underline{X_n(s)},\overline{X_n(s)})ds} d\xi. \end{aligned}$$

Applying the mean value theorem to f_x and g_x again produces

$$\begin{aligned} w\left[P_n\left(X_n(t)\right)\right] &\leq \int_0^t \left[f_{xx}\left(\xi, \mu(\xi)\right) - g_{xx}\left(t, \eta(\xi)\right)\right] \left(\overline{X_n(\xi)} - \underline{X_n(\xi)}\right)^2 \\ &\quad \times e^{\int_\xi^t H_x(s,\underline{X_n(s)},\overline{X_n(s)})ds} d\xi. \end{aligned}$$

Then,

$$w\left[P_n\left(X_n(t)\right)\right] \leq (M+N) \int_0^t w\left[X_n(t)\right]^2 e^{Q(t-\xi)} d\xi$$

where $-g_{xx}(t,x) \leq N$, $f_{xx}(t,x) \leq M$ and $|H_x(t,x,y)| \leq Q$ on J when $\alpha_0(t) \leq y(t) \leq x(t) \leq \beta_0(t)$. Taking the maximum and simplifying, yields

$$\max_J \left[P_n\left(X_n(t)\right)\right] \leq \frac{(M+N)\, e^{QT}}{Q} \max_j w\left[X_n(t)\right]^2.$$

This result indicates that the operator of each P_n produces inequality (2.3.4) in Theorem 2.3.1, thus the sequence converges quadratically. The proof of Theorem 2.3.3 is complete.

2.4 HIGHER ORDER CONVERGENCE

A method to solve a nonlinear equation is to find a sequence of approximate solutions converging to a solution of the given problem. One of the well known techniques in this direction is the method of upper and lower solutions coupled with the monotone iterative technique. It provides a constructive procedure for the solutions and the iterates serve as upper and lower bounds for those solutions.

On the other hand, it is very important to have higher order of convergence of a numerical scheme from the practical point of view. We recall here that for a convergent sequence $\{x_n\} \to x$, the order of convergence is $k = 1, 2, \cdots$ if there exists $\lambda > 0$ such that

$$\|x_{n+1} - x\| \leq \lambda \|x_n - x\|^k .$$

For $k = 2$ we say that the convergence is quadratic.

It is easy to see that the convergence of the sequence of approximate solutions given by the monotone iterative technique is very slow. Indeed, that convergence is linear. As we know, under some convexity conditions, the methods of quasilinearization provides a monotone increasing sequence converging uniformly and quadratically to the solution.

It would be very important to give some general methods that give monotone sequences converging to a solution with order of convergence $k \geq 2$.

To be specific, let us consider the following initial value problem

$$u' = f(t, u), \quad u(0) = u_0, \, t \in [0, T], \quad T > 0, \tag{2.4.1}$$

where $f : I \times R \to R$ is a continuous function.

As usual, we say that $\alpha \in C^1(I)$ is a lower solution for the problem (2.4.1) if

$$\alpha'(t) \leq f(t, \alpha(t)), \quad t \in I, \quad \alpha(0) \leq u_0.$$

Similarly, $\beta \in C^1(I)$ is an upper solution for the problem (2.4.1) if

$$\beta'(t) \geq f(t, \beta(t)), \quad t \in I, \quad \beta(0) \geq u_0.$$

It is well known that $\alpha \leq \beta$ on I implies that there exists at least one solution u of (2.4.1), with $u \in [\alpha, \beta] = \{v \in C(I) : \alpha(t) \leq v(t) \leq \beta(t), t \in I\}$.

Moreover, if there exists $M > 0$ such that

$$f(t, u) + Mu \geq f(t, v) + Mv \tag{2.4.2}$$

for every $t \in I$ and $\alpha(t) \leq v \leq u \leq \beta(t)$, then it is possible to construct monotone sequences $\{\alpha_n(t)\}$ nondecreasing, and $\{\beta_n(t)\}$ nonincreasing, which converge uniformly to the extremal solution ϕ and ψ of (2.4.1). We insist that, in general, the order of convergence of the monotone sequences is at most $k = 1$ in the space $E = C(I)$ with the usual uniform norm.

If there exists $M_1 > 0$ such that $f(t,u) + M_1 u \leq f(t,v) + M_1 v$ whenever $u \geq v$, then (2.4.1) possesses a unique solution.

Note that if $\partial f / \partial u$ exists and it is continuous in

$$\Omega = \{(t,u) \in I \times R;\ \alpha(t) \leq u \leq \beta(t)\},$$

then we have the uniqueness for (2.4.1) and inequality (2.4.2) holds, and it is equivalent to the following condition:

$$\frac{\partial f}{\partial u}(t,u) \geq -M,\ (t,u) \in \Omega. \tag{2.4.3}$$

As we have seen the method of generalized quasilinearization does not demand $f(t,u)$ to be convex in u for $t \in I$ but imposes the following less restrictive condition: there exists an $M > 0$ such that

$$f(t,u) + Mu^2 \text{ is convex in } u \text{ for any } t \in I. \tag{2.4.4}$$

If $\partial^2 f / \partial u^2$ exists and it is continuous for every $(t,u) \in \Omega$, then (2.4.4) holds and it is equivalent to

$$\frac{\partial^2 f}{\partial u^2}(t,u) \geq -2M,\ (t,u) \in \Omega. \tag{2.4.5}$$

If this last condition is satisfied, then there exists a nondecreasing sequence starting at the lower solution and converging uniformly and quadratically to the unique solution of (2.4.1).

In this section, we show that we can generate two monotone sequences of approximate solutions, one nondecreasing starting at the lower solution and the other one nonincreasing starting at the upper solution, that converge uniformly to the unique solution of (2.4.1) and the order of convergence is k provided that there exists $\partial^k f / \partial u^k$, and it is continuous in Ω. Observe that in this case there exists $M > 0$ such that

$$\frac{\partial^k f}{\partial u^k}(t,u) \geq -(k!)M,\ (t,u) \in \Omega. \tag{2.4.6}$$

Note that for $k = 1$ we have (2.4.3) and for $k = 2$ we obtain (2.4.5). Thus, we extend previous results. Moreover, we obtain two sequences of approximation solution. The sequence of lower solutions is provided generalizing the ideas for the case $k = 2$. However, to construct the sequence of upper solution it is necessary to modify the method.

As is known, if the property (2.4.2) is verified then there exist two monotone sequences $\{\alpha_n\}$ and $\{\beta_n\}$ which converge uniformly to the unique solution of (2.4.1) lying in $[\alpha, \beta]$. We recall that $\alpha_0 = \alpha$, $\beta_0 = \beta$, and for any $n \geq 1$, α_n and β_n are given as the unique solution of the following linear problems:

$$\alpha_n' + M\alpha_n = f(t, \alpha_{n-1}) + M\alpha_{n-1}, \quad \alpha_n(0) = u_0$$

and

$$\beta_n' + M\beta_n = f(t, \beta_{n-1}) + M\beta_{n-1}, \quad \beta_n(0) = u_0. \tag{2.4.7}$$

However, this convergence is not rapid in general, as we can see in the following example.

Example 2.4.1 Let us consider the following initial value problem

$$u' = f(u), \quad t \in I; \quad u(0) = 0, \tag{2.4.8}$$

where $f : R \to R$ is defined as follows:

$$f(u) = \begin{cases} -u & \text{if } u \leq -1 \\ -2u - 1/n & \text{if } u \in I_n = (-1/n, -(2n+1)/(2n(n+1))) \quad n = 1, 2, \cdots \\ 1/(n+1) & \text{if } u \in [-(2n+1)/(2n(n+1)), -1/(n+1)] \quad n = 1, 2, \cdots \\ 0 & \text{if } u = 0 \\ -2u + 1/n & \text{if } u \in J_n = ((2n+1)/(2n(n+1)), 1/n) \quad n = 1, 2, \cdots \\ -1/(n+1) & \text{if } u \in [1/(n+1), (2n+1)/(2n(n+1))] \quad n = 1, 2, \cdots \\ -u & \text{if } u \geq 1. \end{cases}$$

Clearly $|f(t,u) - f(t,v)| \leq 2|u - v|$ for all u, $v \in R$, Thus, (2.4.8) has a unique solution $u \equiv 0$, and (2.4.2) holds with $M = 2$.

By the theory of differential inequalities, we have that any lower solution α of (2.4.1) is such that $\alpha \leq 0$ in I, and any upper solution β of (2.4.1) satisfies $\beta \geq 0$ in I. Therefore, if α is a strict lower solution (that is, a lower solution but not a solution), and β is a strict upper solution, we have $\alpha(t) < \beta(t)$ for some $t \in [0, T]$. In consequence, $M = 2$is the smallest positive constant for which (2.4.2) is valid since $f$is linear on the interval I_n and J_n.

For simplicity, let $\alpha < 0$ be a constant lower solution and $\beta > 0$ be a constant upper solution.

Thus, let $\beta_0 = \beta$. For $n \geq 1$, β_n is given by the expression (2.4.7) with $M = 2$:

$$\beta_{n+1}(t) = \int_0^t e^{2(s-t)} \left[f(\beta_n(s)) + 2\beta_n(s)\right] ds.$$

Since $\beta > 0$, we have that $\beta_1(t) > 0$ for all $t \in (0, T]$. By induction, $\beta_n(t) > 0$ for all $t \in (0, T]$ and $n \in N$. In consequence, using that $f(u) \geq -u$ for all $u \geq 0$, we have that

$$\beta_{n+1}(t) \geq e^{-2T} \int_0^t \beta_n(s) ds \equiv f_n(t).$$

Now, let $\lambda > 0$ fixed, and $g_n(t) = \lambda(\beta_n(t))^2$. Using expression (2.4.7), we see that β_n' converges uniformly to $u \equiv 0$ in I. So, there exists $n_0 \in N$ such that

$$2\lambda\beta_n'(t) < e^{-2T}, \quad \text{for all } n \geq n_0 \text{ and } t \in I.$$

But this last expression is equivalent to assure that $f_n'(t) > g_n'(t)$ for all $n \geq n_0$ and $t \in I$. Therefore, since $f_n(0) = g_n(0) = 0$, we conclude that $f_n(t) > g_n(t)$ for all $n \geq n_0$ and $t \in (0, T]$. As a consequence, we have proved that for every $\lambda > 0$ fixed, there exists $n_0(\lambda) \in N$, such that

$$\|\beta_{n+1}\| > \lambda \|\beta_n\|^2, \text{ for all } n \geq n_0.$$

That is, $\{\beta_n\} \to 0$ linearly but not quadratically.

Analogously, $\{\alpha_n\} \to 0$ uniformly and linearly, but not quadratically.

Theorem 2.4.1 *Suppose that there exist $\alpha \leq \beta$, α, β are lower and upper solutions respectively for the problem (2.4.1). If there exists $\frac{\partial^k f}{\partial u^k}$, $k \geq 1$, and it is continuous in Ω, then there exist two monotone sequences $\{\alpha_n\}$ and $\{\beta_n\}$ with $\alpha_0 = \alpha$ and $\beta_0 = \beta$, which converge uniformly to the solution ψ of (2.4.1). This convergence is of order k.*

Proof We first note that of the partial derivatives $\partial^k f/\partial u^k$ implies that (2.4.1) has a unique solution ψ between α and β.

To construct the sequence $\{\alpha_n\}$, let $t \in I$ and $\alpha(t) \leq v \leq u \leq \beta(t)$. We first note that for given $t \in I$:

$$f(t, u) = \sum_{i=0}^{k-1} \frac{\partial^i f}{\partial u^i}(t, v)\frac{(u-v)^i}{i!} + \frac{\partial^k f}{\partial u^k}(t, \xi(t))\frac{(u-v)^k}{k!}, \tag{2.4.9}$$

with $\xi(t) \in [v, u]$.

Now, since there exists $\partial^k f/\partial u^k$ and it is continuous in Ω, (2.4.6) is verified.

Thus, we define

$$g(t,u,v) = \sum_{i=0}^{k-1} \frac{\partial^i f}{\partial u^i}(t,v)\frac{(u-v)^i}{i!} - M(u-v)^k. \tag{2.4.10}$$

As a consequence, using (2.4.6), we obtain that

$$g(t,v,u) \le f(t,u), \text{ for all } t \in I \text{ and } \alpha(t) \le v \le u \le \beta(t). \tag{2.4.11}$$

Now, let us consider the initial value problem

$$u'(t) = g(t,u(t),\alpha(t)),\ t \in I,\ u(0) = u_0. \tag{2.4.12}$$

Now,

$$\beta'(t) \ge f(t,\beta(t)) \ge g(t,\beta(t),\alpha(t)),\ t \in I,\ \beta(0) \ge u_0$$

and

$$\alpha'(t) \le f(t,\alpha(t)) = g(t,\alpha(t),\alpha(t)),\ t \in I\ ,\ \alpha(0) \le u_0,$$

that is, α and β are lower and upper solutions for (2.4.12) respectively.

Theorem 1.1.5 shows that α_1 is the unique solution of (2.4.12), with $\alpha_1 \in [\alpha,\beta]$.

Now, suppose we have constructed $\alpha_0 = \alpha \le \alpha_1 \le \cdots \le \alpha_n \le \beta$. Where α_n is the unique solution of

$$u'(t) = g(t,u(t),\alpha_{n-1}(t)),\ t \in I,\ u(0) = u_0.$$

In this case, we have that

$$\beta'(t) \ge f(t,\beta(t)) \ge g(t,\beta(t),\alpha_n(t)),\ \beta(0) \ge u_0$$

and

$$\alpha_n'(t) = g(t,\alpha_n(t),\alpha_{n-1}(t)) \le f(t,\alpha_n(t)) = g(t,\alpha_n(t),\alpha_n(t)),\ \alpha_n(0) = u_0.$$

We conclude, using again Theorem 1.1.5, that there exists a unique $\alpha_{n+1} \in [\alpha_n,\beta]$, solution of

$$u'(t) = g(t,u(t),\alpha_n(t)),\ t \in I,\ u(0) = u_0. \tag{2.4.13}$$

Thus, $\{\alpha_n\}$ is a nondecreasing sequence, which is bounded in $C^1(I)$. Then $\{\alpha_n\}$ converges in $C(I)$ to some continuous function $\psi \in [\alpha,\beta]$.

Since

$$\alpha_n(t) = u_0 + \int_0^t g(s, \alpha_n(s), \alpha_{n-1}(s))ds,$$

we have that

$$\psi(t) = u_0 + \int_0^t g(s, \psi(s), \psi(s))ds = u_0 + \int_0^t f(s, \psi(s))ds.$$

Therefore, ψ is the unique solution of (2.4.1) in $[\alpha, \beta]$.

Now, we prove that the convergence is of order k. For that, using (2.4.9), we have that

$$\sum_{i=0}^{k-1} \frac{\partial^i f}{\partial u^i}(t, \alpha_n(t))\frac{(\psi(t) - \alpha_n(t))^i}{i!} + \frac{\partial^k f}{\partial u^k}(t, \rho_n(t))\frac{(\psi(t) - \alpha_n(t))^k}{k!}, \ \psi(0) = u_0,$$

$\rho_n \in [\alpha_n, \psi]$.

On the other hand, by (2.4.10) and (2.4.13), it is verified that for $n \geq 0$

$$\begin{aligned} \alpha'_{n+1}(t) &= g(t, \alpha_{n+1}(t), \alpha_n(t)) \\ &= \sum_{i=0}^{k-1} \frac{\partial^i f}{\partial u^i}(t, \alpha_n(t))\frac{(\alpha_{n+1}(t) - \alpha_n(t))^i}{i!} - M\frac{(\alpha_{n+1}(t) - \alpha_n(t))^k}{k!}, \\ \alpha_{n+1}(0) &= u_0. \end{aligned}$$

Let $e_n = \psi - \alpha_n$ and $a_n = \alpha_{n+1} - \alpha_n$. Thus, we have that

$$\begin{aligned} e'_{n+1} &= \sum_{i=1}^{k-1} \frac{\partial^i f}{\partial u^i}(t, \alpha_n(t))\left[\frac{e_n^i(t) - a_n^i(t)}{i!}\right] \\ &\quad + \frac{\partial^k f}{\partial u^k}(t, \rho_n(t))\frac{e_n^k(t)}{k!} + Ma_n^k(t), \ e_{n+1}(0) = 0. \end{aligned}$$

Now, there exists $N > 0$ such that

$$\frac{\partial^k f}{\partial u^k(t, x)} \leq (k!)N \text{ for all } t \in I \text{ and } x \in [\alpha(t), \beta(t)]. \tag{2.4.14}$$

Furthermore, $a_n^k(t) \leq e_n^k(t)$, for all $n \in N$ and $t \in I$. Finally, using that for all A, $B \in R$, $A^i - B^i = (A - B)\sum_{j=0}^{i-1} A^{i-1-j}B^j$ we can write that

$$e'_{n+1}(t) - p_n(t)e_{n+1}(t) \leq ce_n^k(t), \ t \in I,$$

where $c = N + M > 0$ and

$$p_n(t) = \sum_{i=1}^{k-1} \frac{\partial^i f}{\partial u^i}(t, \alpha_n(t)) \left(\frac{1}{i!} \sum_{j=0}^{i-1} e_n^{i-1-j}(t) a_n^j(t) \right).$$

Therefore, we have

$$\begin{aligned} 0 &\leq e_{n+1}(t) \leq c \int_0^t e_n^k(s) \exp\left(\int_s^t p_n(r) dr \right) ds \\ &\leq c \max_{s \in I} \{ e_n^k(s) \} \int_0^{2\pi} \exp\left(\int_0^t p_n(r) dr \right) ds = \lambda \| e_n \|^k, \end{aligned}$$

which implies

$$\| \alpha_{n+1} - \psi \| \leq \lambda \| \alpha_n - \psi \|^k.$$

To construct the sequence $\{\beta_n\}$ we define the following function:

$$h(t, u, v) = \begin{cases} \sum_{i=0}^{k-1} \frac{\partial^i f}{\partial u^i}(t, v) \frac{(u-v)^i}{i!} - M(u - v)^k & \text{if } k \text{ odd} \\ \sum_{i=0}^{k-1} \frac{\partial^i f}{\partial u^i}(t, v) \frac{(u-v)^i}{i!} + N(u - v)^k & \text{if } k \text{ even.} \end{cases}$$

Here M and N are nonnegative constant given by (2.4.6) and (2.4.14) respectively.

Thus, it is easy to see that

$$h(t, u, v) \geq f(t, u), \text{ for all } t \in I \text{ and } \alpha(t) \leq u \leq v \leq \beta(t). \tag{2.4.15}$$

Now, let $\beta_0 = \beta$. For $n \geq 1$ we define β_n by induction, as the unique solution of the following initial value problem

$$u'(t) = h(t, u(t), \beta_{n-1}(t)), \, t \in I, \, u(0) = u_0. \tag{2.4.16}$$

2.5 ANOTHER REFINEMENT OF QUASILINEARIZATION

As we have seen in Section 1.2, while describing the original method of quasilinearization of Bellman and Kalaba, that the relation (1.2.5) provides an explicit analytic representation of the solution of (1.2.1). Since its value lies not so much in the explicit representation as in the pointwise estimates that can be derived from (1.2.3) by simple choices of $y(t)$, we described in several sections the generalizations, extensions and refinements in this direction.

In this section, we shall concentrate on the explicit representation of the solution and by transforming this solution formula into another form, so that it can be applied to a wider class of problems. In other words, we shall refine the original formula to expand its applicability.

Consider the equation

$$\frac{dy}{dt} = h(t, y), \tag{2.5.1}$$

where $h(t, y)$ is continuous and has a bounded first derivative $h'_y(t, y)$ in some domain containing the point $(0, c)$. Moreover, $h(t, y)$ is assumed to be strictly convex in y for any fixed t. The solution $\phi(t)$, $t \in [0, b]$ of equation (2.5.1) through the point $(0, c)$ can be expressed as

$$\begin{aligned}\phi(t) \quad = \quad & \max_v \left[c e^{\int_0^t h'_y(\xi, v(\xi)) d\xi} + \int_0^t e^{\int_\xi^t h'_y(\xi, v(\xi)) d\xi} (h(\xi, v(\xi)) \right. \\ & \left. - v(\xi) h'_y(\xi, v(\xi))) d\xi \right], \ t \in [0, b], \end{aligned} \tag{2.5.2}$$

where $v(t)$, $t \in [0, b]$ is an admissible function, i.e., a function for which all the integrals involved exist. The maximum is attained for $v = \phi$ and that the second derivative $h''_{y^2}(t, y)$ exists and is bounded is not necessary for this result, but in the following considerations we assume that $h'_y(t, y)$ is continuous.

Now, any continuous function $v(t)$, $t \in [0, b]$ is admissible and the maximum value in (2.5.2) is also valid with $v(t)$, $t \in [0, b]$ in the smaller class of continuous functions. In this form it can easily be transformed by the substitution

$$z(t) = e^{-\int_0^t h'_y(\xi, v(\xi)) d\xi}, \ t \in [0, b]. \tag{2.5.3}$$

Notice that $z(t)$, $t \in [0, b]$ is a positive C^1 function. By a straightforward calculation the expression under the max sign in (2.5.2) becomes

$$\frac{1}{z(t)} \left[c + \int_0^t \left[h(\xi, v(\xi)) z(\xi) + v(\xi) z'(\xi) \right] d\xi \right]. \tag{2.5.4}$$

Observe that the function

$$y \to h(t, y) z(t) + y z'(t), \ t \text{ fixed}$$

has the derivative $z(t) \left[h'_y(t, y) - h'_y(t, v(t)) \right]$, which changes sign at $y = v(t)$. Hence

$$\min_y \left(h(t, y) z(t) + y z'(t) \right) = h(t, v(t)) z(t) + v(t) z'(t)$$

for any fixed t. By (2.5.4) the representation (2.5.2) can be rewritten as

$$\phi(t) = \max_z \left[\frac{1}{z(t)} \left(c + \int_0^t \min_y \left(h(\xi, y) z(\xi) + y z'(\xi) \right) d\xi \right) \right], \ t \in [0, b], \tag{2.5.5}$$

where $z(t)$, $t \in [0, b]$ is given by (2.5.3) with $v \in [[0, b], R]$ is continuous. Any such function is in the class of all positive C^1 functions $z(t)$, $t \in [0, b]$ for which $-z'(t)[z(t)]^{-1}$ for every fixed t belongs to the domain of values for the function $y \to h'_y(t, y)$ and for which $z(0) = 1$. It is easy to see that the converse is also true and that the connection between a continuous v and a z in the afore mentioned class is one to one. In other words, the representation (2.5.5) with z in the mentioned class is another form of the Bellman-Kalaba formula (2.5.2).

Suppose now that the convexity assumption is altered into *weak* convexity. An easy inspection of the proof shows that the representation (2.5.2) still holds, and so does the representation (2.5.5) with z given by (2.5.3). However, we may no longer conclude that the connection between a continuous v and a z in the afore-mentioned class is one to one. Nonetheless the representation (2.5.5) still holds. Instead of examining the substitution (2.5.3) with $v(t)$, $t \in [0, b]$ is only admissible, we give a direct proof of this statement. It turns out that this proof holds if z is allowed to belong to a certain larger class than the above mentioned, giving a representation (2.5.5) with inf instead of min. It should be pointed out that the integrand in the representation so obtained will be locally integrable only in extended sense, when $h(t, y)$ is only weakly convex in y.

We wish to consider (2.5.1) with weak convexity in a domain $I_0 \times J_0$, where I_0 and J_0 are open intervals, and prove the representation corresponding to (2.5.5) for the solution $\phi(t)$ through an arbitrary point $(t_0, y_0) \in I_0 \times J_0$. Moreover, it will be shown that this representation under further assumptions remains valid if the solution $\phi(t)$ is characterized by a point (t_0, y_0) on the boundary of $I_0 \times J_0$. All the results also apply mutatis mutandis if $h(t, y)$ instead of being weakly convex is weakly concave in y for any fixed t.

Throughout this section, I and J denote intervals.

Lemma 2.5.1 *Let $u(t, y)$, $(t, y) \in I \times J$ be a continuous function. Assume further that the function $u(t, y)$, $y \in J$ for every fixed $t \in I$ is strictly increasing and attains the value zero. Then there exists a unique continuous function $\psi(t)$, $t \in I$ such that*

$$u(t, \psi(t)) = 0 \text{ for all } t \in I. \tag{2.5.6}$$

Proof The existence and uniqueness of the function $\psi(t)$, $t \in I$ is obvious. We show that it is continuous in every point $t_0 \in I$. Let $\epsilon > 0$ be given. If $\psi(t_0)$ is an interior point in J we may suppose that $\psi(t_0) \pm \varepsilon$ are also interior points. Then we get $u(t_0, \psi(t_0) - \varepsilon) < 0$ and $u(t_0, \psi(t_0) + \varepsilon) > 0$. Hence some $\delta > 0$ can be chosen such that $u(t, \psi(t_0) - \varepsilon) < 0$ and $u(t, \psi(t_0) + \varepsilon) > 0$ hold for all $t \in (t_0 - \delta, t_0 + \delta) \cap I$. We conclude that $\psi(t_0) - \varepsilon < \psi(t) < \psi(t_0) + \varepsilon$

for all $t \in (t_0 - \delta, t_0 + \delta) \cap I$. If $\psi(t_0)$ is an endpoint of J, this reasoning holds with a slight modification.

Clearly, Lemma 2.5.1 remains valid if *increasing* is substituted by *decreasing*. From the example

$$u(t, y) = \begin{cases} t(1-y) & \text{if } t < 0 \\ ty & \text{if } t \geq 0 \end{cases}$$

it follows that Lemma 2.5.1 does not remain valid if the function $u(t, y)$, $y \in J$is allowed to be only weakly monotonic for some $t \in I$.

Next, we mention the well-known fact that if $u(t, y)$, $t \in I \times J$ is continuous and J is *compact* then $\min_{y \in J} u(t, y)$, $t \in J$ and $\max_{y \in J} u(t, y)$, $t \in I$ are continuous functions.

Lemma 2.5.2 *Let $u(t, y)$, $(t, y) \in I \times J$ be a continuous function having a continuous derivative $u'_y(t, y)$, $(t, y) \in I \times J$, which is assumed to be weakly increasing in $y \in J$ for every fixed $t \in I$. Assume further that*

$$m(t) = \inf_{y \in J} u(t, y),\ t \in I \tag{2.5.7}$$

is finite. Then $m(t)$, $t \in I$ is Lebesgue measurable and in extended sense locally Lebesgue integrable. Moreover, $m(t)$ is continuous in a neighborhood of the point $t_0 \in I$ if the function $u'_y(t, y)$, $y \in J$ either assumes both negative and positive values for $t = t_0$ or is identically zero for all t in some neighborhood of t_0.

Proof We note that if y_t is an interior point of J then $m(t) = u(t, y_t) \iff u'_y(t, y_t) = 0$.

Let (J_n) be an increasing sequence of compact subintervals of J with $\bigcup_{n \in \mathbb{N}} J_n = J$. By the remark preceding the lemma we may define the continuous function

$$m(t) = \min_{y \in J_n} u(t, y),\ t \in I,$$

$n \in \mathbb{N}$. Clearly $m_n(t) \to m(t)$ as $n \to +\infty$ pointwise and decreasing for $t \in I$. Hence the function (2.5.7) is measurable. Moreover, if $[a, b]$ is a subinterval of I then the integral

$$\int_a^b m(t)dt$$

either exists as a real number or exists as the value $-\infty$, i.e., $m(t)$, $t \in [a, b]$ is integrable in the extended sense.

Suppose next that there exist interior points y_1, y_2 in J such that $u'_y(t_0, y_1) < 0$ and $u'_y(t_0, y_2) > 0$. By an argument of the same type as in Lemma 2.5.1 we can choose $\delta > 0$ such that

$$\forall t \in I : |t - t_0| < 0 \Longleftrightarrow \exists y \in [y_1, y_2] : u'_y(t, y) = 0.$$

From the first remark of the proof we conclude that

$$m(t) = \min_{y \in [y_1, y_2]} u(t, y),\ t \in (t_0 - \delta, t_0 + \delta) \cap I.$$

By the remark preceding the lemma, $m(t)$ is continuous in $(t_0 - \delta, t_0 + \delta) \cap I$.

If $u'_y(t, y) = 0$ for all $y \in J$ and all t in some neighborhood U of t_0, then for any fixed $y_0 \in J$ we have

$$m(t) = u(t, y_0),\ t \in U,$$

proving the last statement of the lemma.

Remark 2.5.1 Lemma 2.5.2 remains valid with (2.5.7) replaced by

$$M(t) = \sup_{y \in J} u(t, y),\ t \in I, \tag{2.5.8}$$

if $u'_y(t, y)$ is weakly decreasing instead of increasing in $y \in J$ for every fixed $t \in I$.

The example

$$u(t, y) = t^4 y^2 - 2ty,\ (t, y) \in I\!R^2,$$

where

$$m(t) = \begin{cases} -\frac{1}{t^2} & \text{if } t \neq 0 \\ 0 & \text{if } t = 0, \end{cases}$$

shows that the function (2.5.7) may not be locally integrable in the ordinary sense.

We consider the equation

$$\frac{dy}{dt} = h(t, y),\ (t, y) \in I_0 \times J_0, \tag{2.5.9}$$

where I_0 and J_0 are open intervals and $h(t, y)$ a continuous function having a continuous derivative $h'_y(t, y)$, $(t, y) \in I_0 \times J_0$, which is assumed to be weakly increasing in $y \in J_0$ for every fixed $t \in I_0$. In other words, $h(t, y)$ is weakly convex in y for every fixed t. It follows that there exists a unique maximal solution through any given point $(t_0, y_0) \in I_0 \times J_0$.

For every fixed $t \in I_0$ we denote by $A_0(t)$ the domain of values for the function $h'_y(t,y)$, $y \in J_0$, i.e.,

$$A_0(t) = \left\{a \in \mathbb{R} | \exists y \in J_0 : a = h'_y(t,y)\right\}, \; t \in I_0. \tag{2.5.10}$$

Further, we define the set

$$A(t) = \left\{a \in \mathbb{R} | \inf_{y \in J_0} (h(t,y) - ay) > -\infty\right\}, \; t \in I_0. \tag{2.5.11}$$

If $a \in A_0(t)$ then $h(t,y) - ay$, $y \in J_0$ has a minimum. Hence

$$A_0(t) \subseteq A(t) \text{ for all } t \in I_0. \tag{2.5.12}$$

It is not difficult to describe the precise connection between the two sets $A_0(t)$ and $A(t)$. Since it depends on the type of the interval J_0, this description, however, is a little extensive. Apart from the proof of Lemma 2.5.3 it is not necessary to know what follows. Therefore we postpone the description (and the proof of Lemma 2.5.3) to the end of this section.

In the following considerations I denotes an arbitrary subinterval of I_0. Let $z(t)$, $t \in I$ be any positive C^1-function. Equation. (2.5.9) is equivalent to

$$\frac{dy}{dt}(zy) = h(t,y)z(t) + yz'(t), \; (t,y) \in I \times J_0 \tag{2.5.13}$$

in the domain $I \times J_0$. We define the continuous function

$$H_z(t,y) = h(t,y)z(t) + yz'(t), \; (t,y) \in I \times J_0, \tag{2.5.14}$$

having the continuous derivative

$$\frac{\partial}{\partial y} H_z(t,y) = h'_y(t,y)z(t) + z'(t), \; (t,y) \in I \times J_0, \tag{2.5.15}$$

which is weakly increasing in y for any fixed $t \in I$. Suppose further that the function $z(t)$, $t \in I$ satisfies the condition

$$\forall t \in I : -\frac{z'(t)}{z(t)} \in A(t). \tag{2.5.16}$$

Then the function $H_z(t,y)$, $y \in J_0$ has a finite infimum for any fixed $t \in I$ and we may define

$$m_z(t) = \inf_{y \in J_0} H_z(t,y), \; t \in I. \tag{2.5.17}$$

If in particular the function $z(t)$, $t \in I$ satisfies the stronger condition

$$\forall t \in I : -\frac{z'(t)}{z(t)} \in A_0(t). \tag{2.5.18}$$

then the function $\partial/\partial y H_z(t,y)$, $y \in J_0$ has a zero for any fixed $t \in I$. Then inf in (2.5.17) can be replaced by min. We note that (2.5.18) implies the existence of a $y_t \in J_0$ such that $z'(t) + h'_y(t, y_t)z(t) = 0$, by which

$$m_z(t) = H_z(t, y_t),\ t \in I. \tag{2.5.19}$$

Note that the definition (2.5.17) is related to the Fenchel transformation of the function $h(t,y)$, $y \in J_0$, i.e., the function

$$h^*(t,w) = \sup_{y \in J_0} (wy - h(t,y)),\ t \in I_0,\ w \in A(t).$$

It is seen that

$$m_z(t) = -z(t)h^*\left(t, -\frac{z'(t)}{z(t)}\right),\ t \in I.$$

By Lemma 2.5.2 the function (2.5.17) is a Lebesgue measurable function, which is locally Lebesgue integrable in extended sense.

By $\mathcal{M}\{I\}$ we denote the class of all positive C^1-functions $z(t)$, $t \in I$ satisfying condition (2.5.16). By $\mathcal{M}_0\{I\}$ we denote the subclass of functions in $\mathcal{M}\{I\}$ satisfying condition (2.5.18).

If $v : I \to J_0$ is any continuous function, then each of the positive C^1-functions

$$z(t) = Ce^{-\int_{t_0}^{t} h'_y(\xi, v(\xi))d\xi},\ t \in I, \tag{2.5.20}$$

where $t_0 \in I$ is a fixed point of I and C a positive constant, satisfies condition (2.5.18). Hence they will belong to $\mathcal{M}_0\{I\}$, which is consequently nonempty.

Lemma 2.5.3 *Let I and I_1 be subintervals of I_0, $I_1 \subset I$, such that an endpoint of I_1 belonging to I also belongs to I_1. Then the class $\mathcal{M}\{I_1\}$ consists of the restrictions to I_1 of the functions in the class $\mathcal{M}\{I\}$.*

The proof of Lemma 2.5.3 and further discussion of the sets A_0, A_1 will be given at the end of the section to avoid interruption.

Before proceeding we summarize: If $z(t)$, $t \in I$ belongs to the class $\mathcal{M}\{I\}$, then (2.5.9) is equivalent to

$$\frac{d}{dt}(zy) = H_z(t,y),\ (t,y) \in I \times J_0 \tag{2.5.21}$$

in the domain $I \times J_0$. The right hand side of (2.5.21) is given by (2.5.14), and the function (2.5.17) is locally integrable in extended sense.

Theorem 2.5.1 *Let $\phi(t)$, $t \in I$ be a maximal solution of* (2.5.9) *and let t_0 be any point of I. Then*

$$\phi(t) = \min_z \left[\frac{1}{z(t)} \left(z(t_0)\phi(t_0) + \int_{t_0}^{t} m_z(\xi)d\xi \right) \right], \; t \in I, \; t \le t_0 \tag{2.5.22}$$

and

$$\phi(t) = \max_z \left[\frac{1}{z(t)} \left(z(t_0)\phi(t_0) + \int_{t_0}^{t} m_z(\xi)d\xi \right) \right], \; t \in I, \; t \ge t_0 \tag{2.5.23}$$

where $z \in \mathcal{M}\{I\}$ in both formulae and

$$m_z(t) = \inf_{y \in J_0} \left(h(t,y)z(t) + yz'(t) \right), \; t \in I. \tag{2.5.24}$$

Moreover, the minimum in (2.5.22) *and the maximum in* (2.5.23) *are attained for the functions*

$$z(t) = Ce^{-\int_{t_0}^{t} h'_y(\xi,\phi(\xi))d\xi}, \; t \in I, \tag{2.5.25}$$

where C is a constant. On the other hand, if for some function $z \in \mathcal{M}\{I\}$ the minimum in (2.5.22) *(the maximum in* (2.5.23)*) is attained for all t in some interval $[t_1, t_0]$ (some interval $[t_0, t_1]$), where $t_1 \in I \setminus \{t_0\}$, then this function coincides with one of the functions* (2.5.25) *in the interval $[t_1, t_0]$ (the interval $[t_0, t_1]$).*

If in particular $h'_y(t,y)$ is strictly increasing in y for every fixed $t \in I_0$ except for a set of measure zero, the last statement can be sharpened as follows. If for some function $z \in \mathcal{M}_0\{I\}$ the minimum in (2.5.22) *(the maximum in* (2.5.23)*) is attained in some point t_1, $t_1 \neq t_0$, then this function coincides with one of the functions* (2.5.25) *in the interval $[t_1, t_0]$ (the interval $[t_0, t_1]$).*

Remark 2.5.2 If the integral $\int_{t_0}^{t} m_z(\xi)d\xi$ $(t > t_0)$ in formula (2.5.23) is not finite, it has the value $-\infty$. In this case the expression under the max-sign in (2.5.23) of course shall be interpreted as $-\infty$. Analogously the same holds for formula (2.5.22).

Proof of Theorem 2.5.1 Since all the functions (2.5.25) are of the form (2.5.20), they all belong to $\mathcal{M}_0\{I\}$ and hence to $\mathcal{M}\{I\}$. Moreover, by (2.5.19)

$$m_z(t) = H_z(t,\phi(t)) = h(t,\phi(t))z(t) + \phi(t)z'(t), \; t \in I \tag{2.5.26}$$

for any of the functions (2.5.25). It is seen that $m_z(t)$, $t \in I$ in this case is continuous and hence locally integrable.

Let $z \in \mathcal{M}\{I\}$ be arbitrary. For any $t \notin I$ we get from (2.5.21) by integration

$$
\begin{aligned}
\phi(t) &= \frac{1}{z(t)}\left[z(t)\phi(t)\right] \\
&= \frac{1}{z(t)}\left[z(t_0)\phi(t_0) + \int_{t_0}^{t} H_z(\xi, \phi(\xi))d\xi\right] \\
&\begin{cases} \leq \frac{1}{z(t)}\left[z(t_0)\phi(t_0) + \int_{t_0}^{t} m_z(\xi)d\xi\right] & \text{if } t \leq t_0 \\ \geq \frac{1}{z(t)}\left[z(t_0)\phi(t_0) + \int_{t_0}^{t} m_z(\xi)d\xi\right] & \text{if } t \geq t_0. \end{cases}
\end{aligned}
$$

By (2.5.26), equality sign holds if $z(t)$, $t \in I$ is one of the functions of the form (2.5.25). From this the two formulae (2.5.22) and (2.5.23) follow, and the statement thereafter holds in the theorem, too.

Next, suppose that, for example,

$$
\phi(t) = \frac{1}{z(t)}\left[z(t_0)\phi(t_0) + \int_{t_0}^{t} m_z(\xi)d\xi\right], \text{ for all } t \in [t_1, t_0]
$$

holds for some $t_1 \in (-\infty, t_0] \cap I$ and some function $z \in \mathcal{M}\{I\}$. Then

$$
\int_{t_0}^{t} m_z(\xi)d\xi = z(t)\phi(t) - z(t_0)\phi(t_0) \text{ for all } t \in [t_1, t_0],
$$

showing that both sides are C^1-functions. Hence

$$
m_z(t) = \frac{d}{dt}\left[z(t)\phi(t)\right], \text{ a.e. for } t \in [t_1, t_0],
$$

i.e.,

$$
\inf_{y \in J_0} H_z(t, y) = H_z(t, \phi(t)), \text{ a.e. for } t \in [t_1, t_0]
$$

by (2.5.17) and (2.5.21). This implies

$$
\frac{\partial}{\partial y} H_z(t, y) \mid_{y=\phi(t)} = 0, \text{ a.e. for } t \in [t_1, t_0]
$$

giving (see (2.5.15))

$$
h'_y(t, \phi(t))z(t) + z'(t) = 0, \text{ a.e. for } t \in [t_1, t_0].
$$

By continuity this holds everywhere in the interval $[t_1, t_0]$. Solving for $z(t)$ we get

$$
z(t) = z(t_0)e^{-\int_{t_0}^{t} h'_y(\xi, \phi(\xi))d\xi} \text{ for all } t \in [t_1, t_0],
$$

which proves the last statement of the general part of the theorem.

Suppose now that the sharpened condition is fulfilled. It then follows from Lemma 2.5.2 that the function $m_z(t)$, $t \in I$ is continuous in a neighborhood of ξ for almost any $\xi \in I$ and for any $z \in \mathcal{M}_0\{I\}$.

Finally, suppose that, for example,

$$\phi(t_1) = \frac{1}{z(t_1)}\left[z(t_0)\phi(t_0) + \int_{t_0}^{t_1} m_z(t)dt\right]$$

holds for some $t_1 \in (-\infty, t_0) \cap I$ and some $z \in \mathcal{M}_0\{I\}$. Then by (2.5.21)

$$\begin{aligned}\int_{t_0}^{t_1} m_z(t)dt &= z(t_1)\phi(t_1) - z(t_0)\phi(t_0) \\ &= \int_{t_0}^{t_1} \frac{d}{dt}[z(t)\phi(t)]\,dt \\ &= \int_{t_0}^{t_1} H_z(t,\phi(t))dt,\end{aligned}$$

i.e.,

$$\int_{t_1}^{t_0} [H_z(t,\phi(t)) - m_z(t)]\,dt = 0. \tag{2.5.27}$$

From the definition (2.5.17) we derive

$$H_z(t,\phi(t)) - m_z(t) \geq 0 \text{ for all } t \in I. \tag{2.5.28}$$

Since $H_z(t,\phi(t)) - m_z(t)$, $t \in I$ is continuous in a neighborhood of ξ for almost any $\xi \in I$ we conclude from (2.5.27) and (2.5.28) that $H_z(t,\phi(t)) - m_z(t) = 0$ a.e. for $t \in [t_1, t_0]$, i.e.,

$$\inf_{y \in J_0} H_z(t,y) = H_z(t,\phi(t)) \text{ a.e. for } t \in [t_1, t_0].$$

From this the desired property follows as before.

It is a straightforward procedure to see that Theorem 2.5.1 holds with inf in (2.5.24) replaced by min if the class $\mathcal{M}\{I\}$ is replaced by the class $\mathcal{M}_0\{I\}$ throughout.

It follows from the proof that for the formulae to be valid it suffices that the functions $z(t)$ are defined only on the interval $(-\infty, t_0] \cap I$ respectively the interval $[t_0, +\infty) \cap I$. In other words, the two formulae are valid with $z \in \mathcal{M}\{(-\infty, t_0] \cap I\}$ and $z \in \mathcal{M}\{[t_0, +\infty) \cap I\}$, respectively. However, Lemma 2.5.3 shows that the two sets of formulae are equivalent. We have preferred to formulate the theorem with a class of functions z, independent on the point t_0 chosen. Further, it follows from (2.5.14) and (2.5.17) that the expression under the min and max signs is positively homogeneous in z. Hence we can impose the restriction $z(t_0) = 1$ on z, giving, for example, the formula

$$\phi(t) = \max_z \left[\frac{1}{z(t)}\left(\phi(t_0) + \int_{t_0}^{t} m_z(\xi)d\xi\right)\right], \; t \in I, \, t \geq t_0, \tag{2.5.29}$$

where z belongs to the class of all functions $z \in \mathcal{M}\{I\}$ with $z(t_0) = 1$. If in particular $t_0 = 0$ and $\phi(t_0)$ is denoted by c, this gives the representation (2.5.5) with inf instead of min as in the introduction (by (2.5.24) and the remark immediately after the proof.)

Example 2.5.1 We consider the Riccati equation

$$\frac{dy}{dt} = y^2 + g(t), \; (t, y) \in I\!R^2,$$

where $g(t)$, $t \in I\!R$ is a continuous function. Let $\phi(t)$, $t \in I$ be a maximal solution and let $t_0 \in I$ be arbitrary. In order to get the representation (2.5.29) we first find the sets $A(t)$, $t \in I$, next the class $\mathcal{M}\{I\}$ and finally the function $m_z(t)$, $t \in I$.

From the definition (2.5.11) it is easily seen that $A(t) = I\!R$ for any $t \in I\!R$, by which condition (2.5.16) is trivially satisfied. Hence, the class $\mathcal{M}\{I\}$ consists of all positive C^1-functions $z(t)$, $t \in I$. It is a straightforward calculation to derive that

$$m_z(t) = g(t)z(t) - \frac{1}{4}\left[z(t)\right]^{-1}\left[z'(t)\right]^2, \; t \in I$$

for any function $z \in \mathcal{M}\{I\}$. Inserting this in (2.5.29) we obtain the representation

$$\phi(t) = \max_z \left[\frac{1}{z(t)}\left(\phi(t_0) + \int_{t_0}^{t}\left(g(\xi)z(\xi) - \frac{1}{4}\left[z(\xi)\right]^{-1}\left[z'(\xi)\right]^2\right)d\xi\right)\right], \; t \in I, \, t \geq t_0,$$

where $z(t)$, $t \in I$ denotes a positive C^1-function with $z(t_0) = 1$. By (2.5.25) with $C = 1$ the maximizing function is

$$z(t) = \exp\left[-2\int_{t_0}^{t}\phi(\xi)d\xi\right], \; t \in I.$$

If in particular $0,\ b \in I$ and we choose $t_0 = 0$, the representation in the subinterval $[0, b]$ becomes

$$\phi(t) = \max_z \left[\frac{1}{z(t)}\left(C + \int_{t_0}^{t}\left(g(\xi)z(\xi) - \frac{1}{4}\left[z(\xi)\right]^{-1}\left[z'(\xi)\right]^2\right)d\xi\right)\right], \; t \in [0, b].$$

This is formula (2.5.5) for the considered Riccati equation.

It is unimportant that we have chosen the function $g(t)$ to be defined for all $t \in \mathbb{R}$, since any continuous function on the interval $[0, b]$ can be extended to a continuous function on $\mathbb{R}$. Hence, the above representation can be directly compared with the corresponding representation, coming from formula (2.5.2). It is seen that in the above representation the variable function z occurs together with its derivative, whereas the admissible function v in (1.9) (loc. cit.) occurs together with its *primitive*. This difference of course comes from the substitution (2.5.3).

In the next example it is crucial that the function $h(t, y)$ in (2.5.9) may be only *weakly* convex in y.

Example 2.5.2 If in particular

$$h(t, y) = p(t)y + q(t), \ (t, y) \in I_0 \times \mathbb{R},$$

where $p(t)$, $t \in I_0$ and $q(t)$, $t \in I_0$ are continuous, equation. (2.5.9) is a first order *linear* equation. Since $h'_y(t, y) = p(t)$ we have $A_0(t) = A(t) = \{p(t)\}$ for all $t \in I_0$. Hence condition (2.5.16) with $I = I_0$ reads

$$z'(t) + p(t)z(t) = 0, \ t \in I_0.$$

We conclude that

$$z(t) \triangleq z_0(t) = e^{-\int_{t_0}^{t} p(\xi)d\xi}, \ t \in I_0 \tag{2.5.30}$$

is the only function in the class $\mathcal{M}\{I_0\}$ with $z(t_0) = 1$, where t_0 denotes some chosen point in I_0. Hence formula (2.5.29) and its analogue for $t \leq t_0$ coincide with

$$\phi(t) = \frac{1}{z_0(t)} \left(\phi(t_0) + \int_{t_0}^{t} m_{z_0}(\xi)d\xi \right), \ t \in I_0, \tag{2.5.31}$$

where by (2.5.24)

$$m_{z_0}(t) = \inf_{y \in \mathbb{R}} \left[(p(t)y + q(t)) z_0(t) + y z_0'(t) \right] = q(t)z_0(t), \ t \in I_0. \tag{2.5.32}$$

If (2.5.30) and (2.5.32) are inserted in (2.5.31) the well-known solution formula occurs.

Example 2.5.3 In this example we present a classical representation, which also can be derived from Theorem 2.5.1.

Consider a curve in the polar plane given by an equation

$$r = p(\theta), \ \theta \in \left[0, \frac{\pi}{2}\right),$$

where $p(\theta)$, $\theta \in [0, \pi/2)$ is a positive C^1-function with $p(0) = 1$. The length of the arc from the initial point to a variable point $P(\theta)$ on the curve is

$$l(\theta) = \int_0^\theta \left([p(\xi)]^2 + [p'(\xi)]^2\right)^{1/2} d\xi,\ \theta \in \left[0, \frac{\pi}{2}\right).$$

The distance from $P(\theta)$ to the polar axis is

$$d(\theta) = p(\theta)\sin\theta,\ \theta \in \left[0, \frac{\pi}{2}\right).$$

The classical theorem on the shortest distance from a point to a line gives $d(\theta) \le l(\theta)$ for all $\theta \in [0, \pi/2)$, i.e.,

$$\sin\theta \le \frac{1}{p(\theta)} l(\theta),\ \theta \in \left[0, \frac{\pi}{2}\right).$$

Moreover, the equality sign is attained for all $\theta \in [0, \pi/2)$ precisely when the curve is the vertical line through the point $P(0)$, i.e., when $p(\theta) = (\cos\theta)^{-1}$, $\theta \in [0, \pi/2)$. Hence

$$\sin\theta = \min_p \frac{1}{p(\theta)} \int_0^\theta \left([p(\xi)]^2 + [p'(\xi)]^2\right)^{1/2} d\xi,\ \theta \in \left[0, \frac{\pi}{2}\right),$$

where $p(\theta)$, $\theta \in [0, \pi/2)$ denotes a positive C^1-function with $p(0) = 1$, and where the only minimizing function is $p(\theta) = (\cos\theta)^{-1}$, $\theta \in [0, \pi/2)$.

It is interesting that this classical representation actually is a special case of (2.5.29)—after a change of sign— which is shown below.

The equation

$$\frac{dy}{dt} = -\left(1 - y\right)^{1/2},\ (x, y) \in I\!R \times (-1, 1)$$

has the desired properties. By inspection $y = -\sin t$, $t \in (-\pi/2, \pi/2)$ is a maximal solution. We will find the representation (2.5.29) with $t_0 = 0$ and proceed as in Example 2.5.1.

Since $J_0 = (-1, 1)$ is bounded, it is obvious that $A(t) = I\!R$ for any $t \in I\!R$. Again, condition (2.5.16) is trivially satisfied. Hence $\mathcal{M}\{(-\pi/2, \pi/2)\}$ consists of all positive C^1-functions $z(t)$, $t \in (-\pi/2, \pi/2)$. Next,

$$\begin{aligned} m_z(t) &= \inf_{y \in (-1,1)} \left(-\left(1 - y^2\right)^{1./2} z(t) + y z'(t)\right) \\ &= -\left(1 - y_0^2\right)^{1/2} z(t) + y_0 z'(t), \end{aligned}$$

where

$$y_0\left(1-y_0^2\right)^{-1/2} z(t)+z'(t)=0,\ t\in\left(-\frac{\pi}{2},\frac{\pi}{2}\right).$$

By an elementary calculation one derives

$$m_z(t)=-\left([z(t)]^2+[z'(t)]^2\right)^{1/2},\ t\in\left(-\frac{\pi}{2},\frac{\pi}{2}\right)$$

for any function $z\in\mathcal{M}\{(-\pi/2,\pi/2)\}$. Inserting in (2.5.29) we get a representation for $\phi(t)=-\sin t$, $t\in[0,\pi/2)$. Changing sign on both sides of this representation we obtain

$$\sin t=\min_z\int_0^t\left([z(\xi)]^2+[z'(\xi)]^2\right)^{1/2}d\xi,\ t\in\left[0,\frac{\pi}{2}\right),$$

where by Lemma 2.5.3 the variable function can be taken as a positive C^1-function $z(t)$, $t\in[0,\pi/2)$ with $z(0)=1$. From (2.5.25) with $C=1$ we get that the only minimizing function is

$$z(t)=\exp\left[-\int_0^t\frac{-\sin\xi}{\cos\xi}d\xi\right]=\frac{1}{\cos t},\ t\in\left[0,\frac{\pi}{2}\right).$$

Apart from the notations this result is the classical one mentioned above.

In the formulae hitherto mentioned t_0 denotes an *interior* point of the interval I. Under further assumptions it is possible to extend the validity to the case where t_0 is an *endpoint* of I. In order to state this we introduce two subclasses of $\mathcal{M}\{I\}$ corresponding to the cases $\inf I$ finite and $\sup I$ finite, respectively.

First, let I be a subinterval of I_0 with $\alpha=\inf I$ finite. By $\mathcal{N}_1\{I\}$ we denote the class of all functions $z\in\mathcal{M}\{I\}$ for which

(i) $z(\alpha^+)$ exists and is finite

(ii) for some $\alpha_1\in I$ be integral (2.5.33)

$\int_0^{\alpha_1} m_z(t)dt$ exists as a value in $[-\infty,+\infty)$.

Recall that $m_z(t)$, $t\in I$ is locally integrable in extended sense for all $z\in\mathcal{M}\{I\}$ (see (2.5.14) and (2.5.17)). Next, let I be a subinterval of I_0 with $\beta=\sup I$ finite. By $\mathcal{N}_2\{I\}$ we denote the class of all functions $z\in\mathcal{M}\{I\}$ for which

(i) $z(\beta^-)$ exists and is finite

(ii) for some $\beta_1\in I$ be integral (2.5.34)

$\int_\beta^{\beta_1} m_z(t)dt$ exists as a value in $[-\infty,+\infty)$.

Clearly, if $\alpha = \inf I$ belongs to I the class $\mathcal{N}_1\{I\}$ which coincides with the class $\mathcal{M}\{I\}$ and is nonempty. Otherwise $\mathcal{N}_1\{I\}$ may be empty. Analogously as to the class $\mathcal{N}_1\{I\}$.

Theorem 2.5.2 *Let $\phi(t)$, $t \in I$ be a maximal solution of* (2.5.9). *If the following conditions*

$$\begin{array}{l} \text{(i) } \alpha = \inf I \text{ is finite} \\ \text{(ii) } \phi(\alpha^+) \text{ exists and is finite} \\ \text{(iii) for some } \alpha_1 \in I \text{ be integral} \\ \quad \int_0^{\alpha_1} h'_y(t,\phi(t))dt \text{ exists as a value in } [-\infty,+\infty) \end{array} \tag{2.5.35}$$

are satisfied, then

$$\phi(t) = \max_z \left[\frac{1}{z(t)}\left(z(\alpha^+)\phi(\alpha^+) + \int_\alpha^t m_z(\xi)d\xi\right)\right], \ t \in I, \tag{2.5.36}$$

where $z \in \mathcal{N}_1\{I\}$. Moreover, the maximum is attained for the functions

$$z(t) = Ce^{-\int_{\alpha_1}^t h'_y(\xi,\phi(\xi))d\xi}, \ t \in I, \tag{2.5.37}$$

where C is a positive constant. If for some function $z \in \mathcal{N}_2\{I\}$ the maximum in (2.5.36) *is attained for all $t \in (\alpha, t_1)$, where $t_1 \in I$, then this function coincides with one of the functions* (2.5.37) *in the interval (α, t_1).*

If the following conditions

$$\begin{array}{l} \text{(i) } \beta = \sup I \text{ is finite} \\ \text{(ii) } \phi(\beta^-) \text{ exists and is finite} \\ \text{(iii) for some } \beta_1 \in I \text{ be integral} \\ \quad \int_\beta^{\beta_1} h'_y(t,\phi(t))dt \text{ exists as a value in } (-\infty,+\infty] \end{array} \tag{2.5.38}$$

are satisfied, then

$$\phi(t) = \min_z \left[\frac{1}{z(t)}\left(z(\beta^-)\phi(\beta^-) + \int_\beta^t m_z(\xi)d\xi\right)\right], \ t \in I, \tag{2.5.39}$$

where $z \in \mathcal{N}_2\{I\}$. Moreover, the minimum is attained for the functions

$$z(t) = Ce^{-\int_{\beta_1}^t h'_y(\xi,\phi(\xi))d\xi}, \ t \in I, \tag{2.5.40}$$

where C is a positive constant. If for some function $z \in \mathcal{N}_2\{I\}$ the minimum in (2.5.39) *is attained for all $t \in [t_1, \beta)$, where $t_1 \in I$, then this function coincides with one of the functions* (2.5.40) *in the interval $[t_1, \beta)$.*

In both formulae

$$\mathrm{m_z(t)} = \inf_{\mathrm{y \in J_0}} \left(\mathrm{h(t,y)z(t) + yz'(t)}\right), \ \mathrm{t \in I.}$$

Remark 2.5.3 The two formulae (2.5.36) and (2.5.39) are generalizations of the two formulae (2.5.23) and (2.5.22), respectively.

Proof of Theorem 2.5.2 Suppose that conditions (2.5.35) are satisfied. As before, any of the functions (2.5.37) belongs to $\mathcal{M}_0\{I\}$ and hence to $\mathcal{M}\{I\}$. If $z(t)$, $t \in I$ is one of these functions, then $z(\alpha^+)$ exists and is finite by (2.5.35)(iii). Moreover, by (2.5.19) and (2.5.21) for any $t \in I$

$$\begin{aligned}\int_{t_1}^{t} m_z(\xi)d\xi &= \int_{t_1}^{t} H_z(\xi,\phi(\xi))d\xi = z(t)\phi(t) - z(t_1)\phi(t_1) \\ &\rightarrow z(t)\phi(t) - z(\alpha^+)\phi(\alpha^+) \text{ as } t_1 \rightarrow \alpha^+.\end{aligned}$$

i.e.,

$$\int_{\alpha}^{t} m_z(t)dt = z(t)\phi(t) - z(\alpha^+)\phi(\alpha^+), \ t \in I. \tag{2.5.41}$$

Hence any of the functions (2.5.37) belongs to the class $\mathcal{N}_1\{I\}$.

Let $z \in \mathcal{N}_1\{I\}$ be arbitrary. Exactly as in the proof of Theorem 2.5.1 we get for $t \in I$, $t_1 \in (\alpha, t)$

$$\begin{aligned}\phi(t) &= \frac{1}{z(t)}\left[z(t_1)\phi(t_1) + \int_{t_1}^{t} H_z(\xi,\phi(\xi))d\xi\right] \\ &\geq \frac{1}{z(t)}\left[z(t_1)\phi(t_1) + \int_{t_1}^{t} m_z(\xi)d\xi\right].\end{aligned}$$

Passing to the limit $t_1 \rightarrow \alpha^+$ we get by (2.5.33)

$$\phi(t) \geq \frac{1}{z(t)}\left[z(\alpha^+)\phi(\alpha^+) + \int_{\alpha}^{t} m_z(\xi)d\xi\right],$$

where the integral may have the value $-\infty$. By (2.5.41) the equality sign holds for all $t \in I$ if in particular $z(t)$, $t \in I$ is one of the functions (2.5.37). This proves (2.5.36) and the statement thereafter. The proof of the statement after (2.5.37) is exactly as the proof of the last statement in the general part of Theorem 2.5.1.

The second part of the theorem is proved analogously.

Remark 2.5.4 Assume that $h'_y(t,y)$ is strictly increasing in y for every fixed $t \in I_0$ except for a set of measure zero. If for some function $z \in \mathcal{M}_0\{I\} \cap \mathcal{N}_1\{I\}$ the maximum in (2.5.36) is attained in some point $t_1 \in I$, then this function coincides with one of the functions (2.5.37) in the interval $(\alpha, t_1]$. If for some function $z \in \mathcal{M}_0\{I\} \cap \mathcal{N}_2\{I\}$ the minimum in (2.5.39) is attained in some point $t_1 \in I$, then this function coincides with one of the functions (2.5.40) in the interval $[t_1, \beta)$.

Proof Suppose that, e.g.,

$$\phi(t) = \frac{1}{z(t)}\left[z(\alpha^+)\phi(\alpha^+) + \int_\alpha^{t_1} m_z(t)dt\right] \tag{2.5.42}$$

holds for some $t_1 \in I$ and some $z \in \mathcal{M}_0\{I\} \cap \mathcal{N}_1\{I\}$. Then the integral in (2.5.42) must be finite and by (2.5.21)

$$\begin{aligned}\int_\alpha^{t_1} m_z(t)dt &= z(t_1)\phi(t_1) - z(\alpha^+)\phi(\alpha^+)\\ &= \lim_{\xi\to\alpha^+}\int_\xi^{t_1} \frac{d}{dt}\left[z(t)\phi(t)\right]dt\\ &= \lim_{\xi\to\alpha^+}\int_\xi^{t_1} H_z(t,\phi(t))dt.\end{aligned}$$

We conclude that $H_z(t,\phi(t))$, $t \in (\alpha, t_1]$ is integrable and

$$\int_\alpha^{t_1}\left[H_z(t,\phi(t)) - m_z(t)\right]dt = 0.$$

Then we proceed exactly as in the corresponding part of the proof of Theorem 2.5.1. The second part of Remark 2.5.4 which relates to Theorem 2.5.2 is proved analogously.

Example 2.5.4 Let $c \in (1, +\infty)$ be arbitrary but fixed. The function $\phi(t) = [\log t]^{1/c}$, $t \in (1, +\infty)$ is a maximal solution of the equation

$$\frac{dy}{dt} = \frac{1}{cty^{c-1}},\ (t,y) \in (0,+\infty)\times(0,+\infty),$$

which satisfies the conditions of equation. (2.5.9). Since

$$\int_1^2\left(\frac{1}{c} - 1\right)\frac{1}{t}\left[\log t\right]^{-1}dt = -\infty,$$

the requirements (2.5.35) are also satisfied. In order to find the representation (2.5.36) we proceed along the lines of Examples 2.5.1 and 2.5.3.

From (2.5.11) we get that $A(t) = (-\infty, 0]$ for any $t \in (0, +\infty)$, by which condition (2.5.16) with $I = (1, +\infty)$ becomes $z'(t) \geq 0$ for all $t \in (1, +\infty)$. Hence, the class $\mathcal{M}\{(1, +\infty)\}$ consists of all positive C^1-functions $z(t)$, $t \in (1, +\infty)$ fro which $z'(t) \geq 0$ for all $t \in (1, +\infty)$. Next,

$$\begin{aligned} m_z(t) &= \inf_{y \in (0,+\infty)} \left(\frac{z(t)}{cty^{c-1}} + yz'(t) \right) \\ &= \left[\frac{c}{c-1} \right]^{1-1/c} \left[\frac{1}{t} z(t) \left[z'(t)\right]^{c-1} \right]^{1/c}, \ t \in (1, +\infty) \end{aligned}$$

by a straightforward (although tedious) calculation. By (2.5.33) for a function $z \in \mathcal{M}\{(1, +\infty)\}$ to belong to the subclass $\mathcal{N}_1\{[1, +\infty)\}$ it is required that $z(1^+)$ exists and that, e.g., the integral

$$\begin{aligned} I_z &= \int_1^2 \left[\frac{1}{t} z(t) \left[z'(t)\right]^{c-1} \right]^{1/c} dt \\ &= \int_1^2 \left[\frac{z(t)}{t} \right]^{1/c} \left[z'(t)\right]^{-1/c} dt \end{aligned}$$

exists as a value in $[-\infty, \infty)$. From the Hölder inequality we get

$$\begin{aligned} \int_t^2 \left[\frac{z(\xi)}{\xi} \right]^{1/c} \left[z'(\xi)\right]^{1-1/c} d\xi &\leq \left[\int_t^2 \frac{z(\xi)}{\xi} d\xi \right] \left[\int_t^2 z'(\xi) \right]^{1-1/c} d\xi \\ &= \left[\int_t^2 \frac{z(\xi)}{\xi} d\xi \right]^{1/c} (z(2) - z(t))^{1-1/c} \end{aligned}$$

for any $t \in (1, 2]$. When $z(1^+)$ exists, this implies the convergence of the integral I_z. Hence, the class $\mathcal{N}_1\{(1, +\infty)\}$ consists of all positive C^1-functions $z(t)$, $t \in (1, +\infty)$ for which $z(1^+)$ exists and for which $z'(t) \geq 0$ for all $t \in (1, +\infty)$. The representation (2.5.36) becomes

$$[\log t]^{1/c} = \left[\frac{c}{c-1} \right]^{1-1/c} \max_z \frac{1}{z(t)} \int_1^t \left[\frac{z(\xi)}{\xi} \left[z'(\xi)\right]^{c-1} \right]^{1/c} d\xi, \ t \in (1, +\infty),$$

where $z \in \mathcal{N}_1\{(1, +\infty)\}$, Notice that since $z(1^+)$ may be zero we cannot impose the restriction $z(1^+) = 1$. By (2.5.37) the maximizing functions are

$$\begin{aligned} z(t) &= C \exp\left[-\int_2^t \left(\frac{1}{c} - 1 \right) \frac{1}{\xi} [\log \xi]^{-1} d\xi \right] \\ &= C_1 [\log t]^{1-1/c}, \qquad t \in (1, +\infty), \end{aligned}$$

where C and C_1 are arbitrary constants.

We may insert a suitable function z in the representation. Choosing, e.g.,

$$z(t)\,[z'(t)]^{c-1} = 1, \quad t \in (1, +\infty)$$

and $z(1^+) = 0$, by which

$$z(t) = \left[\frac{c}{c-1}(t-1)\right]^{1-1/c}, \; t \in (1, +\infty)$$

belongs to $\mathcal{N}_1\,\{(1, +\infty)\}$, we obtain

$$\begin{aligned} [\log t]^{1/c} &\geq \left[\frac{c}{c-1}\right]^{1-1/c} \left[\frac{c}{c-1}(t-1)\right]^{-1+1/c} \int_1^t \xi^{-1/c} d\xi \\ &= (t-1)^{-1+1/c} \frac{c}{c-1}\left(t^{-1/c+1} - 1\right), \; t \in (1, +\infty). \end{aligned}$$

Substituting $1 - 1/c = k$ we get from this by a simple calculation

$$\log t \geq \left[\frac{1(t^k - 1)}{k(t-1)^k}\right]^{1/(1-k)}, \; t \in (1, +\infty),$$

which is valid for any $k \in (0, 1)$. Since the right hand side of this inequality has a finite limit as $t \to +\infty$, The result of course is only interesting as $t \to 1^+$.

Remark 2.5.5 If the function $h'_y(t, y)$, $y \in J_0$ instead of weakly increasing is weakly decreasing for every fixed $t \in I_0$, the investigation in this section can be carried out with minor modifications. Instead of definition the set $A(t)$, $t \in I_0$ by (2.5.11) we define it by

$$A(t) = \left\{ a \in I\!R : \sup_{y \in J_0} (h(t, y) - ay) < +\infty \right\}, \; t \in I_0. \tag{2.5.43}$$

Theorem 2.5.1, 2.5.2 and Remark 2.5.4 to Theorem 2.5.2 still hold if all the signs $\max_z$ and $\min_z$ are interchanged and if the function (2.5.24) is replaced by the function

$$M_z(t) = \sup_{y \in J_0} H_z(t, y) = \sup_{y \in J_0} \left(h(t, y)z(t) + yz'(t)\right), \; t \in I, \tag{2.5.44}$$

where $z \in \mathcal{M}\{I\}$. If the integral $\int_a^b M_z(t)dt$ (where $[a, b] \subset I$) is not finite, it has the value $+\infty$.

If the class $\mathcal{M}\{I\}$ is replaced throughout by the class $\mathcal{M}_0\{I\}$ then all these results are valid with sup in (2.5.44) replaced by max.

We shall now discuss the relationship of the sets A_0 and A and proof of Lemma 2.5.3.

We consider the two sets $A_0(t)$ and $A(t)$, $t \in I_0$ introduced in the beginning.. Obviously they both are intervals. The discussion of their relationship takes place in terms of the type of the interval J_0.

Case 1. The interval J_0 is bounded. Since the function $h(t, y) - ay$, $y \in J_0$ is weakly convex for any $a \in \mathbb{R}$ and hence bounded below, $A(t) = \mathbb{R}$ for any $t \in I_0$ in this case.

Case 2. The interval J_0 is bounded downwards, but not upwards, i.e., $\sup J_0 = +\infty$. Let $t \in I_0$. If $a \leq \inf A_0(t)$, the function $h(t, y) - ay$, $y \in J_0$ is weakly increasing and weakly convex. Clearly, it is bounded below, i.e., $a \in A(t)$.

If $a > \sup A_0(t)$, the function $h(t, y) - by$, $y \in J_0$ is strictly decreasing for any $b \in (\sup A_0(t), a)$. Hence a cannot belong to $A(x)$, since otherwise

$$h(t, y) - by = h(t, y) - ay + (a - b)y \to +\infty \quad \text{as } y \to +\infty,$$

which contradicts the monotonicity.

From (2.5.12) we conclude that

$$(-\infty, \sup A_0(t)) \subseteq A(t) \subseteq (-\infty, \sup A_0(t)). \tag{2.5.45}$$

Suppose that $k = \sup A_0(t)$ is finite. Then we have for some arbitrarily chosen $y_0 \in J_0$

$$\begin{aligned} h(t, y) &= h(t, y_0) + \int_{y_0}^{y} h'_y(t, \eta) d\eta \\ &= h(t, y_0) + k(y - y_0) - \int_{y_0}^{y} (k - h'_y(t, \eta) d\eta, \end{aligned}$$

by which

$$h(t, y) - ky = (h(t, y_0) - ky_0) - \int_{y_0}^{y} (k - h'_y(t, \eta)) d\eta, \quad y \in J_0. \tag{2.5.46}$$

The integrand in (2.5.46) is non-negative. Hence $k \in A(t)$ if and only if the integral

$$\int_{y_0}^{+\infty} (k - h'_y(t, y)) dy = \int_{y_0}^{+\infty} (\sup A_0(t) - h'_y(t, y)) dy \tag{2.5.47}$$

is convergent. We notice that this in particular is true if $k \in A_0(t)$.

Summarizing we have shown that $A(t)$, $t \in I_0$, is one of the two intervals $(-\infty, \sup A_0(t))$, $(-\infty, \sup A_0(t)]$. Further, $\sup A_0(t)$ belongs to $A(t)$ if and only if it has finite value k and the integral (2.5.47) is convergent for some fixed $y_0 \in J_0$.

Case 3. The interval J_0 is bounded upwards, but not downwards, i.e., $\inf J_0 = -\infty$. This case is treated analogously with the preceding case. The result is

$$(\inf A_0(t), +\infty) \subseteq A(t) \subseteq [\inf A_0(t), +\infty) \tag{2.5.48}$$

for any $t \in I_0$. Moreover, $\inf A_0(t)$ belongs to $A(t)$ if and only if it is finite and the integral

$$\int_{-\infty}^{y_0} (h'_y(t, y) - \inf A_0(t))dy \tag{2.5.49}$$

is convergent for some fixed $y_0 \in J_0$.

Case 4. The interval is unbounded downwards and upwards, i.e., $J_0 = I\!R$. Again the discussion can be carried through as in Case 2. The result is

$$A_0(t) \subseteq A(t) \subseteq \overline{A_0(t)} \tag{2.5.50}$$

for any $t \in I_0$. Moreover, $\sup A_0(t)$ $(\inf A_0(t))$ belongs to $A(t)$ if and only if it is finite and the integral (2.5.47) (the integral (2.5.49)) is convergent for some fixed $y_0 \in I\!R$.

As an introduction to the proof of Lemma 2.5.3 we consider the following problem. Given a continuous function $u(t, y)$, $(t, y) \in I_0 \times J_0$, which is weakly increasing in y for any fixed $t \in I_0$. To any given $\alpha \in I_0$ construct a continuous function $v : I_0 \cap (-\infty, \alpha) \to J_0$ such that

$$u(t, v(t)) \to \sup_{y \in J_0} u(\alpha, y) \quad \text{as } t \to \alpha^-. \tag{2.5.51}$$

To solve this problem we choose a strictly increasing sequence (y_n) in J_0 such that $y_n \to \sup J_0$ as $n \to +\infty$. Next, we choose a strictly increasing sequence (α_n) in $I_0 \cap (-\infty, \alpha)$, such that $\alpha_n \to \alpha$ as $n \to +\infty$ and

$$|u(t, y_n) - u(\alpha, y_n)| < \frac{1}{n} \quad \text{for all } t \in [\alpha_n, \alpha]. \tag{2.5.52}$$

Define $v(t)$, $t \in I_0 \cap (-\infty, \alpha)$ in the following way: $v(t) = y_1$ if $t \leq \alpha_2$ and $v(t)$ linear if $\alpha_n \leq t \leq \alpha_{n+1}$ such that the graph connects the points (α_n, y_{n-1}) and (α_{n+1}, y_n), $n \geq 2$. We have $y_{n-1} \leq v(t) \leq y_n$ for all $t \in [\alpha_n, \alpha_{n+1}]$, by which in view of (2.5.52)

$$\begin{aligned} u(\alpha, y_{n-1}) - \frac{1}{n-1} &< u(t, y_{n-1}) \leq u(t, v(t)) \\ &\leq u(t, y_n) < u(\alpha, y_n) + \frac{1}{n}, \quad n \geq 2. \end{aligned}$$

Since $u(\alpha, y_n) \to \sup_{y \in J_0} u(\alpha, y)$ as $n \to +\infty$, we conclude that (2.5.51) is satisfied. Clearly, $v(t)$, $t \in I_0 \cap (-\infty, \alpha)$ is a continuous function with values in J_0.

Proof of Lemma 2.5.3 If $z \in \mathcal{M}\{I\}$, then obviously the restriction to z to I_1 belongs to $\mathcal{M}\{I_1\}$. On the other hand, let $z_1 \in \mathcal{M}\{I_1\}$ and consider the case $\sup I_1 - \sup I$, $\inf I_1 = \alpha \in I_1$.

In order to construct a function $z \in \mathcal{M}\{I\}$, whose restriction to I_1 is z_1, we consider separately the cases discussed above. If J_0 is bounded we just continue z_1 to the whole interval I as a positive C^1-function. Suppose next that J_0 is bounded downwards, but not upwards, and let $a = -z_1'(\alpha)/z_1\alpha$. Then $a \in A(\alpha)$ by (2.5.16). Recall that $A_0(\alpha)$ is the domain of values for the function $h_y'(\alpha, y)$, $y \in J_0$. We consider the two possible cases.

Case I. If $a \in (-\infty, \sup A_0(\alpha))$, we may choose $y_\alpha \in J_0$ such that $a < h_y'(\alpha, y_\alpha)$. Next, choose some continuous function $v : (-\infty, \alpha] \cap I \to J_0$ since that $v(\alpha) = y_\alpha$ and define

$$z(t) = \begin{cases} z_1(\alpha) e^{-\int_\alpha^t h_y'(\xi, v(\xi)) d\xi} e^{K(t-\alpha)}, & t \in (-\infty, \alpha) \cap I \\ z_1(t), & t \in I_1, \end{cases}$$

where $K = h_y'(\alpha, y_\alpha) - a$ is positive. Clearly, $z(t)$, $t \in I$ is a positive continuous function, which is also a C^1-function, since for $t < \alpha$

$$\begin{aligned} z'(t) &= z(t)(-h_y'(t, v(t)) + K) \to z_1(\alpha)(-h_y'(\alpha, y_\alpha) + K) \\ &= -z_1(\alpha) a = z_1'(\alpha) \quad \text{as } t \to \alpha^-. \end{aligned}$$

Clearly, condition (2.5.45) is satisfied if $t \in I_1$. If $t < \alpha$ we have

$$-\frac{z'(t)}{z(t)} = h_y'(t, v(t)) - K < h_y'(t, v(t)) \le \sup A_0(t),$$

i.e., $-z'(t)/z(t) \in A(t)$ by (2.5.45). Hence $z(t)$, $t \in I$ belongs to the class $\mathcal{M}\{I\}$ and the restriction to I_1 is $z_1(t)$, $t \in I_1$.

Case II. If $a = \sup A_0(\alpha)$, we may choose a continuous function $v : (-\infty, \alpha) \cap I \to J_0$ such that

$$h_y'(t, v(t)) \to a, \quad \text{as } t \to \alpha^-.$$

Define the continuous function

$$z(t) = \begin{cases} z_1(\alpha) e^{-\int_\alpha^t h_y'(\xi, v(\xi)) d\xi}, & t \in (-\infty, \alpha) \cap I \\ z_1(t), & t \in I_1. \end{cases}$$

This is a positive C^1-function, since for $t < \alpha$

$$z'(t) = z(t)(-h_y'(t, v(t))) \to z_1(\alpha)(-a) = z_1'(\alpha) \quad \text{as } t \to \alpha^-.$$

Condition (2.5.16) is fulfilled since for $t < \alpha$

$$-\frac{z'(t)}{z(t)} = h'_y(t, v(t)) \in A_0(t).$$

Hence $z(t)$, $t \in I$ belongs to the class $\mathcal{M}\{I\}$ and the restriction to I_1 is $z_1(t)$, $t \in I_1$.

The two remaining cases as to the type of J_0 can be treated in a similar way.

The two other possibilities $\inf I_1 = \inf I$, $\sup I_1 = \beta \in I_1$ and $\inf I < \inf I_1 = \alpha \in I_1$, $\sup I > \sup I_1 = \beta \in I_1$ are dealt with analogously.

We notice that Lemma 2.5.3 remains valid if the classes $\mathcal{M}\{I\}$ and $\mathcal{M}\{I_1\}$, respectively, are replaced by the classes $\mathcal{M}_0\{I\}$ and $\mathcal{M}_0\{I_1\}$. The proof of this follows from the proof of Case I above with minor modifications (we may choose $a = h'_y(\alpha, y_\alpha)$).

2.6 EXTENSION TO SYSTEM OF DIFFERENTIAL EQUATIONS

We have seen so far how the method of quasilinearization can be generalized, extended and refined for the scalar differential equations relative to initial value problems, periodic boundary value problems, and antiperiodic boundary value problems. We shall devote this section to extend some of the typical results to systems of differential equations with respect to IVPs. Other classes of results still await extension. As we shall see that the extension is more complicated compared to the monotone iterative technique since the situation demands the study of uncoupled systems.

Throughout this section, we assume that the inequalities are understood to be componentwise.

We now consider the IVP:

$$u' = f(t, u), \qquad u(0) = u_0 \tag{2.6.1}$$

where $f \in C[\Omega, R^n]$ and Ω is the closed set defined as

$$\Omega = [(t, u) : \alpha_0(t) \leq u \leq \beta_0(t), t \in J];$$

$\alpha_0, \beta_0 \in C^1[J, R^n]$ being such that $\alpha_0(t) \leq \beta_0(t)$ on $J = [0, T]$.

We begin by proving a result that gives the conditions under which the monotone sequences $\{\alpha_n(t)\}$ and $\{\beta_n(t)\}$ that converge to the unique solution, $u(t)$, of (2.6.1). Further the convergence is quadratic. The results are obtained when $f(t, u)$ is not convex in u. This means that $f_{i,u_j u_k}$ is not nonnegative. However the function $f + \phi$ is convex for some convex function $\phi(t, u)$. That is $(f + \phi)_{i,u_j u_k} \geq 0$ for all $i, j, k = 1, 2, \cdots, n$..

Theorem 2.6.1 *Assume that*

(A_1) $\alpha_0, \beta_0 \in C^1[J, R^n]$ *are lower and upper solutions of* (2.6.1) *such that* $\alpha_0(t) \leq \beta_0(t)$ *on* J.

(A_2) $f, \phi \in C[\Omega, R^n]$, $f(t,u), \phi(t,u)$ *are quasimonotone nondecreasing in* u *for* $t \in J$. *Further* f_x, f_{xx}, ϕ_x, ϕ_{xx} *exist and are continuous on* Ω, *satisfying*

(a) $f_{xx} + \phi_{xx} \geq 0$ *with* $\phi_{xx} \geq 0$;

(b) $a_{ij}(t, \alpha_0, \beta_0) \geq 0$ *for* $i \neq j$, *where* $A(t, \alpha_0, \beta_0) = [a_{ij}(t, \alpha_0, \beta_0)]$ *is an* $n \times n$ *matrix given by*

$$A(t, \alpha_0, \beta_0) \equiv f_u(t, \alpha_0) + \phi_u(t, \alpha_0) - \phi_u(t, \beta_0) = F_u(t, \alpha_0) - \phi_u(t, \beta_0).$$

Here $F_u(t,u) = f_u(t,u) + \phi_u(t,u)$.

Then, there exists monotone sequences $\{\alpha_n(t)\}$, $\{\beta_n(t)\}$ *which converge uniformly to the unique solution,* $u(t)$ *of* (2.6.1) *and the convergence is quadratic.*

Proof Let us observe that (A_2) implies for any $x \geq y$, $x, y \in \Omega$,

$$f(t,x) \geq f(t,y) + [f_x(t,y) + \phi_x(t,y)](x-y) - [\phi(t,x) - \phi(t,y)]. \tag{2.6.2}$$

Also for any $\alpha_0(t) \leq y \leq x \leq \beta_0(t)$, we have

$$a_{ij}(t, y, x) \geq 0, \qquad i \neq j, \tag{2.6.3}$$

since $a_{ij}(t,y,x) \geq a_{ij}(t, \alpha_0, \beta_0)$ for all i, j in view of the fact $f_u(t,u) + \phi_u(t,u)$ and $\phi_u(t,u)$ are nondecreasing in u for each $t \in J$ by our assumption. It is also clear that for $\alpha_0(t) \leq y \leq x \leq \beta_0(t)$,

$$f_i(t, x_1, \cdots, x_n) - f_i(t, y_1, \cdots, y_n) \leq L_i \sum_{i=1}^{n} (x_i - y_i). \tag{2.6.4}$$

Let α_1, β_1 be the solutions of IVP's;

$$\alpha_1' = f(t, \alpha_0) + [A(t, \alpha_0, \beta_0)](\alpha_1 - \alpha_0), \qquad \alpha_1(0) = u_0, \tag{2.6.5}$$

and

$$\beta_1' = f(t, \beta_0) + [A(t, \alpha_0, \beta_0(t)](\beta_1 - \beta_0), \qquad \beta_1(0) = u_0, \tag{2.6.6}$$

where $F(t,x) = f(t,x) + \phi(t,x)$ and $\alpha_0(0) \leq u_0 \leq \beta_0(0)$. We shall prove that $\alpha_0 \leq \alpha_1$ on J. To do this, let $p = \alpha_0 - \alpha_1$, so that $p(0) \leq 0$. Then using (2.6.5), we obtain

$$\begin{aligned} p' &= \alpha_0' - \alpha_1' \\ &\leq f(t,\alpha_0) - [f(t,\alpha_0) + (A(t,\alpha_0,\beta_0)(\alpha_1 - \alpha_0)] \\ &= A(t,\alpha_0,\beta_0)p. \end{aligned}$$

Since $A(t, \alpha_0, \beta_0)$ is quasimonotone nondecreasing by assumption (A_2) and it follows from Theorem 1.1.3 that $p(t) \leq 0$ on J proving $\alpha_0 \leq \alpha_1$ on J. Now we let $p = \alpha_1 - \beta_0$ and note that $p(0) \leq 0$. Also,

$$\begin{aligned} p' &= \alpha_1' - \beta_0' \\ &\leq f(t,\alpha_0) + [A(t,\alpha_0,\beta_0)](\alpha_1 - \alpha_0) - f(t,\beta_0). \end{aligned}$$

Since $\beta_0 \geq \alpha_0$, using (2.6.2), we get

$$f(t,\beta_0) \geq f(t,\alpha_0) + F_u(t,\alpha_0)(\beta_0 - \alpha_0) - [\phi(t,\beta_0) - \phi(t,\alpha_0)].$$

In view of (A_2), we have

$$\phi(t,\beta_0) - \phi(t,\alpha_0) \leq \phi_u(t,\beta_0)(\beta_0 - \alpha_0)$$

which yields

$$f(t,\beta_0) \geq f(t,\alpha_0) + [F_u(t,\alpha_0) - \phi_u(t,\beta_0)](\beta_0 - \alpha_0).$$

Hence, we obtain

$$p' \leq A(t,\alpha_0,\beta_0)p.$$

This implies $\alpha_1(t) \leq \beta_0(t)$ on J using Theorem 1.1.3. As a result, we have $\alpha_0(t) \leq \alpha_1(t) \leq \beta_0(t)$ on J. In a similar way, we can prove that $\alpha_0(t) \leq \beta_1(t) \leq \beta_0(t)$ on J. To show $\alpha_1(t) \leq \beta_1(t)$, we use (2.6.2), (2.6.4), and the fact ϕ_u is nondecreasing in u to obtain

$$\begin{aligned} \alpha_1' &= f(t,\alpha_0) + [F_u(t,\alpha_0) - \phi_u(t,\beta_0)](\alpha_1 - \alpha_0) \\ &\leq f(t,\alpha_1) + \phi(t,\alpha_1) - \phi(t,\alpha_0) - \phi_u(t,\beta_0)(\alpha_1 - \alpha_0) \\ &\leq f(t,\alpha_1) + \left[\int_0^1 \phi_u(t, s\alpha_1 + (1-s)\alpha_0)ds\right] \\ &\quad \times(\alpha_1 - \alpha_0) - \phi_u(t,\beta_0)(\alpha_1 - \alpha_0) \\ &\leq f(t,\alpha_1) + \phi_u(t,\alpha_1)(\alpha_1 - \alpha_0) - \phi_u(t,\beta_0)(\alpha_1 - \alpha_0) \\ &\leq f(t,\alpha_1). \end{aligned}$$

Using a similar argument, it is easy to show that $\beta_1' \geq f(t,\beta_1)$, and therefore by Theorem 1.1.3, it follows that $\alpha_1(t) \leq \beta_1(t)$ on J since $f(t,u)$ is Lipschitzian in u on Ω. This proves that

$$\alpha_0(t) \leq \alpha_1(t) \leq \beta_1(t) \leq \beta_0(t) \text{ on } J.$$

Next, we will assume that for some $k \geq 1$, $\alpha_k' \leq f(t, \alpha_k)$, $\beta_k' \geq f(t, \beta_k)$, and

$$\alpha_0(t) \leq \alpha_k(t) \leq \beta_k(t) \leq \beta_0(t) \text{ on } J.$$

We shall show that

$$\alpha_k(t) \leq \alpha_{k+1}(t) \leq \beta_{k+1}(t) \leq \beta_k(t) \text{ on } J \tag{2.6.7}$$

where $\alpha_{k+1}(t)$ and $\beta_{k+1}(t)$ are the solutions of linear IVP's:

$$\alpha_{k+1}' = f(t, \alpha_k) + [F_u(t, \alpha_k) - \phi_u(t, \beta_k)](\alpha_{k+1} - \alpha_k), \alpha_{k+1}(0) = u_0, \tag{2.6.8}$$

and

$$\beta_{k+1}' = f(t, \beta_k) + [F_u(t, \alpha_k) - \phi_u(t, \beta_k)](\beta_{k+1} - \beta_k), \beta_{k+1}(0) = u_0. \tag{2.6.9}$$

We note that $F_u(t, \alpha_k) - \phi_u(t, \beta_k) = A(t, \alpha_k, \beta_k)$. Now, set $p = \alpha_k - \alpha_{k+1}$ so that

$$\begin{aligned} p' = \alpha_k' - \alpha_{k+1}' &\leq f(t, \alpha_k) - [f(t, \alpha_k) + (F_u(t, \alpha_k) - \phi_u(t, \beta_k))](\alpha_{k+1} - \alpha_k) \\ &= [F_u(t, \alpha_k) - \phi_u(t, \beta_k)]p \\ &= A(t, \alpha_k, \beta_k)p \end{aligned}$$

and $p(0) = 0$. It follows from Theorem 1.1.3 and using (2.6.2) that $\alpha_k(t) \leq \alpha_{k+1}(t)$ on J. On the other hand, letting $p = \alpha_{k+1} - \beta_k$ yields,

$$\begin{aligned} p' &= \alpha_{k+1}' - \beta_k' \\ &\leq f(t, \alpha_k) + [F_u(t, \alpha_k) - \phi_u(t, \beta_k)](\alpha_{k+1} - \alpha_k) - f(t, \beta_k). \end{aligned}$$

Since $\beta_k \geq \alpha_k$, (2.6.2) and (A_2) give, as before,

$$f(t, \beta_k) \geq f(t, \alpha_k) + [F_u(t, \alpha_k) - \phi_u(t, \beta_k)](\beta_k - \alpha_k)$$

which shows that

$$p' \leq [F_u(t, \alpha_k) - \phi_u(t, \beta_k)]p = A(t, \alpha_k, \beta_k)p.$$

This proves that $p(t) \leq 0$ using Theorem 1.1.3 and (2.6.2) since $p(0) = 0$. Hence we get

$$\alpha_k(t) \leq \alpha_{k+1}(t) \leq \beta_k(t) \text{ on } J.$$

Similarly, we can prove that $\alpha_k(t) \leq \beta_{k+1}(t) \leq \beta_k(t)$ on J. Also, by equations (2.6.2), (2.6.8), and the fact $\alpha_{k+1} \geq \alpha_k$, we obtain

$$\begin{aligned} \alpha_{k+1}' &\leq f(t, \alpha_{k+1}) + \phi(t, \alpha_{k+1}) - \phi(t, \alpha_k) - \phi_u(t, \beta_k)(\alpha_{k+1} - \alpha_k) \\ &\leq f(t, \alpha_{k+1}) + \left[\int_0^1 \phi_u(t, s\alpha_{k+1} + (1-s)\alpha_k)ds\right](\alpha_{k+1} - \alpha_k) \\ &\quad -\phi_u(t, \beta_k)(\alpha_{k+1} - \alpha_k) \\ &\leq f(t, \alpha_{k+1}) + \phi_u(t, \alpha_{k+1})(\alpha_{k+1} - \alpha_k) - \phi_u(t, \beta_k)(\alpha_{k+1} - \alpha_k) \\ &\leq f(t, \alpha_{k+1}). \end{aligned}$$

Using a similar argument, we have, as before, $\beta'_{k+1} \geq f(t, \beta_{k+1})$ and hence Theorem 1.1.3 shows that $\alpha_{k+1}(t) \leq \beta_{k+1}(t)$ on J since $f(t,u)$ is Lipschitzian in u on Ω and f is quasimonotone nondecreasing in u. This proves (2.6.7). Therefore, by induction, we have for all n,

$$\alpha_0(t) \leq \alpha_1(t) \leq \cdots \leq \alpha_n(t) \leq \beta_n(t) \leq \cdots \leq \beta_1(t) \leq \beta_0(t) \text{ on } J.$$

Employing the standard procedure, it is now easy to prove that the sequences $\{\alpha_n(t)\}$ and $\{\beta_n(t)\}$ converge uniformly and monotonically to the unique solution, $u(t)$, of (2.6.1) on J.

We will now show that the convergence of $\{\alpha_n(t)\}$ and $\{\beta_n(t)\}$ to $u(t)$ is quadratic. We first set $p_{n+1} = u - \alpha_{n+1}$ and $q_{n+1} = \beta_{n+1} - u$ on J so that $p_{n+1}(0) = 0$ and $q_{n+1}(0) = 0$. Then, using integral mean value theorem and the fact that $u \geq \alpha_{n+1}$ and (A_2) we get

$$\begin{aligned}
p'_{n+1} &= u' - \alpha'_{n+1} \\
&= f(t,u) - [f(t,\alpha_n) + (F_u(t,\alpha_n) - \phi_u(t,\beta_n))(\alpha_{n+1} - \alpha_n)] \\
&= F(t,u) - F(t,\alpha_n) - [F_u(t,\alpha_n) - \phi_u(t,\beta_n)](\alpha_{n+1} - \alpha_n) \\
&\quad -[\phi(t,u) - \phi(t,\alpha_n)] \\
&= \left(\int_0^1 F_u(t, su + (1-s)\alpha_n)ds\right) p_n \\
&\quad - (F_u(t,\alpha_n) - \phi_u(t,\beta_n))\,(-p_{n+1} + p_n) \\
&\quad - \left[\int_0^1 \phi_u(t, su + (1-s)\alpha_n)ds\right) (u - \alpha_n) \\
&\leq (F_u(t,u) - F_u(t,\alpha_n))\,p_n + (\phi_u(t,\beta_n) - \phi_u(t,\alpha_n))\,p_n + f_u(t,\alpha_n)p_{n+1}.
\end{aligned}$$

So, we get, using T for transpose,

$$\begin{aligned}
p'_{n+1} &\leq p_n^T \left[\int_0^1 F_{uu}(t, su + (1-s)\alpha_n)ds\right] p_n \\
&\quad +(q_n + p_n)^T \left[\int_0^1 \phi_{uu}(t, s\beta_n + (1-s)\beta_n ds\right] p_n + f_u(t,\alpha_n)p_{n+1} \\
&\leq M p_{n+1} + L(p_n \cdot p_n) + N(q_n + p_n) \cdot p_n,
\end{aligned}$$

where $f_u \leq M$, $\sum_{i=1}^n f_{uu}^i \leq L$ and $\sum_{i=1}^n \phi_{uu}^i \leq N$ in Ω and L, M, N are $n \times n$ positive matrices such that M^{-1} exists. We note that there is no loss of generality in assuming that M^{-1} exists such that $m_{ij} \geq 0$. Now using Corollary 1.1.1, we can get the estimate

$$p_{n+1} \leq \int_0^T e^{M(t-s)} \left[L p_n^2 + \frac{3}{2} N p_n^2 + \frac{1}{2} N q_n^2\right] ds.$$

This implies

$$\max_J \|p_{n+1}\| \le M^{-1}e^{MT}\left[\left(L+\frac{3N}{2}\right)\max_J \|p_n\|^2 + \frac{N}{2}\max_J \|q_n\|^2\right]$$

where $\|p_j\|^2 = (|p_j^1|^2, \cdots, |p_j^n|^2)$ and $\|q_j\|^2 = (|q_j^1|^2, \cdots, |q_j^n|^2)$. Following a similar procedure, we can also show that

$$\max_J \|q_{n+1}\| \le M^{-1}e^{MT}\left[\frac{L}{2}\max_J \|p_n\|^2 + \left(\frac{3L}{2}+\frac{5N}{2}\right)\max_J \|q_n\|^2\right]$$

which yields the quadratic convergence. Hence the proof is complete.

The next result offers the decomposition of f into convex and concave parts in the special case. The result proved is much more general.

Theorem 2.6.2 *Assume that*

(A_1) $\alpha_0, \beta_0 \in C^1[J, R^n]$ *are lower and upper solutions of* (2.6.1), *respectively such that* $\alpha_0(t) \le \beta(t)$ *on* J.

(A_2) $f \in C[\Omega, R^n], f$ *admits a decomposition* $f = F+G+H$ *where* F_u, G_u, F_{uu}, G_{uu} *exist and are continuous satisfying*

(a) $B_{uu} = F_{uu} + \phi_{uu} \ge 0$, *and* $D_{uu} = G_{uu} + \psi_{uu} \le 0$ *on* Ω, *where* $\phi, \psi \in C[\Omega, R^n]$, $\phi_u, \phi_{uu}, \psi_u, \psi_{uu}$, *exist, are continuous and* $\phi_{uu}(t,u) \ge 0$, $\psi_{uu}(t,u) \le 0$ *on* Ω, *where*

$$B(t,u) = F(t,u) + \phi(t,u),$$
$$J(t,u) = G(t,u) + \psi(t,u).$$

(b) $H(t,u)$ *satisfies*

$$H_i(t, x_1, x_2, ..., x_n) - H_i(t, y_1, y_2, ..., y_n) \le K_i \sum_{i=1}^{n} (x_i - y_i)$$

whenever $x \ge y$, *where* $K_i > 0$ *is the Lipschitz constant.*

Furthermore, $f(t,u), \phi(t,u)$ *are quasimonotone nondecreasing in* u *for* $t \in J = [0,T]$, $\psi(t,u)$ *is quasimonotone nonincreasing in* u *for* $t \in J = [0,T]$.

Then there exist monotone sequences $\{\alpha_n(t)\}$, $\{\beta_n(t)\}$ *which converge uniformly to the unique solution of* (2.6.1) *and the convergence is quadratic.*

Proof Let us observe that for any $x \geq y,\ x, y \in \Omega$

$$\begin{aligned} F(t,x) &\geq F(t,y) + [F_x(t,y) + \phi_x(t,y)](x-y) \\ &\quad -[\phi(t,x) - \phi(t,y)] \\ G(t,x) &\geq G(t,y) + [G_x(t,x) + \psi_x(t,x)](x-y) \\ &\quad -[\psi(t,x) - \psi(t,y)] \end{aligned} \tag{2.6.10}$$

Also,from (A_2), we have

$$f_i(t, x_1, ..., x_n) - f_i(t, y_1, ..., y_n) \leq L_i \sum_{i=1}^{n} (x_i - y_i) \tag{2.6.11}$$

whenever $\alpha_0(t) \leq y \leq x \leq \beta_0(t)$.

Consider the IVP,

$$\begin{aligned} w'(t) &= F(t,\alpha_0) + G(t,\alpha_0) + H(t,w) \\ &\quad +[B_u(t,\alpha_0) + D_u(t,\beta_0) - \phi_u(t,\beta_0) - \psi_u(t,\alpha_0)](w-\alpha_0) \\ &\equiv M(t, w, \alpha_0, \beta_0) \\ w(0) &= u_0 \end{aligned} \tag{2.6.12}$$

We now show that α_0 is a lower solution of (2.6.12) and β_0 is an upper solution of (2.6.12). By (A_1) we have

$$\begin{aligned} \alpha_0' &\leq f(t,\alpha_0) \\ &= F(t,\alpha_0) + G(t,\alpha_0) + H(t,\alpha_0) \\ &\equiv M(t,\alpha_0,\alpha_0,\beta_0) \\ \alpha_0(0) &\leq u_0. \end{aligned}$$

then $\alpha_0(t)$ is a lower solution of (2.6.12) on J.

Similarly, it follows from (2.6.10) and using the Integral Mean Value Theorem for systems and the fact that ϕ_u is nondecreasing and ψ_u is nonincreasing on Ω that

$$\begin{aligned} \beta_0' &\geq f(t,\beta_0) \\ &= F(t,\beta_0) + G(t,\beta_0) + H(t,\beta_0) \\ &\geq F(t,\alpha_0) + G(t,\alpha_0) + H(t,\beta_0) \\ &\quad +[F_u(t,\alpha_0) + \phi_u(t,\alpha_0) + G_u(t,\beta_0) + \psi_u(t,\beta_0)](\beta_0 - \alpha_0) \\ &\quad -[\phi(t,\beta_0) - \phi(t,\alpha_0)] - [\psi(t,\beta_0) - \psi(t,\alpha_0)] \end{aligned}$$

$$
\begin{aligned}
&= F(t,\alpha_0)+G(t,\alpha_0)+H(t,\beta_0)\\
&\quad+[B_u(t,\alpha_0)+D_u(t,\beta_0)](\beta_0-\alpha_0)\\
&\quad-\left[\int_0^1 \phi_u(t,s\beta_0+(1-s)\alpha_0)ds\right](\beta_0-\alpha_0)\\
&\quad-\left[\int_0^1 \psi_u(t,s\beta_0+(1-s)\alpha_0)ds\right](\beta_0-\alpha_0)\\
&\geq F(t,\alpha_0)+G(t,\alpha_0)+H(t,\beta_0)\\
&\quad+[B_u(t,\alpha_0)+D_u(t,\beta_0)-\phi_u(t,\beta_0)-\psi_u(t,\alpha_0)](\beta_0-\alpha_0)\\
&\equiv M(t,\beta_0,\alpha_0,\beta_0).
\end{aligned}
$$

Therefore, $\beta_0(t)$ is an upper solution of (2.6.12) on J and consequently by Theorem 1.1.6, there exists a solution $\alpha_1(t)$ of (2.6.12) such that $\alpha_0(t) \leq \alpha_1(t) \leq \beta_0(t)$ on J. Since M satisfies a Lipschitz condition, it follows that $\alpha_1(t)$ is unique on J.

Consider the IVP,

$$
\begin{aligned}
v' &= F(t,\beta_0)+G(t,\beta_0)+H(t,v)\\
&\quad+[B_u(t,\alpha_0)+D_u(t,\beta_0)-\phi_u(t,\beta_0)-\psi_u(t,\alpha_0)]\\
&\quad\times(v-\beta_0) \qquad\qquad (2.6.13)\\
&\equiv N(t,v,\alpha_0,\beta_0)\\
v(0) &= u_0.
\end{aligned}
$$

Next, we show that α_0 and β_0 are lower and upper solutions of (2.6.13) respectively. By (A_1) we have

$$
\begin{aligned}
\beta_0' &\geq f(t,\beta_0)\\
&= F(t,\beta_0)+G(t,\beta_0)+H(t,\beta_0)\\
&\equiv N(t,\beta_0,\alpha_0,\beta_0).
\end{aligned}
$$

Hence, $\beta_0(t)$ is an upper solution of (2.6.13) on J.

Also, using the Integral Mean Value Theorem for systems and from (2.6.10) and the fact that ϕ_u is nondecreasing and ψ_u is nonincreasing in u, we obtain the following;

$$
\begin{aligned}
\alpha_0' &\leq f(t,\alpha_0)\\
&= F(t,\alpha_0)+G(t,\alpha_0)+H(t,\alpha_0)\\
&\leq F(t,\beta_0)+G(t,\beta_0)+H(t,\alpha_0)\\
&\quad-[B_u(t,\alpha_0)+D_u(t,\beta_0)](\beta_0-\alpha_0)+\phi(t,\beta_0)
\end{aligned}
$$

$$
\begin{aligned}
& -\phi(t,\alpha_0)+\psi(t,\beta_0)-\psi(t,\alpha_0) \\
= \;& F(t,\beta_0)+G(t,\beta_0)+H(t,\alpha_0) \\
& +[B_u(t,\alpha_0)+D_u(t,\beta_0)](\alpha_0-\beta_0) \\
& +\left[\int_0^1 \phi_u(t,s\beta_0+(1-s)\alpha_0)ds\right](\beta_0-\alpha_0) \\
& +\left[\int_0^1 \psi_u(t,s\beta_0+(1-s)\alpha_0)ds\right](\beta_0-\alpha_0) \\
\leq \;& F(t,\beta_0)+G(t,\beta_0)+H(t,\alpha_0) \\
& +[B_u(t,\alpha_0)+D_u(t,\beta_0)-\phi_u(t,\beta_0)-\psi_u(t,\alpha_0)](\alpha_0-\beta_0) \\
\equiv \;& N(t,\alpha_0,\alpha_0,\beta_0).
\end{aligned}
$$

Thus, $\alpha_0(t)$ is a lower solution of (2.6.13) on J. It follows from Theorem 1.1.6 that there exists a solution $\beta_1(t)$ of (2.6.13) such that $\alpha_0(t) \leq \beta_1(t) \leq \beta_0(t)$ on J. Further, since N satisfies a Lipschitz condition, it follows that $\beta_1(t)$ is unique on J.

Next, we need to show that $\alpha_1(t) \leq \beta_1(t)$ on J, by showing that α_0 and β_0 are lower and upper solutions of (2.6.1), respectively. Since $\alpha_0(t) \leq \alpha_1(t) \leq \beta_0(t)$ and from (2.6.10) and the fact that D_u and ψ_u are nonincreasing, and ϕ_u is nondecreasing on Ω, we obtain,

$$
\begin{aligned}
\alpha_1'(t) \;=\;& F(t,\alpha_0)+G(t,\alpha_0)+H(t,\alpha_1) \\
& +[B_u(t,\alpha_0)+D_u(t,\beta_0)-\phi_u(t,\beta_0)-\psi_u(t,\alpha_0)](\alpha_1-\alpha_0) \\
\leq \;& F(t,\alpha_1)+G(t,\alpha_1)+H(t,\alpha_1) \\
& -[B_u(t,\alpha_0)+D_u(t,\alpha_1)](\alpha_1-\alpha_0)+\phi(t,\alpha_1) \\
& -\phi(t,\alpha_0)+\psi(t,\alpha_1)-\psi(t,\alpha_0) \\
& +[B_u(t,\alpha_0)+D_u(t,\beta_0)-\phi_u(t,\beta_0)-\psi_u(t,\alpha_0)](\alpha_1-\alpha_0) \\
\leq \;& F(t,\alpha_1)+G(t,\alpha_1)+H(t,\alpha_1) \\
& -[B_u(t,\alpha_0)+D_u(t,\beta_0)](\alpha_1-\alpha_0) \\
& +\left[\int_0^1 \phi_u(t,s\alpha_1+(1-s)\alpha_0)ds\right](\alpha_1-\alpha_0) \\
& +\left[\int_0^1 \psi_u(t,s\alpha_1+(1-s)\alpha_0)ds\right](\alpha_1-\alpha_0) \\
& +[B_u(t,\alpha_0)+D_u(t,\beta_0)-\phi_u(t,\beta_0)-\psi_u(t,\alpha_0)](\alpha_1-\alpha_0) \\
\leq \;& f(t,\alpha_1)+\phi_u(t,\alpha_1)(\alpha_1-\alpha_0)+\psi_u(t,\alpha_1)(\alpha_1-\alpha_0) \\
& -[\phi_u(t,\beta_0)+\psi_u(t,\alpha_0)](\alpha_1-\alpha_0) \\
\leq \;& f(t,\alpha_1)+\phi_u(t,\beta_0)(\alpha_1-\alpha_0)+\psi_u(t,\alpha_0)(\alpha_1-\alpha_0)
\end{aligned}
$$

$$
\begin{aligned}
&-[\phi_u(t,\beta_0)+\psi_u(t,\alpha_0)](\alpha_1-\alpha_0)\\
\equiv\;& f(t,\alpha_1).
\end{aligned}
$$

Therefore, $\alpha_1(t)$ is a lower solution of (2.6.1). Similarly, we prove $\beta_1(t)$ is an upper solution of (2.6.1). Since $\alpha_0(t) \le \beta_1(t) \le \beta_0(t)$ and using (2.6.10) and the fact that B_u is nondecreasing on Ω and ψ_u is nonincreasing on Ω, we obtain,

$$
\begin{aligned}
\beta_1'(t) &= F(t,\beta_0)+G(t,\beta_0)+H(t,\beta_1)\\
&\quad +[B_u(t,\alpha_0)+D_u(t,\beta_0)-\phi_u(t,\beta_0)-\psi_u(t,\alpha_0)](\beta_1-\beta_0)\\
&\ge F(t,\beta_1)+G(t,\beta_1)+H(t,\beta_1)\\
&\quad +[B_u(t,\beta_1)+D_u(t,\beta_0)](\beta_0-\beta_1)\\
&\quad +\phi(t,\beta_1)-\phi(t,\beta_0)+\psi(t,\beta_1)-\psi(t,\beta_0)\\
&\quad +[B_u(t,\alpha_0)+D_u(t,\beta_0)-\phi_u(t,\beta_0)-\psi_u(t,\alpha_0)](\beta_1-\beta_0)\\
&\ge F(t,\beta_1)+G(t,\beta_1)+H(t,\beta_1)\\
&\quad -[B_u(t,\beta_1)+D_u(t,\beta_0)](\beta_1-\beta_0)\\
&\quad -\left[\int_0^1 \phi_u(t,s\beta_0+(1-s)\beta_1)ds\right](\beta_0-\beta_1)\\
&\quad -\left[\int_0^1 \psi_u(t,s\beta_0+(1-s)\beta_1)ds\right](\beta_0-\beta_1)\\
&\quad +[B_u(t,\alpha_0)+D_u(t,\beta_0)-\phi_u(t,\beta_0)-\psi_u(t,\alpha_0)](\beta_1-\beta_0)\\
&\ge f(t,\beta_1).
\end{aligned}
$$

Thus $\beta_1(t)$ is an upper solution of (2.6.1). Hence by Theorem 1.1.3, we get $\alpha_1(t) \le \beta_1(t)$ on J. Consequently, we have that $\alpha_0(t) \le \alpha_1(t) \le \beta_1(t) \le \beta_0(t)$ on J.

We now proceed to show by induction that

$$\alpha_0(t) \le \alpha_1(t) \le \cdots \le \alpha_n(t) \le \alpha_{n+1}(t) \le \beta_{n+1}(t) \le \beta_n(t) \le \cdots \le \beta_1(t) \le \beta_0(t) \text{ on } J.$$

To this end let's consider the IVP,

$$
\begin{aligned}
w'(t) &= F(t,\alpha_n)+G(t,\alpha_n)+H(t,w)\\
&\quad +[B_u(t,\alpha_n)+D_u(t,\beta_n)-\phi_u(t,\beta_n)-\psi_u(t,\alpha_n)]\\
&\quad \times(w-\alpha_n) \qquad (2.6.14)\\
&\equiv M(t,w,\alpha_n,\beta_n)\\
w(0) &= u_0.
\end{aligned}
$$

Assume that for some $n > 1$, $\alpha_0(t) \le \alpha_n(t) \le \beta_n(t) \le \beta_0(t)$ on J. We need to show that

$$\alpha_n(t) \le \alpha_{n+1}(t) \le \beta_{n+1}(t) \le \beta_n(t) \text{ on } J.$$

From (2.6.10) and applying the Integral Mean Value Theorem for systems and the fact that ϕ_u is nondecreasing on Ω, and ψ_u and D_u are nonincreasing on Ω, we obtain

$$\begin{aligned}
\alpha_n'(t) &= F(t,\alpha_{n-1}) + G(t,\alpha_{n-1}) + H(t,\alpha_n) \\
&\quad +[B_u(t,\alpha_{n-1}) + D_u(t,\beta_{n-1}) - \phi_u(t,\beta_{n-1}) - \psi_u(t,\alpha_{n-1})] \\
&\quad \times(\alpha_n - \alpha_{n-1}) \\
&\leq F(t,\alpha_n) + G(t,\alpha_n) + H(t,\alpha_n) \\
&\quad -[B_u(t,\alpha_{n-1}) + D_u(t,\alpha_n)](\alpha_n - \alpha_{n-1}) + \phi(t,\alpha_n) \\
&\quad -\phi(t,\alpha_{n-1}) + \psi(t,\alpha_n) - \psi(t,\alpha_{n-1}) + [B_u(t,\alpha_{n-1}) + D_u(t,\beta_{n-1}) \\
&\quad -\phi_u(t,\beta_{n-1}) - \psi_u(t,\alpha_{n-1})](\alpha_n - \alpha_{n-1}) \\
&\leq F(t,\alpha_n) + G(t,\alpha_n) + H(t,\alpha_n) \\
&\quad -[B_u(t,\alpha_{n-1}) + D_u(t,\beta_{n-1})](\alpha_n - \alpha_{n-1}) \\
&\quad + \left[\int_0^1 \phi_u(t, s\alpha_n + (1-s)\alpha_{n-1})ds\right](\alpha_n - \alpha_{n-1}) \\
&\quad + \left[\int_0^1 \psi_u(t, s\alpha_n + (1-s)\alpha_{n-1})ds\right](\alpha_n - \alpha_{n-1}) \\
&\quad +[B_u(t,\alpha_{n-1}) + D_u(t,\beta_{n-1}) - \phi_u(t,\beta_{n-1}) - \psi_u(t,\alpha_{n-1})] \\
&\quad \times(\alpha_n - \alpha_{n-1}) \\
&\leq f(t,\alpha_n) + [\phi_u(t,\alpha_n) + \psi_u(t,\alpha_n)](\alpha_n - \alpha_{n-1}) - [\phi_u(t,\beta_{n-1}) \\
&\quad +\psi_u(t,\alpha_{n-1})](\alpha_n - \alpha_{n-1}) \\
&\leq f(t,\alpha_n) \\
&\equiv M(t,\alpha_n,\alpha_n,\beta_n).
\end{aligned}$$

Thus $\alpha_n(t)$ is a lower solution of (2.6.14). In a similar fashion we can show $\beta_n(t)$ is an upper solution of (2.6.14). Using (2.6.10) and applying the Integral Mean Value Theorem for systems and the fact that B_u and ϕ_u are nondecreasing on Ω and D_u and ψ_u are nonincreasing on Ω we obtain the following,

$$\begin{aligned}
\beta_n'(t) &= F(t,\beta_{n-1}) + G(t,\beta_{n-1}) + H(t,\beta_n) \\
&\quad +[B_u(t,\alpha_{n-1}) + D_u(t,\beta_{n-1}) - \phi_u(t,\beta_{n-1}) \\
&\quad -\psi_u(t,\alpha_{n-1})](\beta_n - \beta_{n-1}) \\
&\geq F(t,\alpha_n) + G(t,\alpha_n) + H(t,\beta_n) \\
&\quad +[B_u(t,\alpha_n) + D_u(t,\beta_{n-1})](\beta_{n-1} - \alpha_n) \\
&\quad +\phi(t,\alpha_n) - \phi(t,\beta_{n-1}) + \psi(t,\alpha_n) - \psi(t,\beta_{n-1}) \\
&\quad +[B_u(t,\alpha_{n-1}) + D_u(t,\beta_{n-1}) - \phi_u(t,\beta_{n-1})
\end{aligned}$$

$$
\begin{aligned}
& -\psi_u(t,\alpha_{n-1})](\beta_n-\beta_{n-1}) \\
\geq\ & F(t,\alpha_n)+G(t,\alpha_n)+H(t,\beta_n) \\
& +B_u(t,\alpha_n)(\beta_{n-1}-\alpha_n+\beta_n-\beta_n) \\
& -\left[\int_0^1 \phi_u(t,s\beta_{n-1}+(1-s)\alpha_n)ds\right](\beta_{n-1}-\alpha_n+\beta_n-\beta_n) \\
& +D_u(t,\beta_{n-1})(\beta_{n-1}-\alpha_n+\beta_n-\beta_{n-1}) \\
& -\left[\int_0^1 \psi_u(t,s\beta_{n-1}+(1-s)\alpha_n)ds\right](\beta_{n-1}-\alpha_n+\beta_n-\beta_n) \\
& +[B_u(t,\alpha_{n-1})+D_u(t,\beta_{n-1})-\phi_u(t,\beta_{n-1}) \\
& -\psi_u(t,\alpha_{n-1})](\beta_n-\beta_{n-1}) \\
\geq\ & F(t,\alpha_n)+G(t,\alpha_n)+H(t,\beta_n) \\
& +B_u(t,\alpha_n)(\beta_n-\alpha_n)-B_u(t,\alpha_{n-1})(\beta_n-\beta_{n-1}) \\
& -\phi_u(t,\beta_n)(\beta_n-\alpha_n)+\phi_u(t,\beta_{n-1})(\beta_n-\beta_{n-1}) \\
& +D_u(t,\beta_n)(\beta_n-\alpha_n)-D_u(t,\beta_{n-1})(\beta_n-\beta_{n-1}) \\
& -\psi_u(t,\alpha_n)(\beta_n-\alpha_n)+\psi_u(t,\alpha_{n-1})(\beta_n-\beta_{n-1}) \\
& +[B_u(t,\alpha_{n-1})+D_u(t,\beta_{n-1})-\phi_u(t,\beta_{n-1})-\psi_u(t,\alpha_{n-1})] \\
& \times(\beta_n-\beta_{n-1}) \\
\equiv\ & M(t,\beta_n,\alpha_n,\beta_n).
\end{aligned}
$$

Therefore, $\beta_n(t)$ is an upper solution of (2.6.14). Now, by Theorem 1.1.6 there exists a solution $\alpha_{n+1}(t)$ of (2.6.14) such that $\alpha_n(t)\leq\alpha_{n+1}(t)\leq\beta_n(t)$ on J. Furthermore, $\alpha_{n+1}(t)$ is unique since M satisfies a Lipschitz condition. Similarly, it can be shown that $\alpha_n(t)\leq\beta_{n+1}(t)\leq\beta_n(t)$ on J. We need to show $\alpha_{n+1}(t)\leq\beta_{n+1}(t)$ on J. From (2.6.10) and applying the Integral Mean Value Theorem for systems and the fact that ϕ_u is nondecreasing and ψ_u and D_u are nonincreasing on Ω, we obtain that

$$
\begin{aligned}
\alpha'_{n+1}(t) =\ & F(t,\alpha_n)+G(t,\alpha_n)+H(t,\alpha_{n+1}) \\
& +[B_u(t,\alpha_n)+D_u(t,\beta_n)-\phi_u(t,\beta_n)-\psi_u(t,\alpha_n)](\alpha_{n+1}-\alpha_n) \\
\leq\ & F(t,\alpha_{n+1})+G(t,\alpha_{n+1})+H(t,\alpha_{n+1}) \\
& -[B_u(t,\alpha_n)+D_u(t,\alpha_{n+1})](\alpha_{n+1}-\alpha_n)+\phi(t,\alpha_{n+1})-\phi(t,\alpha_n) \\
& +\psi(t,\alpha_{n+1})-\psi(t,\alpha_n) \\
& +[B_u(t,\alpha_n)+D_u(t,\beta_n)-\phi_u(t,\beta_n)-\psi_u(t,\alpha_n)](\alpha_{n+1}-\alpha_n) \\
\leq\ & F(t,\alpha_{n+1})+G(t,\alpha_{n+1})+H(t,\alpha_{n+1}) \\
& -[B_u(t,\alpha_n)+D_u(t,\beta_n)](\alpha_{n+1}-\alpha_n)
\end{aligned}
$$

$$
\begin{aligned}
&+\left[\int_0^1 \phi_u(t, s\alpha_{n+1} + (1-s)\alpha_n)ds\right](\alpha_{n+1} - \alpha_n) \\
&+\left[\int_0^1 \psi_u(t, s\alpha_{n+1} + (1-s)\alpha_n)ds\right](\alpha_{n+1} - \alpha_n) \\
&+[B_u(t,\alpha_n) + D_u(t,\beta_n) - \phi_u(t,\beta_n) - \psi_u(t,\alpha_n)](\alpha_{n+1} - \alpha_n) \\
\leq\ & f(t,\alpha_{n+1}) + [\phi_u(t,\alpha_{n+1}) + \psi_u(t,\alpha_{n+1})](\alpha_{n+1} - \alpha_n) \\
&-[\phi_u(t,\beta_n) + \psi_u(t,\alpha_n)](\alpha_{n+1} - \alpha_n) \\
\leq\ & f(t,\alpha_{n+1}) + [\phi_u(t,\beta_n) + \psi_u(t,\alpha_n)](\alpha_{n+1} - \alpha_n) \\
&-[\phi_u(t,\beta_n) + \psi_u(t,\alpha_n)](\alpha_{n+1} - \alpha_n) \\
\equiv\ & f(t,\alpha_{n+1}).
\end{aligned}
$$

Therefore, $\alpha_{n+1}(t)$ is a lower solution of (2.6.1) on J. Similarly we can show that $\beta_{n+1}(t)$ is an upper solution of (2.6.1) on J. Hence by Theorem 1.1.3, $\alpha_{n+1}(t) \leq \beta_{n+1}(t)$ on J. Thus by induction we have shown that for all n,

$$\alpha_0(t) \leq \alpha_1(t) \leq \cdots \leq \alpha_{n-1}(t) \leq \alpha_n(t) \leq \beta_n(t) \leq \beta_{n-1}(t) \leq \cdots \leq \beta_1(t) \leq \beta_0(t) \text{ on } J.$$

Employing standard arguments, it can be shown that the sequences $\{\alpha_n(t)\}$ and $\{\beta_n(t)\}$ converge uniformly and monotonically to the unique solution $u(t)$ of (2.6.1) on J.

It remains to show quadratic convergence. To this end, we let $p_{n+1}(t) = u(t) - \alpha_{n+1}(t)$, and $q_{n+1}(t) = \beta_{n+1}(t) - u(t)$. Then $p_{n+1}(0) = q_{n+1}(0) = 0$.

$$
\begin{aligned}
p'_{n+1}(t) &= u'(t) - \alpha'_{n+1}(t) \\
&= F(t,u) + G(t,u) + H(t,u) \\
&\quad -[F(t,\alpha_n) + G(t,\alpha_n) + H(t,\alpha_{n+1})] \\
&\quad -[B_u(t,\alpha_n) + D_u(t,\beta_n) - \phi_u(t,\beta_n) - \psi_u(t,\alpha_n)](\alpha_{n+1} - \alpha_n) \\
&\quad +\phi(t,\alpha_n) - \phi(t,\alpha_n) + \phi(t,u) - \phi(t,u) \\
&\quad +\psi(t,\alpha_n) - \psi(t,\alpha_n) + \psi(t,u) - \psi(t,u) \\
&= \left[\int_0^1 B_u(t, su + (1-s)\alpha_n)ds\right](u - \alpha_n) \\
&\quad -\left[\int_0^1 \phi_u(t, su + (1-s)\alpha_n)ds\right](u - \alpha_n) \\
&\quad +\left[\int_0^1 D_u(t, su + (1-s)\alpha_n)ds\right](u - \alpha_n) \\
&\quad -\left[\int_0^1 \psi_u(t, su + (1-s)\alpha_n)ds\right](u - \alpha_n) + H(t,u) - H(t,\alpha_{n+1})
\end{aligned}
$$

$$
\begin{aligned}
&-[B_u(t,\alpha_n)+D_u(t,\beta_n)-\phi_u(t,\beta_n)-\psi_u(t,\alpha_n)](\alpha_{n+1}-\alpha_n)\\
\le\ & B_u(t,u)(u-\alpha_n)-\phi_u(t,u)(u-\alpha_n)\\
&+D_u(t,u)(u-\alpha_n)-\psi_u(t,u)(u-\alpha_n)+K(u-\alpha_{n+1})\\
&-B_u(t,\alpha_n)(\alpha_{n+1}-\alpha_n)+\phi_u(t,\beta_n)(\alpha_{n+1}-\alpha_n)\\
&-D_u(t,\beta_n)(\alpha_{n+1}-\alpha_n)+\psi_u(t,\alpha_n)(\alpha_{n+1}-\alpha_n)\\
=\ & B_u(t,u)p_n-B_u(t,\alpha_n)(p_n-p_{n+1})\\
&+D_u(t,u)p_n-D_u(t,\beta_n)(p_n-p_{n+1})\\
&-\phi_u(t,u)p_n+\phi_u(t,\beta_n)(p_n-p_{n+1})\\
&-\psi_u(t,u)p_n+\psi_u(t,\alpha_n)(p_n-p_{n+1})+Kp_{n+1}\\
=\ & p_n^T\left[\int_0^1 B_{uu}(t,su+(1-s)\alpha_n)ds\right]p_n+B_u(t,\alpha_n)p_{n+1}\\
&-q_n^T\left[\int_0^1 D_{uu}(t,s\beta_n+(1-s)u)ds\right]p_n+D_u(t,\beta_n)p_{n+1}\\
&+q_n^T\left[\int_0^1 \phi_{uu}(t,s\beta_n+(1-s)u)ds\right]p_n-\phi_u(t,\beta_n)p_{n+1}\\
&-p_n^T\left[\int_0^1 \psi_{uu}(t,su+(1-s)\alpha_n)ds\right]p_n-\psi_u(t,\alpha_n)p_{n+1}\\
&+Kp_{n+1}\\
\le\ & p_n^T\left[\int_0^1 B_{uu}(t,su+(1-s)\alpha_n)ds\right]p_n+F_u(t,\alpha_n)p_{n+1}\\
&-q_n^T\left[\int_0^1 D_{uu}(t,s\beta_n+(1-s)u)ds\right]p_n+G_u(t,\beta_n)p_{n+1}\\
&+q_n^T\left[\int_0^1 \phi_{uu}(t,s\beta_n+(1-s)u)ds\right]p_n\\
&-p_n^T\left[\int_0^1 \psi_{uu}(t,su+(1-s)\alpha_n)ds\right]p_n+Kp_{n+1}\\
\le\ & (A+C+K)p_{n+1}+Op_n^tp_n-Qq_n^tp_n+Rq_n^tp_n-Sp_n^tp_n\\
=\ & (A+C+K)p_{n+1}+(O-S)p_n^tp_n+(R-Q)q_n^tp_n
\end{aligned}
$$

where $\sum_{i=1}^{n} B^i_{uu}\le O$, $\sum_{i=1}^{n} D^i_{uu}\le Q$, $\sum_{i=1}^{n}\phi^i_{uu}\le R$, $\sum_{i=1}^{n}\psi^i_{uu}\le S$, $F_u\le A$, $G_u\le C$ on Ω, and O,Q,R,S,A,C,K are $n\times n$ positive matrices such that $(A+C+K)^{-1}$ exists. There is no loss of generality assuming $(A+C+K)^{-1}$ exists such that $a_{ij}\ge 0$, where a_{ij} represents the components of the matrix, Thus we have

$$p'_{n+1} \le (A+C+K)p_{n+1}+(O-S)p_n^Tp_n$$

$$+\frac{1}{2}(R-Q)p_n^t p_n + \frac{1}{2}(R-Q)q_n^T q_n,$$

T being the transpose. And by Corollary 1.1.1, we obtain the estimate

$$p_{n+1} \le \int_0^1 \exp[(A+C+K)(t-s)][(O-S+\frac{1}{2}(R-Q))p_n^2 + \frac{1}{2}(R-Q)q_n^2]ds$$

This implies

$$\begin{aligned}\max\|p_{n+1}\| &\le (A+C+K)^{-1}\exp[(A+C+K)t][(O-S+\tfrac{1}{2}(R-Q))\max\|p_n^2\| \\ &\quad +\tfrac{1}{2}(R-Q)\max\|q_n^2\|]\end{aligned}$$

which yields quadratic convergence. Similarly we obtain for $q_{n+1}(t)$ the estimate

$$q_{n+1} \le \int_0^1 \exp[(A+C+K)(t-s)][\frac{1}{2}(O-S)p_n^2 + \frac{3}{2}(O-S)q_n^2]ds.$$

This implies

$$\max\|q_{n+1}\| \le (A+C+K)^{-1}\exp[(A+C+K)t][\frac{1}{2}(O-S)\max\|p_n^2\| + \frac{3}{2}(O-S)\max\|q_n^2\|$$

which yields quadratic convergence. Here p_n, q_n are as in Theorem 2.6.1.

The final result of this section deals with only semi-quadratic convergence under less restrictive assumptions.

Theorem 2.6.3 *Assume that (A_1) and (A_2) of Theorem 2.6.2 hold. Assume further that*

(A_3) (a) *$H(t,x) - H(t,y) \ge -K(x-y)$, whenever $x \ge y, x, y \in \Omega$, where K is an $n \times n$ matrix with components $k_{ii} > 0$.*

(b) *$a_{ij}(t,\alpha_0,\beta_0) \ge 0$ for $i \ne j$, where $A(t,\alpha_0,\beta_0) = [a_{ij}(t,\alpha_0,\beta_0)]$ is an $n \times n$ matrix given by*

$$A(t,\alpha_0,\beta_0) = B_u(t,\alpha_0) + D_u(t,\beta_0) - \phi_u(t,\beta_0) - \psi_u(t,\alpha_0) - K$$

Then there exists monotone sequences $\{\alpha_n(t)\}$ and $\{\beta_n(t)\}$, where $\alpha_n(t)$ and $\beta_n(t)$ are solutions of linear differential equations, which converge uniformly and monotonically to the unique solution $u(t)$ of (2.6.1), and the convergence is semi-quadratic.

Proof In view of (A_2), $f(t,u)$ is Lipschitzian in u on Ω. Consider the IVP's

$$\begin{aligned} \alpha_1' &= f(t,\alpha_0) + [B_u(t,\alpha_0) + D_u(t,\beta_0) \\ &\quad -\phi_u(t,\beta_0) - \psi_u(t,\alpha_0) - K](\alpha_1 - \alpha_0) \\ \alpha_1(0) &= u_0 \\ \beta_1' &= f(t,\beta_0) + [B_u(t,\alpha_0) + D_u(t,\beta_0) \\ &\quad -\phi_u(t,\beta_0) - \psi_u(t,\alpha_0) - K](\beta_1 - \beta_0) \\ \beta_1(0) &= u_0. \end{aligned} \tag{2.6.15}$$

We need to show that $\alpha_0(t) \le \alpha_1(t)$ on J. Let $p(t) = \alpha_0(t) - \alpha_1(t)$ so that $p(0) \le 0$. Then

$$\begin{aligned} p' &= \alpha_0' - \alpha_1' \\ &\le f(t,\alpha_0) - f(t,\alpha_0) - [B_u(t,\alpha_0) + D_u(t,\beta_0) \\ &\quad -\phi_u(t,\beta_0) - \psi_u(t,\alpha_0) - K](\alpha_1 - \alpha_0) \\ &\equiv A(t,\alpha_0,\beta_0)p. \end{aligned}$$

Since $A(t,\alpha_0,\beta_0)$ is quasimonotone nondecreasing by assumption (A_2) it follows from using special case of Theorem 1.1.3 for the linear case with $p(t) = \alpha(t)$ and $0 = \beta(t)$ that $p(t) \le 0$ on $J = [0,T]$, proving $\alpha_0 \le \alpha_1$ on J.

Now we show $\alpha_1(t) \le \beta_0(t)$. Let $p(t) = \alpha_1(t) - \beta_0(t)$ and note that $p(0) \le 0$. Then

$$\begin{aligned} p' &= \alpha_1' - \beta_0' \\ &\le f(t,\alpha_0) + [B_u(t,\alpha_0) + D_u(t,\beta_0) - \phi_u(t,\beta_0) \\ &\quad -\psi_u(t,\alpha_0) - K](\alpha_1 - \alpha_0) - f(t,\beta_0) \\ &= F(t,\alpha_0) + G(t,\alpha_0) + H(t,\alpha_0) - H(t,\beta_0) - F(t,\beta_0) - G(t,\beta_0) \\ &\quad +[F_u(t,\alpha_0) + \phi_u(t,\alpha_0) + G_u(t,\beta_0) + \psi_u(t,\beta_0) \\ &\quad -\phi_u(t,\beta_0) - \psi_u(t,\alpha_0) - K](\alpha_1 - \alpha_0) \end{aligned}$$

Using (2.6.10) we obtain

$$\begin{aligned} p' &\le B_u(t,\alpha_0)(\alpha_1 - \beta_0) + D_u(t,\beta_0)(\alpha_1 - \beta_0) \\ &\quad +\phi(t,\beta_0) - \phi(t,\alpha_0) + \psi(t,\beta_0) - \psi(t,\alpha_0) \\ &\quad -[\phi_u(t,\beta_0) + \psi_u(t,\alpha_0) + K](\alpha_1 - \alpha_0) + H(t,\alpha_0) - H(t,\beta_0). \end{aligned}$$

By applying the integral Mean Value Theorem on ϕ and ψ we obtain

$$p' \le B_u(t,\alpha_0)(\alpha_1 - \beta_0) + D_u(t,\beta_0)(\alpha_1 - \beta_0)$$

$$\begin{aligned}
& -\phi_u(t,\beta_0)(\alpha_1-\beta_0)-\psi_u(t,\alpha_0)(\alpha_1-\beta_0)-K(\alpha_1-\alpha_0) \\
& +H(t,\alpha_0)-H(t,\alpha_0)+K(\beta_0-\alpha_0) \\
= & [B_u(t,\alpha_0)+D_u(t,\beta_0)-\phi_u(t,\beta_0)-\psi_u(t,\alpha_0)-K]p \\
\equiv & [A(t,\alpha_0,\beta_0)]p.
\end{aligned}$$

This implies as before, using special case of Theorem 1.1.3 that $\alpha_1(t) \le \beta_0(t)$ on J. We thus have $\alpha_0(t) \le \alpha_1(t) \le \beta_0(t)$ on $J = [0,T]$. In a similar way we can prove that $\alpha_0(t) \le \beta_1(t) \le \beta_0(t)$ on $J = [0,T]$. Next we show that $\alpha_1(t) \le \beta_1(t)$, by showing that $\alpha_1(t)$ and $\beta_1(t)$ are lower and upper solutions of (2.6.1) respectively. Using (2.6.10), the fact that ϕ_u is nondecreasing on Ω and both D_u and ψ_u are nonincreasing on Ω, and applying the Integral Mean Value Theorem for systems, we have

$$\begin{aligned}
\alpha_1' = & f(t,\alpha_0)+[B_u(t,\alpha_0)+D_u(t,\beta_0) \\
& -\phi_u(t,\beta_0)-\psi_u(t,\alpha_0)-K](\alpha_1-\alpha_0) \\
= & [F(t,\alpha_0)+B_u(t,\alpha_0)(\alpha_1-\alpha_0)] \\
& +[G(t,\alpha_0)+D_u(t,\beta_0)(\alpha_1-\alpha_0)] \\
& +H(t,\alpha_0)-K(\alpha_1-\alpha_0) \\
& -\phi_u(t,\beta_0)(\alpha_1-\alpha_0)-\psi_u(t,\alpha_0)(\alpha_1-\alpha_0) \\
\le & F(t,\alpha_1)-B_u(t,\alpha_0)(\alpha_1-\alpha_0)+B_u(t,\alpha_0)(\alpha_1-\alpha_0) \\
& +G(t,\alpha_1)-D_u(t,\alpha_1)(\alpha_1-\alpha_0)+D_u(t,\beta_0)(\alpha_1-\alpha_0) \\
& +H(t,\alpha_0)+H(t,\alpha_1)-H(t,\alpha_0) \\
& +[\phi(t,\alpha_1)-\phi(t,\alpha_0)]+[\psi(t,\alpha_1)-\psi(t,\alpha_0)] \\
& -\phi_u(t,\beta_0)(\alpha_1-\alpha_0)-\psi_u(t,\alpha_0)(\alpha_1-\alpha_0) \\
\le & f(t,\alpha_1)
\end{aligned}$$

Similarly we can show that $\beta_1'(t) \ge f(t,\beta_1)$, $\beta_1(0) = u_0$. Since $f(t,u)$ is Lipschitzian in u on Ω and f is quasimonotone nondecreasing in u, we can conclude by Theorem 1.1.3 that $\alpha_1(t) \le \beta_1(t)$ on $J = [0,T]$. This proves that

$$\alpha_0(t) \le \alpha_1(t) \le \beta_1(t) \le \beta_0(t) \text{ on } J = [0,T].$$

Next, we will assume that for some $k \ge 1$, $\alpha_{k+1}' \le f(t,\alpha_k)$, $\beta_{k+1}' \ge f(t,\beta_k)$ and $\alpha_0(t) \le \alpha_k(t) \le \beta_k(t) \le \beta_0(t)$ on J. We shall show that

$$\alpha_k(t) \le \alpha_{k+1}(t) \le \beta_{k+1}(t) \le \beta_k(t) \text{ on } J$$

where $\alpha_{k+1}(t)$ and $\beta_{k+1}(t)$ are the solutions of linear IVP's

$$\alpha_{k+1}' \;=\; f(t,\alpha_k)+[B_u(t,\alpha_k)+D_u(t,\beta_k)$$

$$
\begin{aligned}
&\quad -\phi_u(t,\beta_k) - \psi_u(t,\alpha_k) - K](\alpha_{k+1} - \alpha_k) \\
\alpha_{k+1}(0) &= u_0 \qquad\qquad (2.6.16)\\
\beta'_{k+1} &= f(t,\beta_k) + [B_u(t,\alpha_k) + D_u(t,\beta_k) \\
&\quad -\phi_u(t,\beta_k) - \psi_u(t,\alpha_k) - K](\beta_{k+1} - \beta_k) \\
\beta_{k+1}(0) &= u_0.
\end{aligned}
$$

We note that $A(t,\alpha_k,\beta_k) = B_u(t,\alpha_k) + D_u(t,\beta_k) - \phi_u(t,\beta_k) - \psi_u(t,\alpha_k) - K$ and $a_{ij}(t,\alpha_k,\beta_k) \geq 0$ for $i \neq j$. We need to show that $\alpha_k(t) \leq \alpha_{k+1}(t)$ on J. As before, set $p(t) = \alpha_k(t) - \alpha_{k+1}(t)$. Then $p(0) = 0$, and

$$
\begin{aligned}
p' &= \alpha'_k - \alpha'_{k+1} \\
&\leq f(t,\alpha_k) - f(t,\alpha_k) - [B_u(t,\alpha_k) + D_u(t,\beta_k) \\
&\quad -\phi_u(t,\beta_k) - \psi_u(t,\alpha_k) - K](\alpha_{k+1} - \alpha_k) \\
&\equiv A(t,\alpha_k,\beta_k)p.
\end{aligned}
$$

It follows from using linear case of Theorem 1.1.3 that $\alpha_k(t) \leq \alpha_{k+1}(t)$ on J. We now show $\alpha_{k+1}(t) \leq \beta_k(t)$ on J. Let $p = \alpha_{k+1}(t) - \beta_k(t)$ and note that $p(0) = 0$. Then

$$
\begin{aligned}
p' &= \alpha'_{k+1} - \beta'_k \\
&\leq f(t,\alpha_k) + [B_u(t,\alpha_k) + D_u(t,\beta_k) - \phi_u(t,\beta_k) \\
&\quad -\psi_u(t,\alpha_k) - K](\alpha_{k+1} - \alpha_k) - f(t,\beta_k) \\
&= F(t,\alpha_k) + G(t,\alpha_k) + H(t,\alpha_k) - H(t,\beta_k) - F(t,\beta_k) - G(t,\beta_k) \\
&\quad +[F_u(t,\alpha_k) + \phi_u(t,\alpha_k) + G_u(t,\beta_k) + \psi_u(t,\beta_k) \\
&\quad -\phi_u(t,\beta_k) - \psi_u(t,\alpha_k) - K](\alpha_{k+1} - \alpha_k).
\end{aligned}
$$

Using (2.6.10) we obtain

$$
\begin{aligned}
p' &\leq B_u(t,\alpha_k)(\alpha_{k+1} - \beta_k) + D_u(t,\beta_k)(\alpha_{k+1} - \beta_k) \\
&\quad +\phi(t,\beta_k) - \phi(t,\alpha_k) + \psi(t,\beta_k) - \psi(t,\alpha_k) \\
&\quad -[\phi_u(t,\beta_k) + \psi_u(t,\alpha_k) + K](\alpha_{k+1} - \alpha_k) + H(t,\alpha_k) - H(t,\beta_k).
\end{aligned}
$$

By applying the integral Mean Value Theorem on ϕ and ψ we obtain

$$
\begin{aligned}
p' &\leq B_u(t,\alpha_k)(\alpha_{k+1} - \beta_k) + D_u(t,\beta_k)(\alpha_{k+1} - \beta_k) \\
&\quad -\phi_u(t,\beta_k)(\alpha_{k+1} - \beta_k) - \psi_u(t,\alpha_k)(\alpha_{k+1} - \beta_k) - K(\alpha_{k+1} - \alpha_k) \\
&\quad +H(t,\alpha_k) - H(t,\alpha_k) + K(\beta_k - \alpha_k) \\
&= [B_u(t,\alpha_k) + D_u(t,\beta_k) - \phi_u(t,\beta_k) - \psi_u(t,\alpha_k) - K]p \\
&\equiv [A(t,\alpha_k,\beta_k)]p
\end{aligned}
$$

This implies as before, using special case of Theorem 1.1.3 that $\alpha_{k+1}(t) \leq \beta_k(t)$ on J. Next, we prove that $\alpha_{k+1}(t) \leq \beta_{k+1}(t)$ on J, by showing $\alpha_{k+1}(t)$, $\beta_k(t)$ are lower and upper solutions of (2.6.1), respectively. Using (2.6.10), the fact that ϕ_u is nondecreasing on Ω and both D_u and ψ_u are nonincreasing on Ω, and applying the Integral Mean Value Theorem for systems, we have

$$\begin{aligned}
\alpha'_{k+1} &= f(t,\alpha_k) + [B_u(t,\alpha_k) + D_u(t,\beta_k) \\
&\quad -\phi_u(t,\beta_k) - \psi_u(t,\alpha_k) - K](\alpha_{k+1} - \alpha_k) \\
&= [F(t,\alpha_k) + B_u(t,\alpha_k)(\alpha_{k+1} - \alpha_k)] \\
&\quad +[G(t,\alpha_k) + D_u(t,\beta_k)(\alpha_{k+1} - \alpha_k)] \\
&\quad +H(t,\alpha_k) - K(\alpha_{k+1} - \alpha_k) \\
&\quad -\phi_u(t,\beta_k)(\alpha_{k+1} - \alpha_k) - \psi_u(t,\alpha_k)(\alpha_{k+1} - \alpha_k) \\
&\leq F(t,\alpha_{k+1}) - B_u(t,\alpha_k)(\alpha_{k+1} - \alpha_k) + B_u(t,\alpha_k)(\alpha_{k+1} - \alpha_k) \\
&\quad +G(t,\alpha_{k+1}) - D_u(t,\alpha_{k+1})(\alpha_{k+1} - \alpha_k) + D_u(t,\beta_k)(\alpha_{k+1} - \alpha_k) \\
&\quad +H(t,\alpha_k) + H(t,\alpha_{k+1}) - H(t,\alpha_k) \\
&\quad +[\phi(t,\alpha_{k+1}) - \phi(t,\alpha_k)] + [\psi(t,\alpha_{k+1}) - \psi(t,\alpha_k)] \\
&\quad -\phi_u(t,\beta_k)(\alpha_{k+1} - \alpha_k) - \psi_u(t,\alpha_k)(\alpha_{k+1} - \alpha_k) \\
&\leq f(t,\alpha_{k+1}).
\end{aligned}$$

Similarly, we can show that $\beta'_{k+1}(t) \geq f(t,\beta_{k+1}), \beta_{k+1}(0) = u_0$. Since $f(t,u)$ is Lipschitzian in u on Ω and f is quasimonotone nondecreasing in u, we can conclude by Theorem 1.1.3 that $\alpha_1(t) \leq \beta_1(t)$ on $J = [0,T]$. This proves that

$$\alpha_0(t) \leq \alpha_{k+1}(t) \leq \beta_{k+1}(t) \leq \beta_0(t) \text{ on } J = [0,T].$$

By induction, we have that for all n,

$$\alpha_0(t) \leq \alpha_1(t) \leq \cdots \leq \alpha_{n-1}(t) \leq \alpha_n(t) \leq \beta_n(t) \leq \beta_{n-1}(t) \leq \cdots \leq \beta_1(t) \leq \beta_0(t) \text{ on } J.$$

Employing standard arguments, it can be shown that the sequences $\{\alpha_n(t)\}$ and $\{\beta_n(t)\}$ converge uniformly and monotonically to the unique solution $u(t)$ of (2.6.1) on J. Finally, we prove the convergence is semi-quadratic. Let $p_{n+1}(t) = u(t) - \alpha_{n+1}(t)$ and $q_{n+1}(t) = \beta_{n+1}(t) - u(t)$ on J. Then $p_{n+1}(0) = q_{n+1}(0) = 0$ and by the Integral Mean Value Theorem for systems and the increasing and decreasing natures of B_u, ϕ_u, D_u, ψ_u, we obtain the following,

$$\begin{aligned}
p'_{n+1} &= u' - \alpha_{n+1}(t) \\
&= f(t,u) - f(t,\alpha_n) \\
&\quad -[B_u(t,\alpha_n) + D_u(t,\beta_n) - \phi_u(t,\beta_n) - \psi_u(t,\alpha_n) - K](\alpha_{n+1} - \alpha_n) \\
&\leq [B(t,u) - B(t,\alpha_n)] + [D(t,u) - D(t,\alpha_n)] + K(u - \alpha_n) \\
&\quad -[\phi(t,u) - \phi(t,\alpha_n)] - [\psi(t,u) - \psi(t,\alpha_n)] \\
&\quad -[B_u(t,\alpha_n) + D_u(t,\beta_n) - \phi_u(t,\beta_n) - \psi_u(t,\alpha_n) - K](\alpha_{n+1} - \alpha_n).
\end{aligned}$$

Then,

$$\begin{aligned} p'_{n+1} \leq\ & B_u(t,u)(u-\alpha_n) - B_u(t,\alpha_n)(u-\alpha_n-(u-\alpha_{n+1})) \\ & +D_u(t,u)(u-\alpha_n) - D_u(t,\beta_n)(u-\alpha_n-(u-\alpha_{n+1})) \\ & -\phi_u(t,u)(u-\alpha_n)] + \phi_u(t,\beta_n)(u-\alpha_n-(u-\alpha_{n+1})) \\ & -\psi_u(t,u)(u-\alpha_n) + \psi_u(t,\alpha_n)(u-\alpha_n-(u-\alpha_{n+1})) \\ & +K(u-\alpha_n) + K(u-\alpha_n-(u-\alpha_{n+1})). \end{aligned}$$

Here we apply the Integral Mean Value Theorem for systems to get,

$$\begin{aligned} p'_{n+1} &\leq p_n^T[\textstyle\int_0^1 B_{uu}(t,su+(1-s)\alpha_n)ds]p_n + F_u(t,\alpha_n)p_{n+1} \\ & \quad -q_n^T[\textstyle\int_0^1 D_{uu}(t,s\beta_n+(1-s)u)ds]p_n + G_u(t,\beta_n)p_{n+1} \\ & \quad +q_n^T[\textstyle\int_0^1 \phi_{uu}(t,s\beta_n+(1-s)u)ds]p_n \\ & \quad -p_n^T[\textstyle\int_0^1 \psi_{uu}(t,su+(1-s)\alpha_n)ds]p_n + 2Kp_n - Kp_{n+1} \\ &\leq (N+O-K)p_{n+1} + 2Kp_n + Lp_n^Tp_n - Mq_n^Tp_n + Qq_n^Tp_n - Rp_n^Tp_n \\ &= (N+O-K)p_{n+1} + 2Kp_n + (L-R)p_n^Tp_n + (Q-M)q_n^Tp_n \end{aligned}$$

where $\sum_{i=1}^n B_{uu}^i \leq L$, $\sum_{i=1}^n D_{uu}^i \leq M$, $\sum_{i=1}^n \phi_{uu}^i \leq Q$, $\sum_{i=1}^n \psi_{uu}^i \leq R$, $F_u \leq N$, $G_u \leq O$ on Ω, and K, L, M, N, O, Q, R are $n \times n$ positive matrices such that $(N+O-K)^{-1}$ exists. There is no loss of generality assuming $(N+O-K)^{-1}$ exists such that $a_{ij} \geq 0$, where a_{ij} represents the components of the matrix. Thus we have

$$\begin{aligned} p'_{n+1} \leq\ & (N+O-K)p_{n+1} + 2Kp_n \\ & +(L-R+\tfrac{1}{2}(Q-M))p_n^tp_n + \tfrac{1}{2}(Q-M)q_n^tq_n. \end{aligned}$$

And by Corollary 1.1.1, we obtain the estimate

$$p_{n+1} \leq \int_0^1 \exp[(N+O-K)(t-s)][2Kp_n + (L-R+\frac{1}{2}(Q-M))p_n^2 + \frac{1}{2}(Q-M)q_n^2]ds.$$

This implies

$$\begin{aligned} \max\|p_{n+1}\| \leq\ & (N+O-K)^{-1}\exp[(N+O-K)t][2K\max\|p_n\| \\ & +(L-R+\tfrac{1}{2}(Q-M))\max\|p_n^2\| + \tfrac{1}{2}(Q-M)\max\|q_n^2\|] \end{aligned}$$

which yields semi-quadratic convergence. Here p_n, q_n are as in the proof of Theorem 2.6.1.

2.7 NOTES

The results on periodic boundary value problems developed in Section 2.1 are taken successively from Lakshmikantham and Leela [29], Lakshmikantham [22],

Lakshmikantham and Nieto [35], Lakshmikantham, Shahzad and Nieto [40], Shahzad and Malek [58], and Vatsala and Stutson [71]. All the results on antiperiodic boundary value problems are taken from Yin [72, 73, 74]. Section 2.3 contains the work of Kovach [18]. For details on Interval Analysis, see Moore [52]. Section 2.4 deals with the work of Cabada and Nieto [5]. See also for results on higher order convergence, Mohapatra, Vajravelu and Yin [49, 50]. Section 2.5 contains the work of Sandqvist and Anderson [56, 55]. Section 2.6, devoted to systems of differential equations is adapted from Vatsala, Shahzad and Köksal [70], and Chong and Vatsala [7].

3 SECOND ORDER DIFFERENTIAL EQUATIONS

3.0 INTRODUCTION

In this chapter, we shall extend the method of generalized quasilinearization to second order boundary value problems (BVPs). We shall only consider some typical results.

In section 3.1, we shall develop the method of lower and upper solutions for the BVPs and include the necessary comparison principles and an existence result in the closed set generated by the upper and lower solutions. Section 3.2 is devoted to the extension of the method of generalized quasilinearization directly, avoiding several stages of development as was done for IVPs. While section 3.3 investigates typical results that are generated due to the splitting of the function involved into suitable choices.

General second order BVPs are considered in section 3.4 and 3.5. Section 3.4 obtains results which are consequences of requiring the generated sequences and their derivatives to be monotone. The approach that is employed here is the generalized Nagumo's condition and estimates in terms of Lyapunov-like functions. Section 3.5, on the other hand, deals with only constructing monotone sequences, as is usual, and even this demands some additional conditions to guarantee quadratic convergence.

Finally, section 3.6 is utilized to formulate and prove results to obtain higher order convergence of monotone sequences. This is discussed only for a simple BVP since general results are not yet available.

3.1 METHOD OF UPPER AND LOWER SOLUTIONS.

We shall, in this section, develop the method of upper and lower solutions relative to second order boundary value problems (BVP in short). We shall also give necessary known results for our later use.

Consider the scalar BVP

$$-u'' = f(t, u, u'), \quad Bu(\mu) = b_\mu, \quad \mu = 0, 1, \tag{3.1.1}$$

where $Bu(\mu) = \alpha_\mu u(\mu) + (-1)^{\mu+1}\beta_\mu u'(\mu) = b_\mu$, $\alpha_0, \alpha_1 \geq 0$, $\beta_0, \beta_1 > 0$, $b_\mu \in R$, $f \in C[I \times R \times R, R]$ and $I = [0, 1]$. Let I^0 denote the interior of I.

A function $v \in C^2[I, R]$ is said to be a lower solution of (3.1.1) if

$$-v'' \leq f(t, v, v'), \quad Bv(\mu) \leq b_\mu, \quad \mu = 0, 1, \tag{3.1.2}$$

and an upper solution of (3.1.1) if the reversed inequalities hold in (3.1.2).

Concerning the upper and lower solutions of (3.1.1), we can prove the following basic results:

Theorem 3.1.1 *Assume that*

(i) v, w *are lower and upper solutions of* (3.1.1)*;*

(ii) $f_u, f_{u'}$ *exist, continuous,* $f_u < 0$ *and* $f_u \neq 0$ *on* $\Omega = [(t, u, \bar{u}) : t \in [0, 1]$, $w(t) \leq u \leq v(t)$ *and* $\bar{u} = v'(t) = w'(t)]$.

Then we have $v(t) \leq w(t)$ *on* I.

Proof First we prove the result by assuming one of the inequalities to be strict. In this case we can show that $v(t) < w(t)$ on I. If not, $m(t) = v(t) - w(t)$ will have a nonnegative maximum at some $t \in I$. If $t_0 \in I^0$, then $m(t_0) \geq 0$, $m'(t_0) = 0$ and $m''(t_0) \leq 0$. Consequently, we have $v(t_0) \geq w(t_0)$, $v'(t_0) = w'(t_0)$ and $v''(t_0) \leq w''(t_0)$. Since $f_u \leq 0$, this implies that at $t = t_0$,

$$\begin{aligned}
f(t_0, v, v') \geq -v'' \geq -w'' &> f(t_0, w, w') \\
&= f(t_0, w, v') \\
&= f_u(t_0, \xi, v')(w - v) + f(t_0, v, v') \\
&\geq f(t_0, v, v')
\end{aligned}$$

where ξ lies between $v(t_0)$ and $w(t_0)$. This is a contradiction because of the assumed strict inequality. If $t_0 = 0$, then we only have $v'(0) \leq w'(0)$. However, using boundary condition, we get $\alpha_0 v(0) - \beta_0 v'(0) \leq \alpha_0 w(0) - \beta_0 w'(0)$, which implies

$$\beta_0(w'(0) - v'(0)) \leq 0.$$

Since $\beta_0 > 0$, we get $v'(0) \geq w'(0)$ and therefore $v'(0) = w'(0)$. Now we can get the contradiction arguing as before. A similar argument holds good if $t_0 = 1$.

If one of the inequalities is not strict, then we can show $v(t) \leq w(t)$ on I. If the conclusion is not true, then $v(t) - w(t)$ has a positive maximum, say M, at some $t_0 \in I$. We prove that if such a t_0 exists, then $v(t) - w(t) \equiv M$ for all $t \in I$. If it is not true, there exists a $\tilde{t}$ in I such that $v(\tilde{t}) - w(\tilde{t}) < M$. We shall show that this leads to a contradiction. Suppose first that $\tilde{t} > t_0$. Define the function

$$z(t) = e^{\alpha(t-t_0)} - 1,$$

where α is a positive constant to be determined. We note that $z(t) < 0$ for $0 \leq t < t_0$ and $z(t) > 0$ for $t_0 \leq t \leq 1$, and $z(t_0) = 0$.

Let $m(t) = v(t) - w(t) + \epsilon z(t)$ where ϵ is a positive constant to be chosen. Then we have

$$\begin{aligned}
m''(t) &= v''(t) - w''(t) + \epsilon z''(t) \\
&\geq f_u(t,\xi,w')(w-v) + f_{u'}(t,v,\eta)(w'-v') + \epsilon z''(t) \\
&= -f_u(t,\xi,w')m - f_{u'}(t,v,\eta)m' \\
&\quad +\epsilon\left[f_u(t,\xi,w')\left\{e^{\alpha(t-t_0)} - 1\right\} + f_{u'}(t,v,\eta)\alpha e^{\alpha(t-t_0)} + \alpha^2 e^{\alpha(t-t_0)}\right] \\
&= -f_u(t,\xi,w')m - f_{u'}(t,v,\eta)m' \\
&\quad +\epsilon e^{\alpha(t-t_0)}\left(f_u(t,\xi,w') + \alpha f_{u'}(t,v,\eta) + \alpha^2\right) - \epsilon f_u
\end{aligned}$$

where ξ, η lie between v, w and v', w' respectively. Since f_u, $f_{u'}$ are bounded on $[0,1]$, we can choose an α such that

$$f_u(t,\xi,w') + \alpha f_{u'}(t,v,\eta) + \alpha^2 > 0.$$

This proves that

$$-m'' < f_u(t,\xi,w')m + f_{u'}(t,v,\eta)m' \tag{3.1.3}$$

which implies that $m(t)$ cannot attain its maximum on $(0,1)$. However, now choosing $0 < \epsilon < (M - v(\tilde{t}) + w(\tilde{t}))/z(\tilde{t})$ when $\tilde{t} > t_0$, we get

$$m(\tilde{t}) < v(\tilde{t}) - w(\tilde{t}) + \frac{M - v(\tilde{t}) + w(\tilde{t})}{z(\tilde{t})} z(\tilde{t}) < M.$$

Also, $m(t_0) = M$, which implies by our choice of $z(t)$ that $m(t)$ has a positive maximum greater than or equal to M at some $t^* \in (t_0, \tilde{t})$. This leads to a contradiction because of (3.1.3). If $\tilde{t} < t_0$, we can arrive at a similar conclusion by considering

$$z(t) = e^{-\alpha(t-t_0)} - 1.$$

Here $z(t) > 0$ on $[0, t_0)$ and $z(t) < 0$ on $(t_0, 1]$. If $t_0 = 0$, then certainly $\tilde{t} > 0$. Also if $t^* > 0$, then, as before, we can arrive at a contradiction. If $t^* = 0$, then we only have $m(0) = M$, $m'(0) \le 0$. But then using boundary conditions, we get $m'(0) \ge 0$ since $\beta_0 > 0$. Hence it follows that $m'(0) = 0$ and we can arrive at a contradiction as before. A similar proof holds good if $t_0 = 1$. Thus we have $v(t) - w(t) \equiv M$ on $[0, 1]$. As a consequence of this, we get

$$0 = v'' - w'' \ge -f(t, v, w') + f(t, w, v') = f_u(t, \xi, w')(-M) \ge 0$$

for $t \in (0, 1)$, where ξ lies between v and w. This implies that $f_u(t, \xi, w') \equiv 0$ on Ω, which contradicts the assumption (ii). Hence the proof of the theorem is complete.

Remark 3.1.1 Instead of assumption (ii) in Theorem 3.1.1, it would be enough to suppose a stronger requirement that f is strictly decreasing in u for each (t, u') to prove Theorem 3.1.1, in which case the proof would be simpler. We note that (ii) implies that f is nonincreasing in u. Also, instead of existence of $f_{u'}$, it is enough to suppose

$$|f(t, \eta(t), u_1) - f(t, \eta(t), u_2)| \le L|u_1 - u_2|,$$

with $\eta = v$ or w.

The following special case of Theorem 3.1.1 is known as the maximum principle.

Corollary 3.1.1 *Let q, $r \in C[I, R]$ with $r(t) \ge 0$ on I. Suppose further that $p \in C^2[I, R]$ and*

$$-p'' \le q|p'| - rp, \quad Bp(\mu) \le 0.$$

Then $p(t) \le 0$ on I. If the inequalities are reversed, then $p(t) \ge 0$ on I.

The monotonic character of f in Theorem 3.1.1 which is a result of the assumption (ii) can be dispensed with under a different set of assumptions. The next result deals with this situation.

Theorem 3.1.2 *Suppose that v, w are lower and upper solutions of* (3.1.1) *respectively. Assume further that $z \in C^2[I,R]$, $z(t) > 0$ on I and for every $\lambda > 0$,*

$$-\lambda z'' > f(t, w + \lambda z, w' + \lambda z') - f(t, w, w'),\ t \in I^0,\ Bz(\mu) > 0,\ \mu = 0, 1.$$

Then $v(t) \le w(t)$ on I.

Proof Suppose that the conclusion of the theorem is false. Then there exists a minimal $\lambda > 0$ such that

$$v(t) \le w(t) + \lambda z(t),\quad t \in I,$$

and for some $t_0 \in I^0$, we have $v(t_0) = w(t_0) + \lambda z(t_0)$. If $t_0 \in I^0$, we will also have

$$v'(t_0) = w'(t_0) + \lambda z'(t_0),$$

and

$$v''(t_0) \le w''(t_0) + \lambda z''(t_0).$$

Using the assumptions of the theorem, we get

$$\begin{aligned} -v''(t_0) &\ge -w''(t_0) - \lambda z''(t_0) \\ &> -w''(t_0) + f(t_0, w(t_0) + \lambda z(t_0), w'(t_0) + \lambda z'(t_0)) \\ &\quad - f(t_0, w(t_0), w'(t_0)) \\ &\ge f(t_0, w(t_0) + \lambda z(t_0), w'(t_0) + \lambda z'(t_0)) \\ &= f(t_0, v(t_0), v'(t_0)), \end{aligned}$$

which is a contradiction.

If $t_0 = 0$, we have $v'(0) \le w'(0) + \lambda z'(0)$. From the boundary conditions, this implies $\lambda(\alpha_0 z(0) - \beta_0 z'(0)) \le 0$. Since $\lambda > 0$, this contradicts $Bz(0) > 0$. Similarly if $t_0 = 1$, we can arrive at a contradiction and this completes the proof.

The following special case of Theorem 3.1.2 is known as the generalized maximum principle.

Corollary 3.1.2 *Let the hypothesis of Corollary* 3.1.1 *hold without the requirement $r(t) \ge 0$ on I. Suppose that $z \in C^2[I,R]$, $z(t) > 0$ on I, $-z'' > qz' - rz$ and $Bz(\mu) > 0$ for $\mu = 0, 1$. Then $p(t) \le 0$ on I. If the inequalities are reversed, then $p(t) \ge 0$ on I.*

If we know the existence of upper and lower solutions w, v such that $v \le w$ on I, we can prove the existence of a solution u of BVP (3.1.1) such that $v \le u \le w$ on I. The content of the next result is the following:

Theorem 3.1.3 *Assume that*

(*i*) *v, w are lower and upper solutions of* (3.1.1) *respectively such that $v(t) \le w(t)$ on I;*

(*ii*) *f satisfies a Nagumo condition relative to (v, w); that is, there exists an $h \in C[R_+, (0,\infty)]$ such that*

$$|f(t, u, u')| \le h(|u'|)$$

for $t \in I$, whenever $v(t) \le u \le w(t)$ and

$$\int^{\infty} \frac{sds}{h(s)} = \infty.$$

Then there exists a solution u for the BVP (3.1.1) *such that $v(t) \le u(t) \le w(t)$ and $|u'(t)| \le N$ on I, where N depends on v, w and h.*

Proof of this well-known result can be found in [4].

We remark for future use that when (ii) holds, there exists an $N > 0$ which depends only on v, w and h, such that

$$\int_{\lambda}^{N} \frac{sds}{h(s)} > \max_I w(t) - \min_I v(t)$$

where $\lambda = \max\,[|v(0) - w(1)|, |v(1) - w(0)|]$.

3.2 EXTENSION OF QUASILINEARIZATION

We consider the second order boundary value problem (BVP)

$$-x'' = f(t, x), \qquad Bx(\mu) = b_\mu \tag{3.2.1}$$

where $\mu = 0, 1$,

$$Bx(\mu) = v_\mu x(\mu) + (-1)^{\mu+1} w_\mu x'(\mu),\ v_0, v_1 \ge 0,\ w_0, w_1 > 0 \tag{3.2.2}$$

and $v_0 + v_1 > 0$, $b_\mu \in R$, $f \in C[I \times R, R]$ and $I = [0, 1]$.

Without going through the several stages of development of the method of generalized quasilinearization as was done in Chapter 1 relative to the IVPs of first order differential equations, we shall, once for all, extend to the second order BVP (3.2.1) a result parallel to Theorem 1.3.1. Since the function f in BVP (3.2.1) does not contain the x' term, this generalization would exhibit more clearly the ideas involved in the extension of the method of quasilinearization. We shall develop later the results for the general BVP which show the difficulties that arise when the function f depends on the x' term.

Theorem 3.2.1 *Assume that*

(A_1) $\alpha_0, \beta_0 \in C^2[I, \mathbb{R}]$ *be lower and upper solutions of* (3.2.1) *such that* $\alpha_0 \leq \beta_0$ *on* I;

(A_2) $f \in C[\Omega, \mathbb{R}]$, f *admits a decomposition* $f = F + G$ *where* F_x, G_x, F_{xx}, G_{xx} *exist and are continuous satisfying* $F_x(t,x) \leq 0$, $G_x(t,x) \leq 0$, $F_{xx}(t,x) \geq 0$ *and* $G_{xx}(t,x) \leq 0$ *for* $(t,x) \in \Omega$.

Then there exist monotone sequences $\{\alpha_n(t)\}$, $\{\beta_n(t)\}$ *which converge uniformly and monotonically to the unique solution of* (3.2.1) *and the convergence is quadratic.*

Proof In view of the assumptions on F and G, we have

$$\begin{aligned} F(t,x) &\geq F(t,y) + F_x(t,y)(x-y), \\ G(t,x) &\geq G(t,y) + G_x(t,x)(x-y), \text{ for } x \geq y. \end{aligned} \tag{3.2.3}$$

Also, it is clear whenever $\alpha_0(t) \leq x_2 \leq x_1 \leq \beta_0(t)$,

$$f(t,x_1) - f(t,x_2) \leq L(x_1 - x_2), \quad \text{for some} \quad L > 0. \tag{3.2.4}$$

Consider the linear BVP's

$$\left.\begin{aligned} -u'' &= F^*(t,\alpha_0;\beta_0,u) \\ &= f(t,\alpha_0) + [F_x(t,\alpha_0) + G_x(t,\beta_0)](u-\alpha_0), \\ Bu(\mu) &= b_\mu \end{aligned}\right\} \tag{3.2.5}$$

and

$$\left.\begin{aligned} -v'' &= G^*(t,\alpha_0;\beta_0,v) \\ &= f(t,\beta_0) + [F_x(t,\alpha_0) + G_x(t,\beta_0)](v-\beta_0), \\ Bv(\mu) &= b_\mu \end{aligned}\right\} \tag{3.2.6}$$

where $B\alpha_0(\mu) \le b_\mu \le B\beta_0(\mu)$.

The inequalities (3.2.3) and (A_1) imply

$$\begin{aligned}-\alpha_0'' &\le f(t,\alpha_0) \equiv F^*(t,\alpha_0;\beta_0,\alpha_0),\\ -\beta_0'' &\ge f(t,\beta_0)\\ &\ge F(t,\alpha_0) + F_x(t,\alpha_0)(\beta_0-\alpha_0) + G(t,\alpha_0) + G_x(t,\beta_0)(\beta_0-\alpha_0)\\ &\equiv F^*(t,\alpha_0;\beta_0,\beta_0).\end{aligned}$$

Hence by Theorem 3.1.5, there exists a solution α_1 of (3.2.5) such that $\alpha_0 \le \alpha_1 \le \beta_0$ on I. In view of assumptions on F and G, α_1 is unique by Theorem 3.1.1.

Similarly, we obtain

$$\begin{aligned}-\alpha_0'' &\le f(t,\alpha_0)\\ &\le F(t,\beta_0) + F_x(t,\alpha_0)(\alpha_0-\beta_0) + G(t,\beta_0) + G_x(t,\beta_0)](\alpha_0-\beta_0)\\ &\equiv G^*(t,\alpha_0;\beta_0,\alpha_0)\\ -\beta_0'' &\ge f(t,\beta_0) \equiv G^*(t,\alpha_0;\beta_0,\beta_0),\end{aligned}$$

and therefore, as before, there exists a unique solution β_1 of (3.2.6) such that $\alpha_0 \le \beta_1 \le \beta_0$ on I.

Since $\alpha_0 \le \alpha_1 \le \beta_0$ using (3.2.3) and (3.2.5), we see that

$$\begin{aligned}-\alpha_1'' &= f(t,\alpha_0) + [F_x(t,\alpha_0) + G_x(t,\beta_0)](\alpha_1-\alpha_0)\\ &\le F(t,\alpha_1) + G(t,\alpha_1) + G_x(t,\alpha_1)(\alpha_0-\alpha_1) + G_x(t,\beta_0)(\alpha_1-\alpha_0)\\ &= f(t,\alpha_1) + [G_x(t,\beta_0) - G_x(t,\alpha_1)](\alpha_1-\alpha_0)\\ &\le f(t,\alpha_1)\end{aligned}$$

because of the fact $G_x(t,x)$ is nonincreasing in x.

Similarly we can prove that

$$-\beta_1'' \ge f(t,\beta_1).$$

It then follows from Theorem 3.1.1 that $\alpha_1 \le \beta_1$. As a result we get $\alpha_0 \le \alpha_1 \le \beta_1 \le \beta_0$ on I.

Assume that for some $k > 1$, $-\alpha_k'' \le f(t,\alpha_k)$, $-\beta_k'' \ge f(t,\beta_k)$ and $\alpha_k \le \beta_k$ on I. We shall show that

$$\alpha_k \le \alpha_{k+1} \le \beta_{k+1} \le \beta_k \text{ on } I, \tag{3.2.7}$$

where α_{k+1} and β_{k+1} are the solutions of linear BVP's

$$\begin{aligned}-u'' &= f(t,\alpha_k) + [F_x(t,\alpha_k) + G_x(t,\beta_k)](u-\alpha_k)\\ &\equiv F^*(t,\alpha_k;\beta_k,u),\ Bu(\mu) = b_\mu,\end{aligned} \tag{3.2.8}$$

and

$$\begin{aligned} -v'' &= f(t,\beta_k) + [F_x(t,\alpha_k) + G_x(t,\beta_k)](v - \beta_k) \\ &\equiv G^*(t,\alpha_k;\beta_k,v), \ Bv(\mu) = b_\mu. \end{aligned} \tag{3.2.9}$$

Using (3.2.3), (3.2.8) and (3.2.9), we get

$$-\alpha_k'' \le f(t,\alpha_k) \equiv F^*(t,\alpha_k;\beta_k,\alpha_k),$$

and

$$\begin{aligned} -\beta_k'' &\ge f(t,\beta_k) \\ &\ge F(t,\alpha_k) + F_x(t,\alpha_k)(\beta_k - \alpha_k) + G(t,\alpha_k) + G_x(t,\beta_k)(\beta_k - \alpha_k) \\ &\equiv F^*(t,\alpha_k;\beta_k,\beta_k). \end{aligned}$$

Consequently, by Theorem 3.1.5, there exists a solution α_{k+1} of (3.2.8) such that $\alpha_k \le \alpha_{k+1} \le \beta_k$ on I. In view of the assumptions on F and G, α_{k+1} is a unique solution of (3.2.8) by Theorem 3.1.1.

Similarly, since

$$-\alpha_k'' \le f(t,\alpha_k) \le G^*(t,\alpha_k;\beta_k,\alpha_k)$$

and

$$-\beta_k'' \ge f(t,\beta_k) \equiv G^*(t,\alpha_k;\beta_k,\beta_k),$$

we find that there exists a unique solution β_{k+1} of (3.2.9) satisfying $\alpha_k \le \beta_{k+1} \le \beta_k$ on I. In view of the fact

$$\begin{aligned} -\alpha_{k+1}'' &= F^*(t,\alpha_k;\beta_k,\alpha_{k+1}), \\ -\beta_{k+1}'' &= G^*(t,\alpha_k;\beta_k,\beta_{k+1}), \end{aligned}$$

we have, as before, using (3.2.3)

$$\begin{aligned} -\alpha_{k+1}'' &\le f(t,\alpha_{k+1}), \\ -\beta_{k+1}'' &\ge f(t,\beta_{k+1}), \end{aligned}$$

and hence Theorem 3.1.1 shows that $\alpha_{k+1} \le \beta_{k+1}$ on I, which proves (3.2.7) is true.

Hence by induction we have for all n,

$$\alpha_0 \le \alpha_1 \le \alpha_2 \le \cdots \le \alpha_n \le \beta_n \le \cdots \le \beta_2 \le \beta_1 \le \beta_0 \quad \text{on} \quad I.$$

Employing standard argument (see [21]), it is easy to conclude that the sequences $\{\alpha_n(t)\}$, $\{\beta_n(t)\}$ converge uniformly and monotonically to the unique solution $x(t)$ of (3.2.1) on I.

We shall now prove that the convergence of the sequences $\{\alpha_n(t)\}$, $\{\beta_n(t)\}$ to the unique solution $x(t)$ is quadratic. To do this, consider $p_{n+1} = x - \alpha_{n+1}$, $q_{n+1} = \beta_{n+1} - x$, and note that $Bp_{n+1}(\mu) = 0$, $Bq_{n+1}(\mu) = 0$.

Therefore we can write

$$p_{n+1}(t) = \int_0^1 G_0(t,s)[f(s,x) - F^*(s,\alpha_n;\beta_n,\alpha_{n+1})]ds$$

and

$$q_{n+1}(t) = \int_0^1 G_0(t,s)[G^*(s,\alpha_n;\beta_n,\beta_{n+1}) - f(s,x)]ds$$

where $G_0(t,s)$ is the Green function given by

$$G_0(t,s) = \begin{cases} \frac{1}{c}x(s)y(t), & 0 \le s \le t \le 1 \\ c & \\ \frac{1}{c}x(t)y(s), & 0 \le t \le s \le 1. \end{cases}$$

Here $x(t) = \frac{v_0}{w_0}t + 1$, $y(t) = \frac{v_1}{w_1}(1-t) + 1$ are two linearly independent solutions of $-u'' = 0$.

Now

$$\begin{aligned} 0 \le p_{n+1} &= \int_0^1 G_0(t,s)[f(s,x) - \{f(s,\alpha_n) + [F_x(s,\alpha_n) \\ &\quad + G_x(s,\beta_n)](\alpha_{n+1} - \alpha_n)\}]ds \\ &\le \int_0^1 G_0(t,s)[\{F(s,x) - F(s,\alpha_n) - F_x(s,\alpha_n)(x-\alpha_n)\} \\ &\quad + \{G(s,x) - G(s,\alpha_n) - G_x(s,\beta_n)(x-\alpha_n)\} \\ &\quad + \{F_x(s,\alpha_n) + G_x(s,\beta_n)\}(x - \alpha_{n+1})]ds \\ &\le \int_0^1 G_0(t,s)\left[F_{xx}(s,\xi)\frac{p_n^2}{2} - \{G_x(s,\beta_n) - G_x(s,\alpha_n)\}(x-\alpha_n)\right]ds \\ &\le \int_0^1 G_0(t,s)\left[F_{xx}(s,\xi)\frac{p_n^2}{2} - G_{xx}(s,\sigma)(\beta_n - \alpha_n)(x-\alpha_n)\right]ds \\ &\le \int_0^1 G_0(t,s)\left[F_{xx}(s,\xi)\frac{p_n^2}{2} - G_{xx}(s,\sigma)(q_n + p_n)p_n\right]ds \end{aligned}$$

where $\alpha_n < \xi, \sigma < x$, here we have used the mean value theorem and the facts $F_x(t,x)$ is nondecreasing, $G_x(t,x)$ is nonincreasing, $F_x(t,x) \le 0$ and $G_x(t,x) \le 0$ on Ω. Thus using the inequality $p_n q_n \le \frac{1}{2}\left[p_n^2 + q_n^2\right]$ we get

$$p_{n+1} \le \int_0^1 G_0(t,s)\left[F_{xx}(s,\xi)\frac{p_n^2}{2} - G_{xx}(s,\sigma)\left(\frac{3}{2}p_n^2 + \frac{1}{2}q_n^2\right)\right]ds \text{ on } I$$

which yields the desired result

$$\max_I |x(t) - \alpha_{n+1}(t)|$$
$$\leq D\left[\left(\tfrac{A}{2} + \tfrac{3}{2}B\right)\max_I |x(t) - \alpha_n(t)|^2 + \tfrac{B}{2}\max_I |\beta_n(t) - x(t)|^2\right]$$

where $|F_{xx}(t,x)| \leq A$, $|G_{xx}(t,x)| \leq B$ and $|G_0(t,x)| \leq D$..

Similarly, we can prove that

$$\max_I |\beta_{n+1}(t) - x(t)|$$
$$\leq D\left[\left(\tfrac{3}{2}A + \tfrac{B}{2}\right)\max_I |\beta_n(t) - x(t)|^2 + \tfrac{A}{2}\max_I |x(t) - \alpha_n(t)|^2\right],$$

and the proof is complete.

Remark 3.2.1 Theorem 3.2.1 is parallel to Theorem 1.2.1 when $G(t,x) \equiv 0$, and corresponds to Theorem 1.3.1 as it is in the present setup.

3.3 GENERALIZED QUASILINEARIZATION

We shall continue to consider the BVP (3.2.1), in this section, and discuss the method of generalized quasilinearization in which we relax the assumption of convexity of the function involved. In fact, we shall prove much more. We shall admit a decomposition of f into a sum of two functions F and G, as before, but assume F satisfies only one-sided Lipschitz condition, and G is not convex but can be made convex by adding a convex function ψ. If $F(t,x) \equiv 0$, the result is the generalized quasilinearization.

We shall now prove our main result.

Theorem 3.3.1 *Assume that*

(A_1) α_0, $\beta_0 \in C^2[I, \mathbb{R}]$ *be lower and upper solutions of* (3.2.1) *respectively such that* $\alpha_0 \leq \beta_0$ *on* I;

(A_2) $f \in C[\Omega, \mathbb{R}]$, f *admits a decomposition* $f = F + G$ *where*

(*i*) $F(t,u_1) - F(t,u_2) \leq L(u_1 - u_2)$, $L \leq 0$, *whenever* $\alpha_0(t) \leq u_2 \leq u_1 \leq \beta_0(t)$.

(*ii*) $G_{xx}(t,x) + \psi_{xx}(t,x) \geq 0$ *on* Ω, *where* G, $\psi \in C[\Omega, \mathbb{R}]$, G_x, ψ_x, G_{xx}, ψ_{xx} *exist, are continuous and* $\psi_{xx} \geq 0$ *on* Ω;

(A_3) $z \in C^2[I, \mathbb{R}]$, $z(t) > 0$ *on* I *and for every* $\lambda > 0$,

$$-\lambda z'' > f(t, \beta + \lambda z) - f(t, \beta), \quad t \in I^0, \quad Bz(\mu) > 0, \quad \nu = 0, 1.$$

holds for every upper solution β *of* (3.2.1).

Then there exist monotone sequences $\{\alpha_n(t)\}$, $\{\beta_n(t)\}$ *which converge uniformly and monotonically to the unique solution of* (3.2.1) *and the convergence is quadratic.*

Proof In view of assumptions on $G_{xx}(t,x) + \psi_{xx}(t,x) \geq 0$, we obtain the inequality

$$G(t,x) \geq G(t,y) + [G_x(t,x) + \psi_x(t,y)](x-y) - [\psi(t,x) - \psi(t,y)] \quad (3.3.1)$$

for $x \geq y$, x, $y \in \Omega$.

Consider The BVPs

$$\begin{cases} \text{(a)} \quad -u'' = F(t,u) + G(t,\alpha_0) + [G_x(t,\alpha_0) + \psi_x(t,\alpha_0)](u-\alpha_0) \\ \qquad\qquad - [\psi(t,u) - \psi(t,\alpha_0)]. \\ \text{(b)} \quad -v'' = F(t,v) + G(t,\beta_0) + [G_x(t,\alpha_0) + \psi_x(t,\alpha_0)](v-\beta_0) \\ \qquad\qquad - [\psi(t,v) - \psi(t,\beta_0)]. \end{cases} \quad (3.3.2)$$

which we shall denote for convenience in the following form:

$$\begin{aligned} -u'' &= F^*(t,u:\alpha_0) \equiv H(t,u) + M(t,\alpha_0) + M_x(t,\alpha_0)(u-\alpha_0). \\ -v'' &= G^*(t,v:\alpha_0,\beta_0) \equiv H(t,v) + M(t,\beta_0) + M_x(t,\alpha_0)(v-\beta_0), \end{aligned}$$

where $H(t,x) = F(t,x) - \psi(t,x)$, $M(t,x) = G(t,x) + \psi(t,x)$. $Bu(\mu) = b_\mu = Bv(\mu)$ and $B\alpha_0(\mu) \leq b_\mu \leq B\beta_0(\mu)$.

It is clear that for $\alpha_0(t) \leq u_2 \leq u_1 \leq \beta_0(t)$,

$$H(t,u_1) - H(t,u_2) \leq L_1(u_1 - u_2), \text{ for some } L_1 > 0. \quad (3.3.3)$$

The assumption (A_1) and the inequality (3.3.1) imply

$$-\alpha_0'' \leq f(t,\alpha_0) = F(t,\alpha_0) + G(t,\alpha_0) \equiv F^*(t,\alpha_0:\beta_0),$$

$$\begin{aligned} -\beta_0'' &\geq f(t,\beta_0) = F(t,\beta_0) + G(t,\beta_0) \\ &\geq F(t,\beta_0) + G(t,\alpha_0) + [G_x(t,\alpha_0) + \psi_x(t,\alpha_0)](\beta_0 - \alpha_0) \\ &\quad - [\psi(t,\beta_0) - \psi(t,\alpha_0)] \\ &= H(t,\beta_0) + M(t,\alpha_0) + M_x(t,\alpha_0)(\beta_0 - \alpha_0) \\ &\equiv F^*(t,\beta_0:\alpha_0). \end{aligned}$$

Hence by Theorem 3.1.5, it follows that there exists a solution α_1 of (3.3.2) (a) such that $\alpha_0 \le \alpha_1 \le \beta_0$. Similarly, using (A_1) and (3.3.1), we get

$$\begin{aligned}
-\alpha_0'' &\le f(t,\alpha_0) = F(t,\alpha_0) + G(t,\alpha_0)\\
&\le F(t,\alpha_0) + G(t,\beta_0) - [G_x(t,\alpha_0) + \psi_x(t,\alpha_0)]\,(\beta_0 - \alpha_0)\\
&\quad - [\psi(t,\beta_0) - \psi(t,\alpha_0)]\\
&= H(t,\alpha_0) + M(t,\beta_0) + M_x(t,\alpha_0)(\alpha_0 - \beta_0)\\
&\equiv G^*(t,\beta_0 : \alpha_0,\beta_0)
\end{aligned}$$

$$-\beta_0'' \ge f(t,\beta_0) = F(t,\beta_0) + G(t,\beta_0) \equiv G^*(t,\beta_0 : \alpha_0,\beta_0)$$

and therefore, as before, there exists a solution β_1 of (3.3.2) (b).

Since $-\alpha_1'' = F^*(t,\alpha_1 : \alpha_0)$, using (3.3.1), we get

$$\begin{aligned}
-\alpha_1'' &= H(t,\alpha_1) + M(t,\alpha_0) + M_x(t,\alpha_0)(\alpha_1 - \alpha_0)\\
&= \{F(t,\alpha_1) - \psi(t,\alpha_1)\} + \{G(t,\alpha_0) + \psi(t,\alpha_0)\}\\
&\quad + \{G(t,\alpha_0) + \psi_x(t,\alpha_0)\}\,(\alpha_1 - \alpha_0)\\
&= F(t,\alpha_1) + G(t,\alpha_0) + \{G_x(t,\alpha_0) + \psi_x(t,\alpha_0)\}\,(\alpha_1 - \alpha_0)\\
&\quad - \{\psi(t,\alpha_1) - \psi_x(t,\alpha_0)\}\\
&\le F(t,\alpha_1) + G(t,\alpha_1)\\
&= f(t,\alpha_1), \qquad B\alpha_1(\mu) = b_\mu.
\end{aligned}$$

Similarly, since $-\beta_1'' = G^*(t,\beta_1 : \alpha_0,\beta_0)$, using (3.3.2) again

$$\begin{aligned}
-\beta_1'' &= H(t,\beta_1) + M(t,\beta_0) + M_x(t,\alpha_0)(\beta_1 - \beta_0)\\
&= \{F(t,\beta_1) - \psi(t,\beta_1)\} + \{G(t,\beta_0) + \psi(t,\beta_0)\}\\
&\quad + \{G(t,\alpha_0) + \psi_x(t,\alpha_0)\}\,(\beta_1 - \beta_0)\\
&= F(t,\beta_1) + G(t,\beta_0) + \{G_x(t,\alpha_0) + \psi_x(t,\alpha_0)\}\,(\beta_1 - \beta_0)\\
&\quad + \{\psi(t,\beta_0) - \psi_x(t,\beta_1)\}\\
&\ge F(t,\beta_1) + G(t,\beta_0) - \{G_x(t,\beta_1) + \psi_x(t,\beta_1)\}\,(\beta_0 - \beta_1)\\
&\quad + \{\psi(t,\beta_0) - \psi(t,\beta_1)\}\\
&\ge F(t,\beta_1) + G(t,\beta_1)\\
&= f(t,\beta_1).
\end{aligned}$$

Here we have used the fact that $G_x(t,x) + \psi_x(t,x)$ is nondecreasing in x. Therefore, by Theorem 3.1.2, we have $\alpha_1 \le \beta_1$. As a result we get $\alpha_0 \le \alpha_1 \le \beta_1 \le \beta_0$ on I. Assume that for some $n > 1$,

$$-\alpha_n'' \le f(t,\alpha_n), \ -\beta_n'' \ge f(t,\beta_n) \text{ and } \alpha_n \le \beta_n \text{ on } I. \tag{3.3.4}$$

We shall show that $\alpha_n \le \alpha_{n+1} \le \beta_{n+1} \le \beta_n$ on I, where α_{n+1} and β_{n+1} are respectively the solutions of BVPs

$$\begin{aligned}
-u'' &= H(t,u) + M(t,\alpha_n) + M_x(t,\alpha_n)(u - \alpha_n)\\
&\equiv F^*(t,u : \alpha_n), \ Bu(\mu) = b_\mu,
\end{aligned} \tag{3.3.5}$$

and

$$\begin{aligned} -v'' &= H(t,v) + M(t,\beta_n) + M_x(t,\alpha_n)(v-\beta_n) \\ &\equiv G^*(t,v:\alpha_n,\beta_n)\ Bv(\mu) = b_\mu. \end{aligned} \tag{3.3.6}$$

Using (A_1) and (3.3.1), we get

$$-\alpha_n'' \le f(t,\alpha_n) = F(t,\alpha_n) + G(t,\alpha_n) \equiv F^*(t,\alpha_n:\alpha_n)$$

and

$$\begin{aligned} -\beta_n'' &\ge f(t,\beta_n) = F(t,\beta_n) + G(t,\beta_n) \\ &\ge F(t,\beta_n) + G(t,\alpha_n)\left[G_x(t,\alpha_n) + \psi_x(t,\alpha_n)\right](\beta_n - \alpha_n) \\ &\quad - \left[\psi(t,\beta_n) - \psi(t,\alpha_n)\right] \\ &\equiv F^*(t,\beta_n:\alpha_n). \end{aligned}$$

Consequently, by Theorem 3.1.5, there exists a solution α_{n+1} of (3.3.5) such that

$$\alpha_n \le \alpha_{n+1} \le \beta_n \text{ on } I.$$

Similarly, using (3.3.1) again,

$$-\alpha_n'' \le f(t,\alpha_n) \le G^*(t,\alpha_n:\alpha_n,\beta_n)$$

and

$$-\beta_n'' \ge f(t,\beta_n) \equiv G^*(t,\beta_n:\alpha_n,\beta_n).$$

Hence, as before, there exists a unique solution β_{n+1} of (3.3.6) such that

$$\alpha_n \le \beta_{n+1} \le \beta_n \text{ on } I.$$

In view of the facts

$$\begin{aligned} -\alpha_{n+1}'' &= F^*(t,\alpha_{n+1}:\alpha_n), \\ -\beta_{n+1}'' &= G^*(t,\beta_{n+1}:\alpha_n,\beta_n), \end{aligned}$$

we have, as before, using (3.3.1)

$$\begin{aligned} -\alpha_{n+1}'' &\le f(t,\alpha_{n+1}), \\ -\beta_{n+1}'' &\ge f(t,\beta_{n+1}), \end{aligned}$$

which implies by Theorem 3.1.2 that $\alpha_{n+1} \le \beta_{n+1}$ on I and consequently $\alpha_n \le \alpha_{n+1} \le \beta_{n+1} \le \beta_n$ on I.

Hence by induction we have for all n,

$$\alpha_0 \le \alpha_1 \le \alpha_2 \le \cdots \le \alpha_n \le \beta_n \le \cdots \le \beta_2 \le \beta_1 \le \beta_0 \quad \text{on} \quad I.$$

Employing standard argument [21], it can be shown that the sequences $\{\alpha_n(t)\}$, $\{\beta_n(t)\}$ converge uniformly and monotonically to the unique solution $x(t)$ of (3.2.1) on I.

We shall now prove that the convergence is quadratic. For this purpose, consider $p_n = x - \alpha_n$, $q_n = \beta_n - x$. Then it is easy to observe that $Bp_n(\mu) = 0$, $Bq_n(\mu) = 0$.

Hence we can write $p_n(t) = \int_0^1 K(t,s)[f(s,x) - F^*(s,\alpha_n : \alpha_{n-1})]ds$ and

$$q_n(t) = \int_0^1 K(t,s)[G^*(s,\beta_n : \alpha_{n-1},\beta_{n-1}) - f(s,x)]ds$$

where $K(t,s)$ is the Green's function given by

$$K(t,s) = \begin{cases} \frac{1}{c}x(s)y(t), & 0 \le s \le t \le 1 \\ c & \\ \frac{1}{c}x(t)y(s), & 0 \le t \le s \le 1. \end{cases}$$

Here $x(t) = \frac{v_0}{w_0}t + 1$, $y(t) = \frac{v_1}{w_1}(1-t) + 1$ are two linearly independent solutions of

$$-u'' = 0 \text{ and } c = y(t)x'(t) - x(t)y'(t).$$

It is easy to observe that $K(t,s)$ is nonnegative.

Now

$$\begin{aligned} 0 \le p_n &= \int_0^1 K(t,s)[f(s,x) - \{H(s,\alpha_n) + M(s,\alpha_{n-1}) \\ &\quad + M_x(s,\alpha_{n-1})(\alpha_n - \alpha_{n-1})\}]ds \\ &= \int_0^1 K(t,s)[F(s,x) + G(s,x) - H(s,\alpha_n) - M(s,\alpha_{n-1}) \\ &\quad - M_x(s,\alpha_{n-1})(\alpha_n - \alpha_{n-1})]ds \\ &= \int_0^1 K(t,s)[H(s,x) + M(s,x) - H(s,\alpha_n) - M(s,\alpha_{n-1}) \\ &\quad - M_x(s,\alpha_{n-1})(\alpha_n - \alpha_{n-1})]ds. \end{aligned}$$

By the mean value theorem, we have

$$M(t,x) - M(t,\alpha_{n-1}) - M_x(t,\alpha_{n-1})p_{n-1} = \frac{M_{xx}(t,\zeta)p_{n-1}^2}{2!}$$

where $\alpha_{n-1} \leq \zeta \leq x$. Then we get

$$\begin{aligned} 0 \leq p_n &= \int_0^1 K(t,s)\left[L_1(x-\alpha_n) + \frac{M_{xx}(s,\zeta)p_{n-1}^2}{2!} + M_x(s,\alpha_{n-1})p_n\right] ds \\ &= \int_0^1 K(t,s)\left[(L_1 + M_x(s,\alpha_{n-1}))\, p_n + \frac{M_{xx}(s,\zeta)p_{n-1}^2}{2!}\right] ds, \end{aligned}$$

which yields

$$\max_I |p_n(t)| \leq \frac{B_1}{(1-A_1)} \max_I |p_{n-1}(t)|^2, \tag{3.3.7}$$

where $|K(t,s)| \leq D_1$, $|g_x(t,x)| \leq A$, $|\psi_x(t,x)| \leq B$, $|g_{xx}(t,x)| \leq D$, $|\psi_{xx}(t,x)| \leq C$, $A_1 = D_1(L_1 + A + B)$ and $B_1 = D_1(D+C)/2$.

In obtaining (3.3.7) we have supposed that $A_1 < 1$.

Similarly

$$\begin{aligned} 0 \leq q_n &= \int_0^1 K(t,s)[H(s,\beta_n) + M(s,\beta_{n-1}) \\ &\quad + M_x(s,\alpha_{n-1})(\beta_n - \beta_{n-1}) - f(s,x)]ds \\ &= \int_0^1 K(t,s)[H(s,\beta_n) + M(s,\beta_{n-1}) \\ &\quad + M_x(s,\alpha_{n-1})(\beta_n - \beta_{n-1}) - H(s,x) - M(s,x)]ds. \end{aligned}$$

We know that

$$M(t,\beta_{n-1}) = M(t,x) + M_x(t,x)q_{n-1} + \frac{M_{xx}(t,\theta)q_{n-1}^2}{2!}$$

and

$$M_x(t,x) - M_x(t,\alpha_{n-1}) = M_{xx}(t,\sigma)p_{n-1},$$

where $x < \theta < \beta_{n-1}$ and $\alpha_{n-1} < \sigma < x$. Thus we obtain

$$\begin{aligned} 0 \leq q_n &= \int_0^1 K(t,s)\left[L_1 q_n + M_{xx}(s,\sigma)p_{n-1}q_{n-1} + \frac{M_{xx}(s,\theta)p_{n-1}^2}{2!}\right. \\ &\quad \left. + M_x(s,\alpha_{n-1})q_n\right] ds \\ &= \int_0^1 K(t,s)\left[(L_1 + M_x(s,\alpha_{n-1}))\, q_n + M_{xx}(s,\sigma)\left\{\frac{q_{n-1}^2 + p_{n-1}^2}{2}\right\}\right. \\ &\quad \left. + \frac{M_{xx}(s,\theta)p_{n-1}^2}{2!}\right] ds \\ &= \int_0^1 K(t,s)\left[(L_1 + M_x(s,\alpha_{n-1}))\, q_n + \frac{M_{xx}(s,\sigma) + M_{xx}(s,\theta)}{2} q_{n-1}^2\right. \\ &\quad \left. + \frac{M_{xx}(s,\sigma)p_{n-1}^2}{2}\right] ds. \end{aligned}$$

which yields the desired estimate

$$\max_I |q_n(t)| \leq \frac{1}{(1-A_1)}\left[2B_1 \max_I |q_{n-1}(t)|^2 + B_1 \max_I |p_{n-1}(t)|^2\right] \tag{3.3.8}$$

where A_1 and B_1 are as in (3.3.7).

3.4 GENERAL SECOND ORDER BVP

In this section, we shall extend the method of quasilinearization to the general second order BVP

$$-x'' = f(t, x, x') \quad x(0) = c,\ x(1) = d$$

under the assumption that f is uniformly convex in x and x'. This is achieved in the framework of the method of upper and lower solutions, so that one can obtain monotone iterates which are the solution of the given BVP. To derive the necessary monotone iterates, a more flexible variant of Nagumo's condition is suitable. We shall first prove for convenience, a known existence result in terms of a variant of the Nagumo's condition, whose proof as given in [4] is not decisive and then employing such a result, we shall develop a monotone iterative method in the same set up which is appropriate for our purpose. Since our approach offers monotone properties for the sequence of iterates, which are solutions of linear BVPs, as well as for its derivatives, the result can be used in applications to the real world problems.

We shall employ a variant of Nagumo's condition which is more flexible than Nagumo's condition in the existence results. Since such a condition is not popular, we shall describe the general existence result for convenience. Moreover, there is an error of this result in [4] and consequently, our presentation here would be useful.

Consider the BVP [4, p. 83]

$$-x'' = f(t, x, x'), \quad x(0) = c, \quad x(1) = d, \quad t \in I = [0, 1] \tag{3.4.1}$$

where $f \in C[I \times \mathbb{R} \times R, \mathbb{R}]$. Let α_0, $\beta_0 \in C^2[I, \mathbb{R}]$ be such that $\alpha_0(t) \leq \beta_0(t)$ on I. We define the set

$$\omega = \{(t, x) : \alpha_0(t) \leq x \leq \beta_0(t), \quad t \in I\}.$$

Let ϕ, $\psi \in C^1[\omega, \mathbb{R}]$ with $\phi(t, x) \leq \psi(t, x)$ on ω. Define a modified function $F(t, x, x')$ on $\omega \times \mathbb{R}$ by

$$F(t, x, x') = \begin{cases} f(t, x, \phi(t, x)) & \text{if } x' < \phi(t, x) \\ f(t, x, x') & \text{if } (t, x, x') \in \Omega \\ f(t, x, \phi(t, x)) & \text{if } x' > \psi(x, t) \end{cases}$$

where $\Omega = \{(t,x,x') : (t,x) \in \Omega \text{ and } \phi(t,x) \leq x' \leq \psi(t,x)\}$, and extend $F(t,x,x')$ to $J \times I\!R \times R$ by setting

$$F_0(t,x,x') = \begin{cases} F(t,\alpha_0(t),x') & \text{if } x > \alpha_0(t) \\ F(t,\beta_0(t),x') & \text{if } x > \beta_0(t). \end{cases}$$

Let us consider now the modified BVP

$$-x'' = F_0(t,x,x'), \quad x(0) = c, \quad x(1) = d. \tag{3.4.2}$$

We shall first prove the variant of Nagumo's condition.

Theorem 3.4.1 *Suppose that*

(i) $\alpha_0, \beta_0 \in C^2(I, I\!R)$ *such that* $\alpha_0(t) \leq \beta_0(t)$ *on* I*;*

(ii) $\phi, \psi \in C^1[\omega, I\!R]$ *with* $\phi(t,x) \leq \psi(t,x)$ *on* ω *and on* ω

$$\begin{aligned} \phi_t(t,x) + \phi_x(t,x)\phi(t,x) + f(t,x,\phi(t,x)) &\leq 0 \\ \psi_t(t,x) + \psi_x(t,x)\psi(t,x) + f(t,x,\psi(t,x)) &\geq 0 \end{aligned}$$

where $f \in C[I \times I\!R \times R, I\!R]$.

Then for any solution $x \in C^2[I, I\!R]$ *of* (3.4.2) *such that* $\alpha_0(t) \leq x(t) \leq \beta_0(t)$ *on* I *and* $\phi(0,x(0)) \leq x'(0) \leq \psi(0,x(0))$*, we have* $\phi(t,x(t)) \leq x'(t) \leq \psi(t,x(t))$*,* $t \in I$.

Proof Let $x \in C^2[I, I\!R]$ be any solution of (3.4.2) such that $\phi(0,x(0)) \leq x'(0) \leq \psi(0,x(0))$ and $\alpha_0(t) \leq x(t) \leq \beta_0(t)$, $t \in I$. Suppose that there exists a $t_0 \in (0,1]$ such that $x'(t_0) > \psi(t_0, x(t_0))$. Define

$$V(t) = [-x'(t) + \psi(t,x(t))] \exp\left[-\int_t^{t_0} \psi_x(s,x(s))ds\right].$$

Then

$$\begin{aligned} V'(t)\exp\left[\int_t^{t_0} \psi_x(s,x(s))ds\right] &= -x''(t) + \psi_t(t,x(t)) + \psi_x(t,x(t))x'(t) \\ &\quad + [-x'(t) + \psi(t,x(t))]\,\psi_x(t,x(t)) \\ &= -x''(t) + [\psi_t(t,x(t)) + \psi(t,x(t))\psi_x(t,x(t))] \\ &\geq F_0(t,x,x') - f(t,x,\psi(t,x)) = 0. \end{aligned}$$

This implies that $V'(t) \geq 0$, $0 \leq t \leq t_0$. Thus $V(t) \leq V(t_0) < 0$ and hence $V(0) < 0$ or equivalently $\psi(0, x(0)) < x'(0)$, which is a contradiction.. Hence $x'(t) \leq \psi(t, x(t))$ on I.

Now suppose that there exists a $t_1 \in (0, 1]$ such that $x'(t) < \phi(t_1, x(t_1))$. Define

$$V(t) = [x'(t) - \phi(t, x(t))] \exp\left[-\int_t^{t_1} \phi_x(s, x(s))ds\right].$$

Then,

$$\begin{aligned} V'(t) \exp\left[\int_t^{t_1} \phi_x(s, x(s))ds\right] &= x''(t) - [\phi_t(t, x(t)) + \phi_x(t, x(t))\phi(t, x(t))] \\ &\geq -F_0(t, x, x') + f(t, x, \phi(t, x)) = 0, \end{aligned}$$

in some interval of the left of t_1. Thus we have $V(0) < 0$. But since $\phi(0, x(0)) \leq x'(0)$, it means $V(0) \geq 0$, which is a contradiction. Hence $\phi(t, x(t)) \leq x'(t)$.

Now we are in a position to prove general existence result in a sector generated by the lower and upper solutions.

Theorem 3.4.2 *Suppose that assumptions (i) and (ii) of Theorem* 3.4.1 *holds. Let α_0, β_0 be lower and upper solutions of* (3.4.1), *respectively, on I, such that $\alpha_0(0) = \beta_0(0)$ and $\phi(t, \gamma(t)) \leq \gamma'(t) \leq \psi(t, \gamma(t))$, for $\gamma(t) = \alpha_0(t)$, $\beta_0(t)$, $t \in I$. Then, for any $\alpha_0(1) \leq d \leq \beta_0(1)$ and $\alpha_0(0) = c = \beta_0(0)$, the BVP* (3.4.1) *has a solution $x \in C^2[I, \mathbb{R}]$ such that $(t, x(t), x'(t)) \in \Omega$, $t \in I$.*

Proof Observe that $F_0(t, x, x')$ of (3.4.2) is bounded on $I \times \mathbb{R} \times R$ and therefore satisfies Nagumo's condition. Further (because of the assumption on α_0', β_0'), α_0, β_0 are also lower and upper solutions relative to

$$-x'' = F_0(t, x, x').$$

Consequently, by Theorem 3.1.5, there exists a solution $x \in C^2[I, \mathbb{R}]$ to the modified BVP (3.4.2) such that $\alpha_0(t) \leq x(t) \leq \beta_0(t)$, $t \in I$ for $\alpha_0(0) = c = \beta_0(0)$ and $\alpha_0(1) \leq d \leq \beta_0(1)$. Now apply Theorem 3.4.1 to conclude that this $x(t)$ is actually a solution of the BVP (3.4.1) satisfying $(t, x(t), x'(t)) \in \Omega$, $t \in I$.

We shall use this result several times in the discussion of our main result which is given below.

Theorem 3.4.3 *Assume that*

(*i*) $\alpha_0, \beta_0 \in C^2[I, I\!R]$ *are lower and upper solution of* (3.4.1) *such that* $\alpha_0 \le \beta_0$ *on* I *and* $\alpha_0(0) = \beta_0(0)$;

(*ii*) $f \in C[I \times I\!R \times R, I\!R]$, $f_x(t,x,y)$, $f_y(t,x,y)$, $f_{xx}(t,x,y)$, $f_{yy}(t,x,y)$ *and* $f_{xy}(t,x,y)$ *exist, are continuous on* Ω *satisfying* $f_{xx}(t,x,y) \ge 0$, $f_{xy}(t,x,y) \ge 0$ *and* $f_{yy}(t,x,y) \ge 0$ *on* Ω;

(*iii*) $\phi, \psi \in C^2[\omega, I\!R]$ *with* $\phi(t,x) \le \psi(t,x)$, $\phi_{xx}(t,x) \ge 0$, $\psi_{xx}(t,x) \ge 0$ *on* ω *and on* ω

$$\begin{aligned}\phi_t(t,x) + \phi_x(t,x)\phi(t,x) + f(t,y,u) + & f_x(t,z,v)(x-y) \\ + f_x(t,v,z)(\phi(t,x) - u) \;\; & \le \;\; 0,\end{aligned}$$

$$\begin{aligned}\psi_t(t,x) + \psi_x(t,x)\psi(t,x) + f(t,y,u) + & f_x(t,z,v)(x-y) \\ + f_x(t,v,z)(\psi(t,x) - u) \;\; & \ge \;\; 0,\end{aligned}$$

for every $\alpha_0 \le z \le y \le \beta_0$ *and* $\phi(t,\alpha_0) \le v \le u \le \psi(t,\beta_0)$;

(*iv*) $\phi(t,\gamma(t)) \le \gamma'(t) \le \psi(t,\gamma(t))$, *for* $\gamma(t) = \alpha_0(t), \beta_0(t)$, $t \in I$ *and for any* (α_0, β_0) *which are lower and upper solutions of* (3.4.1) *respectively, we have*

$$\begin{aligned}\alpha_0' & \le \phi(t,\alpha_0) + \phi_x(t,\alpha_0)(x - \alpha_0) \\ \beta_0' & \ge \psi(t,\beta_0) + \psi_x(t,\alpha_0)(x - \beta_0)\end{aligned}$$

for $\alpha_0 \le x \le \beta_0$;

(*v*) $z \in C^2[I, I\!R]$, $z(t) > 0$ *on* I *and for every* $\lambda > 0$,

$$\begin{aligned}-\lambda'' & > f(t, \beta_0 + \lambda z, \beta_0' + \lambda z') - f(t, \beta_0, \beta_0'), \\ t & \in I^0,\ z(0) > 0 \text{ and } z(1) > 0,\end{aligned}$$

holds for each upper solution β_0 *of* (3.4.1). *Here* I^0 *is the interior of* I.

Then there exist sequences $\{\alpha_n(t)\}$, $\{\beta_n(t)\}$ *which converge uniformly and monotonically to the unique solution of* (3.4.1) *and the convergence is quadratic.*

Proof Let us first note that the assumptions (ii) and (iv) yield the following inequalities

$$\begin{cases} \phi(t,x) \ge \phi(t,y) + \phi_x(t,y)(x-y), \\ \psi(t,x) \ge \phi(t,y) + \psi_x(t,y)(x-y) \end{cases} \text{for } x \ge y, \tag{3.4.3}$$

and

$$f(t,x,u) \geq f(t,y,v) + f_x(t,y,v)(x-y) + f_{x'}(t,y,v)(u-v), \tag{3.4.4}$$

for $x \geq y$, $u \geq v$.

We find, from (iv) and (3.4.3), that $\alpha_0' \leq \beta_0'$. It is also clear that f satisfies, for any $\alpha_0(t) \leq y \leq x \leq \beta_0(t)$,

$$f(t,x,u) - f(t,y,u) \leq L_1(x-y), \tag{3.4.5}$$

and for any $\alpha_0'(t) \leq v \leq u \leq \beta_0'(t)$,

$$f(t,x,u) - f(t,x,v) \leq L_2(u-v), \tag{3.4.6}$$

for some $L_i > 0$ for $i = 1, 2$.

Consider the linear BVPs

$$\begin{aligned} -x'' &= F(t,x,x';\alpha_0,\alpha_0') \qquad (3.4.7)\\ &\equiv f(t,\alpha_0,\alpha_0') + f_x(t,\alpha_0,\alpha_0')(x-\alpha_0) + f_{x'}(t,\alpha_0,\alpha_0')(x'-\alpha_0'), \end{aligned}$$

$x(0) = c$, $x(1) = d$ and

$$\begin{aligned} -x'' &= G(t,x,x';\alpha_0,\alpha_0',\beta_0,\beta_0') \qquad (3.4.8)\\ &\equiv f(t,\beta_0,\beta_0') + f_x(t,\alpha_0,\alpha_0')(x-\beta_0) + f_{x'}(t,\alpha_0,\alpha_0')(x'-\beta_0'), \end{aligned}$$

$x(0) = c$, $x(1) = d$.

The inequality (3.4.4) implies

$$\begin{aligned} -\alpha_0'' &\leq f(t,\alpha_0,\alpha_0') \equiv F(t,\alpha_0,\alpha_0';\alpha_0,\alpha_0')\\ -\beta_0'' &\geq f(t,\beta_0,\beta_0')\\ &\geq f(t,\alpha_0,\alpha_0') + f_x(t,\alpha_0,\alpha_0')(\beta_0-\alpha_0) + f_{x'}(t,\alpha_0,\alpha_0')(\beta_0'-\alpha_0')\\ &\equiv F(t,\beta_0,\beta_0';\alpha_0,\alpha_0'). \end{aligned}$$

Also, from assumption (iii), we get, setting $y = z = \alpha_0$, $u = v = \alpha_0'$,

$$\begin{aligned} \phi_t(t,x) + \phi_x(t,x)\phi(t,x) + F(t,x,\phi(t,x);\alpha_0,\alpha_0') &\geq 0,\\ \psi_t(t,x) + \psi_x(t,x)\psi(t,x) + F(t,x,\psi(t,x);\alpha_0,\alpha_0') &\geq 0, \end{aligned}$$

on γ. Consequently, by Theorem 3.4.2, there exists a solution α_1 of (3.4.7) such that $\alpha_0 \leq \alpha_1 \leq \beta_0$ and $\phi(t,\alpha_1) \leq \alpha_1' \leq \psi(t,\alpha_1)$ on I. But then from (iv) and (3.4.3), we find that

$$\begin{aligned} \alpha_0' &\leq \phi(t,\alpha_0) + \phi_x(t,\alpha_0)(\alpha_1-\alpha_0) \leq \phi(t,\alpha_1) \leq \alpha_1'\\ \beta_0' &\geq \psi(t,\beta_0) + \psi_x(t,\alpha_0)(\alpha_1-\beta_0) \geq \psi(t,\alpha_1) \geq \alpha_1' \end{aligned}$$

and hence $\phi(t,\alpha_0) \le \alpha_0' \le \alpha_1' \le \beta_0' \le \psi(t,\beta_0)$ on I.

Similarly, using (3.4.4), we obtain

$$\begin{aligned}
-\alpha_0'' &\le f(t,\alpha_0,\alpha_0') \\
&\le f(t,\beta_0,\beta_0') - f_x(t,\alpha_0,\alpha_0')(\beta_0-\alpha_0) - f_{x'}(t,\alpha_0,\alpha_0')(\beta_0'-\alpha_0') \\
&= f(t,\beta_0,\beta_0') + f_x(t,\alpha_0,\alpha_0')(\alpha_0-\beta_0) + f_{x'}(t,\alpha_0,\alpha_0')(\alpha_0'-\beta_0') \\
&\equiv G(t,\alpha_0,\alpha_0';\alpha_0,\alpha_0',\beta_0,\beta_0') \\
-\beta_0'' &\ge f(t,\beta_0,\beta_0') \equiv G(t,\beta_0,\beta_0';\alpha_0,\alpha_0',\beta_0,\beta_0').
\end{aligned}$$

Using (iii), with $y=\beta_0$, $z=\alpha_0$, $u=\beta_0'$ and $v=\alpha_0'$, we have

$$\begin{aligned}
\phi_t(t,x) + \phi_x(t,x)\phi(t,x) + G(t,x,\phi(t,x);\alpha_0,\alpha_0',\beta_0,\beta_0') &\le 0, \\
\psi_t(t,x) + \psi_x(t,x)\psi(t,x) + G(t,x,\psi(t,x);\alpha_0,\alpha_0',\beta_0,\beta_0') &\ge 0,
\end{aligned}$$

on Ω and therefore, as before, there exists a solution β_1 of (3.4.8) such that $\alpha_0 \le \beta_1 \le \beta_0$, $\phi(t,\beta_1) \le \beta_1' \le \psi(t,\beta_1)$ on I and hence $\phi(t,\alpha_0) \le \alpha_0' \le \beta_1' \le \beta_0' \le \psi(t,\beta_0)$ on I.

Now since $-\alpha_1'' = F(t,\alpha_1,\alpha_1';\alpha_0,\alpha_0')$, we get, using (3.4.4),

$$\begin{aligned}
-\alpha_1'' &= f(t,\alpha_0,\alpha_0') + f_x(t,\alpha_0,\alpha_0')(\alpha_1-\alpha_0) + f_{x'}(t,\alpha_0,\alpha_0')(\alpha_1'-\alpha_0') \\
&\le f(t,\alpha_1,\alpha_1'), \qquad \alpha_1(0)=c,\ \alpha_1(0)=d.
\end{aligned}$$

Since $-\beta_1'' = G(t,\beta_1,\beta_1';\alpha_0,\alpha_0',\beta_0,\beta_0')$, we obtain, from (ii),

$$\begin{aligned}
-\beta_1'' &= f(t,\beta_0,\beta_0') + f_x(t,\alpha_0,\alpha_0')(\beta_1-\beta_0) + f_{x'}(t,\alpha_0,\alpha_0')(\beta_1'-\beta_0') \\
&= f(t,\beta_0,\beta_1') - f_x(t,\alpha_0,\alpha_0')(\beta_0-\beta_1) - f_{x'}(t,\alpha_0,\alpha_0')(\beta_0'-\beta_1') \\
&\ge f(t,\beta_0,\beta_0') - f_x(t,\beta_1,\beta_1')(\beta_0-\beta_1) - f_{x'}(t,\beta_0,\beta_0')(\beta_0'-\beta_1') \\
&\ge f(t,\beta_1,\beta_1'), \qquad \beta_1(0)=c,\ \beta_1(0)=d.
\end{aligned}$$

It then follows from Theorem 3.1.2 that $\alpha_1 \le \beta_1$. Since α_1, β_1 are lower and upper solutions of f, we find, from (iv) and (3.4.3), that

$$\alpha_1' \le \phi(t,\alpha_1) + \phi_x(t,\alpha_1)(\beta_1-\alpha_1) \le \phi(t,\beta_1) \le \beta_1',$$

and hence $\alpha_1' \le \beta_1'$ on I. This can be proved from β_1' as follows,

$$\begin{aligned}
\beta_1' &\le \psi(t,\beta_1) + \psi_1(t,\alpha_0)(\alpha_1-\beta_1) \\
&= \psi(t,\beta_1) - \psi_x(t,\alpha_0)(\beta_1-\alpha_1) \\
&\ge \psi(t,\beta_1) - \psi_x(t,\alpha_1)(\beta_1-\alpha_1) \\
&\ge \psi(t,\alpha_1) \ge \alpha_1'.
\end{aligned}$$

Here we have used the fact that $\psi_x(t,x)$ is nondecreasing in x. As a result we get $\alpha_0 \leq \alpha_1 \leq \beta_1 \leq \beta_0$ and $\phi(t,\alpha_0) \leq \alpha_0' \leq \alpha_1' \leq \beta_1' \leq \beta_0' \leq \psi(t,\beta_0)$ on I.

Assuming that for $k > 1$, $-\alpha_k'' \leq f(t,\alpha_k,\alpha_k')$, $-\beta_k'' \geq f(t,\beta_k,\beta_k')$, $\alpha_k \leq \beta_k$ on I. It follows, from (iv) and (3.4.3), that $\alpha_k' \leq \beta_k'$ on I. We shall show that

$$\begin{cases} \alpha_k \leq \alpha_{k+1} \leq \beta_{k+1} \leq \beta_k, \\ \alpha_k' \leq \alpha_{k+1}' \leq \beta_{k+1}' \leq \beta_k' \end{cases} \tag{3.4.9}$$

where α_{k+1} and β_{k+1} are the solutions of linear BVPs

$$\begin{aligned} -x'' &= f(t,\alpha_k,\alpha_k') + f_x(t,\alpha_k,\alpha_k')(x-\alpha_k) + f_{x'}(t,\alpha_k,\alpha_k')(x'-\alpha_k') \\ &\equiv F(t,x,x';\alpha_k,\alpha_k'), \qquad x(0)=c,\ x(1)=d, \end{aligned} \tag{3.4.10}$$

and

$$\begin{aligned} -x'' &= f(t,\beta_k,\beta_k') + f_x(t,\alpha_k,\alpha_k')(x-\beta_k) + f_{x'}(t,\alpha_k,\alpha_k')(x'-\beta_k') \\ &\equiv G(t,x,x';\alpha_k,\alpha_k',\beta_k,\beta_k'), \qquad x(0)=c,\ x(1)=d. \end{aligned} \tag{3.4.11}$$

Observe that

$$-\alpha_k'' \leq f(t,\alpha_k,\alpha_k') \equiv F(t,\alpha_k,\alpha_k';\alpha_k,\alpha_k')$$

and using (3.4.4)

$$\begin{aligned} -\beta_k'' &\geq f(t,\beta_k,\beta_k') \\ &\geq f(t,\alpha_k,\alpha_k') + f_x(t,\alpha_k,\alpha_k')(\beta_k-\alpha_k) + f_{x'}(t,\alpha_k,\alpha_k')(\beta_k'-\alpha_k') \\ &\equiv F(t,\beta_k,\beta_k';\alpha_k,\alpha_k'). \end{aligned}$$

We find, from (iii) with $y = \alpha_k = z$ and $u = \alpha_k' = v$, that

$$\begin{aligned} \phi_t(t,x) + \phi_x(t,x)\phi(t,x) + F(t,x,\phi(t,x);\alpha_k,\alpha_k') &\leq 0 \\ \psi_t(t,x) + \psi_x(t,x)\psi(t,x) + F(t,x,\psi(t,x);\alpha_k,\alpha_k') &\geq 0 \end{aligned}$$

on I. Thus, by Theorem 3.4.2, there exists a solution α_{k+1} of (3.4.10) such that $\alpha_k \leq \alpha_{k+1} \leq \beta_k$ and $\phi(t,\alpha_{k+1}) \leq \alpha_{k+1}' \leq \psi(t,\alpha_{k+1})$ on I. Now (iv) and (3.4.3) imply that

$$\begin{aligned} \alpha_k' &\leq \phi(t,\alpha_k) + \phi_x(t,\alpha_k)(\alpha_{k+1}-\alpha_k) \leq \phi(t,\alpha_{k+1}) \leq \alpha_{k+1}', \\ \beta_k' &\geq \psi(t,\beta_k) + \psi_x(t,\alpha_k)(\alpha_{k+1}-\beta_k) \geq \psi(t,\alpha_{k+1}) \geq \alpha_{k+1}'. \end{aligned}$$

Similarly, using (3.4.4) again, we have

$$\begin{aligned} -\alpha_k'' &\leq f(t,\alpha_k,\alpha_k') \leq G(t,\alpha_k,\alpha_k';\alpha_k,\alpha_k',\beta_k,\beta_k') \\ -\beta_k'' &\geq f(t,\beta_k,\beta_k') \equiv G(t,\beta_k,\beta_k';\alpha_k,\alpha_k',\beta_k,\beta_k'). \end{aligned}$$

Also from (iii),

$$\begin{aligned}
\phi_t(t,x) + \phi_x(t,x)\phi(t,x) + G(t,x,\phi(t,x);\alpha_k,\alpha_k',\beta_k,\beta_k') &\leq 0 \\
\psi_t(t,x) + \psi_x(t,x)\psi(t,x) + G(t,x,\psi(t,x);\alpha_k,\alpha_k',\beta_k,\beta_k') &\leq 0.
\end{aligned}$$

We find that, there exists a β_{k+1} which is a solution of (3.4.11) satisfying $\alpha_k \leq \beta_{k+1} \leq \beta_k$ and $\phi(t,\beta_{k-1}) \leq \beta_{k+1}' \leq \psi(t,\beta_{k+1})$ on I and hence $\alpha_k' \leq \beta_{k+1}' \leq \beta_k'$ on I.

We have as before, using (3.4.4) and (ii)

$$\begin{aligned}
-\alpha_{k+1}'' &\leq f(t,\alpha_{k+1},\alpha_{k+1}') \\
-\beta_{k+1}'' &\geq f(t,\beta_{k+1},\beta_{k+1}')
\end{aligned}$$

and therefore Theorem 3.1.2 shows that $\alpha_{k+1} \leq \beta_{k+1}$ on I. Since α_{k+1} and β_{k+1} are lower and upper solutions of f respectively, it follows from (iv) that $\alpha_{k+1}' \leq \beta_{k+1}'$, which proves (3.4.9) is true. Hence by induction we have for all n,

$$\alpha_0 \leq \alpha_1 \leq \alpha_2 \leq \cdots \leq \alpha_n \leq \beta_n \leq \cdots \leq \beta_2 \leq \beta_1 \leq \beta_0$$

and

$$\alpha_0' \leq \alpha_1' \leq \alpha_2' \leq \cdots \leq \alpha_n' \leq \beta_n' \leq \cdots \leq \beta_2' \leq \beta_1' \leq \beta_0'$$

on I. Therefore employing standard arguments, it can be shown that the sequences $\{\alpha_n(t)\}$, $\{\beta_n(t)\}$ converge uniformly and monotonically to the unique solution $x(t)$ of 3.4.1 on I, which implies that $\alpha_n \leq x \leq \beta_n$, $\alpha_n' \leq x' \leq \beta_n'$ on I for all n.

Finally we show that the convergence of $\{\alpha_n\}$, $\{\beta_n\}$ to x is quadratic. For this purpose, consider $p_n = x - \alpha_n$ and $q_n = \beta_n - x$. Then it is easy to observe that $p_n(\delta) = 0$ and $q_n(\delta) = 0$ for $\delta = 0, 1$. Therefore we can write

$$p_n(t) = \int_0^1 K(t,s)\left[f(s,x,x') - F(s,\alpha_n,\alpha_n';\alpha_{n-1},\alpha_{n-1}'\right] ds$$

and

$$q_n(t) = \int_0^1 K(t,s)\left[G(s,\beta_n,\beta_n';\alpha_{n-1},\alpha_{n-1}',\beta_{n-1},\beta_{n-1}') - f(s,x,x')\right] ds, \tag{3.4.12}$$

where $K(t,s)$ is the Green's function given by

$$K(t,s) = \begin{cases} t(1-s), & 0 \leq s \leq t \leq 1 \\ s(1-t), & 0 \leq t \leq s \leq 1. \end{cases}$$

It is easy to see that $K(t,s)$ is nonnegative. From (3.4.12), we can get

$$p_n'(t) = \int_0^1 K_t(t,s)\left[f(s,x,x') - F(s,\alpha_n,\alpha_n';\alpha_{n-1},\alpha_{n-1}'\right] ds$$

and

$$q_n'(t) = \int_0^1 K_t(t,s)\left[G(s,\beta_n,\beta_n';\alpha_{n-1},\alpha_{n-1}',\beta_{n-1},\beta_{n-1}') - f(s,x,x')\right] ds. \tag{3.4.13}$$

Now

$$\begin{aligned}
0 \le p_n(t) &= \int_0^1 K(t,s)\left[f(s,x,x') - \{f(s,\alpha_{n-1},\alpha_{n-1}')\right.\\
&\quad \left. + f_x(s,\alpha_{n-1},\alpha_{n-1}')(\alpha_n - \alpha_{n-1}) + f_{x'}(s,\alpha_{n-1},\alpha_{n-1}')(\alpha_n' - \alpha_{n-1}')\}\right] ds\\
&= \int_0^1 K(t,s)\left[f(s,x,x') - f(s,\alpha_{n-1},x') + f(s,\alpha_{n-1},x') - f(s,\alpha_{n-1},\alpha_{n-1}')\right.\\
&\quad \left. - f_x(s,\alpha_{n-1},\alpha_{n-1}')(\alpha_n - \alpha_{n-1}) - f_{x'}(s,\alpha_{n-1},\alpha_{n-1}')(\alpha_n' - \alpha_{n-1}')\right] ds\\
&\le \int_0^1 K(t,s)\left[f_x(s,x,x')(x - \alpha_{n-1}) + f_x(s,\alpha_{n-1},x')(x' - \alpha_{n-1}')\right.\\
&\quad - f_x(s,\alpha_{n-1},\alpha_{n-1}')(x - \alpha_{n-1}) + f_x(s,\alpha_{n-1},\alpha_{n-1}')(x - \alpha_n)\\
&\quad \left. - f_{x'}(s,\alpha_{n-1},\alpha_{n-1}')(x' - \alpha_{n-1}') + f_{x'}(s,\alpha_{n-1},\alpha_{n-1}')(x' - \alpha_n')\right] ds\\
&= \int_0^1 K(t,s)\left[\{f_x(s,x,x') - f_x(s,\alpha_{n-1},x') + f_x(s,\alpha_{n-1},x')\right. \qquad (3.4.14)\\
&\quad - f_x(s,\alpha_{n-1},\alpha_{n-1}')\}\, p_{n-1} + \left\{f_{x'}(s,\alpha_{n-1},x') - f_{x'}(s,\alpha_{n-1},\alpha_{n-1}')\right\} p_{n-1}\\
&\quad \left. + f_x(s,\alpha_{n-1},\alpha_{n-1}')p_n + f_{x'}(s,\alpha_{n-1},\alpha_{n-1}')p_n'\right] ds\\
&= \int_0^1 K(t,s)\left[f_{xx}(s,\sigma_1,x')p_{n-1}^2 + f_{xx'}(s,\alpha_{n-1},\sigma_2)p_{n-1}'p_{n-1}\right.\\
&\quad + f_{x'x'}(s,\alpha_{n-1},\sigma_3)(p_{n-1}')^2 + f_x(s,\alpha_{n-1},\alpha_{n-1}')p_n\\
&\quad \left. + f_{x'}(s,\alpha_{n-1},\alpha_{n-1}')p_n'\right] ds\\
&= \int_0^1 K(t,s)\left[f_{xx}(s,\sigma_1,x')p_{n-1}^2 + f_{xx'}(s,\alpha_{n-1},\sigma_2)\left\{(p_{n-1}')^2 + (p_{n-1})^2\right\}\right.\\
&\quad + f_{x'x'}(s,\alpha_{n-1},\sigma_3)(p_{n-1}')^2 + f_x(s,\alpha_{n-1},\alpha_{n-1}')p_n\\
&\quad \left. + f_{x'}(s,\alpha_{n-1},\alpha_{n-1}')p_n'\right] ds,
\end{aligned}$$

where $\alpha_{n-1} < \sigma_1 < x$ and $\alpha_{n-1}' < \sigma_2, \sigma_3 < x'$.

Denote $A_1 = (L+M)/4$, $A_2 = (L+M)/2$, $B_1 = (M+N)/4$, $B_2 = (M+N)/2$, $C_1 = P/2$, $C_2 = P/2$, $D_1 = Q/4$ and $D_2 = Q/2$ where $|K(t,s)| \le 1/4$,

$|K_t(t,s)| \le 1/2$ and on Ω, $|f_{xx}(t,x,y)| \le L$, $|f_{x'x}(t,x,y)| = |f_{xx'}(t,x,y)| \le M$, $|f_{x'x'}(t,x,y)| \le N$, $|f_x(t,x,y)| \le P$ and $|f_{x'}(t,x,y)| \le Q$.

Now we obtain from (3.4.14)

$$\begin{aligned} \max_I |p_n(t)| &\le A_1 \max_I |p_{n-1}(t)|^2 + B_1 \max_I |p'_{n-1}(t)|^2 \\ &\quad + C_1 \max_I |p_n(t)| + D_1 \max_I |p'_n(t)| . \end{aligned} \tag{3.4.15}$$

Similarly from (3.4.13), we can get

$$\begin{aligned} \max_I |p'_n(t)| &\le A_2 \max_I |p_{n-1}(t)|^2 + B_2 \max_I |p'_{n-1}(t)|^2 \\ &\quad + C_2 \max_I |p_n(t)| + D_2 \max_I |p'_n(t)| . \end{aligned} \tag{3.4.16}$$

We find, from (3.4.15) and (3.4.16), that

$$\begin{pmatrix} \max_I |p_n(t)| \\ \max_I |p'_n(t)| \end{pmatrix} \le (I^* - X)^{-1} Y \begin{pmatrix} \max_I |p_{n-1}(t)|^2 \\ \max_I |p'_{n-1}(t)|^2 \end{pmatrix}, \tag{3.4.17}$$

where $X = \begin{pmatrix} C_1 & D_1 \\ C_2 & D_2 \end{pmatrix}$, $Y = \begin{pmatrix} A_1 & B_1 \\ A_2 & B_2 \end{pmatrix}$ and I^* is the identity matrix.

Similarly,

$$\begin{aligned} 0 \le q_n(t) &= \int_0^1 K(t,s) \left[f(s,\beta_{n-1},\beta'_{n-1}) + f_x(s,\alpha_{n-1},\alpha'_{n-1})(\beta_n - \beta_{n-1}) \right. \\ &\quad \left. + f_{x'}(s,\alpha_{n-1},\alpha'_{n-1})(\beta'_n - \beta'_{n-1}) - f_x(s,x,x') \right] ds \\ &= \int_0^1 K(t,s) \left[\{ f(s,\beta_{n-1},\beta'_{n-1}) - f(s,x,\beta'_{n-1}) \} \right. \\ &\quad + \{ f(s,x,\beta'_{n-1}) - f(s,x,x') \} \\ &\quad \left. - f_x(s,\alpha_{n-1},\alpha'_{n-1})(\beta_n - \beta_{n-1}) + f_{x'}(s,\alpha_{n-1},\alpha'_{n-1})(\beta'_n - \beta'_{n-1}) \right] ds \\ &\le \int_0^1 K(t,s) \left[f_x(s,\beta_{n-1},\beta'_{n-1})(\beta_{n-1} - x) + f_{x'}(s,x,\beta'_{n-1})(\beta'_{n-1} - x') \right. \\ &\quad \left. - f_x(s,\alpha_{n-1},\alpha'_{n-1})(\beta_n - \beta_{n-1}) + f_{x'}(s,\alpha_{n-1},\alpha'_{n-1})(\beta'_n - \beta'_{n-1}) \right] ds \\ &= \int_0^1 K(t,s) \left[\{ f_x(s,\beta_{n-1},\beta'_{n-1}) - f_x(s,\alpha_{n-1},\beta'_{n-1}) \right. \\ &\quad + f_x(s,\alpha_{n-1},\beta'_{n-1}) \\ &\quad - f_x(s,\alpha_{n-1},\alpha'_{n-1}) \} q_{n-1} + \{ f_{x'}(s,x,\beta'_{n-1}) - f_{x'}(s,\alpha_{n-1},\beta'_{n-1}) \\ &\quad + f_{x'}(s,\alpha_{n-1},\beta'_{n-1}) - f_{x'}(s,\alpha_{n-1},\alpha'_{n-1}) \} q'_{n-1} \end{aligned}$$

$$
\begin{aligned}
& +f_x(s,\alpha_{n-1},\alpha'_{n-1})q_n + f_{x'}(s,\alpha_{n-1},\alpha'_{n-1})q'_n\big]\,ds \\
= & \int_0^1 K(t,s)\,\big[f_{xx}(s,\sigma_1,\beta'_{n-1})(p_{n-1}+q_{n-1})q_{n-1} \\
& +f_{xx'}(s,\alpha_{n-1},\sigma_2)(q'_{n-1}+p'_{n-1})q_{n-1} + f_{x'x}(s,\sigma_3,\beta'_{n-1})p_{n-1}q'_{n-1} \\
& +f_{x'x'}(s,\alpha_{n-1},\sigma_4)(q'_{n-1}+p'_{n-1})q'_{n-1} + f_x(s,\alpha_{n-1},\alpha'_{n-1})q_n \\
& +f_{x'}(s,\sigma_{n-1},\sigma'_{n-1})q_n\big]\,ds \\
\leq & \int_0^1 K(t,s)\,\big[f_{xx}(s,\sigma_1,\beta'_{n-1})\left\{p_{n-1}^2+2q_{n-1}^2\right\} \\
& +f_{xx'}(s,\alpha_{n-1},\sigma_2)\left\{(q'_{n-1})^2+2q_{n-1}^2+(p'_{n-1})^2\right\} \\
& +f_{x'x}(s,\sigma_3,\beta'_{n-1})\left\{p_{n-1}^2+(q'_{n-1})^2\right\} + f_{x'x'}(s,\alpha_{n-1},\sigma_4)\left\{2(q'_{n-1})^2\right. \\
& \left.+(p'_{n-1})^2\right\} + f_x(s,\alpha_{n-1},\alpha'_{n-1})q_n + f_{x'}(s,\alpha_{n-1},\alpha'_{n-1})q'_n\big]\,ds \\
= & \int_0^1 K(t,s)\,\big[\left\{f_{xx}(s,\sigma_1,\beta'_{n-1})+2f_{xx'}(s,\alpha_{n-1},\sigma_2)\right\}q_{n-1}^2 \\
& +\left\{f_{xx}(s,\sigma_1,\beta'_{n-1})+f_{x'x}(s,\sigma_3,\beta'_{n-1})\right\}p_{n-1}^2 + \left\{f_{xx'}(s,\alpha_{n-1},\sigma_2)\right. \\
& \left.+f_{x'x}(s,\sigma_3,\beta'_{n-1})+2f_{x'x'}(s,\alpha_{n-1},\sigma_4)\right\}(q'_{n-1})^2 + \left\{f_{xx'}(s,\alpha_{n-1},\sigma_2)\right. \\
& \left.+f_{x'x'}(s,\alpha_{n-1},\sigma_4)\right\}(p'_{n-1})^2 + f_x(s,\alpha_{n-1},\alpha'_{n-1})q_n \\
& +f_{x'}(s,\alpha_{n-1},\alpha'_{n-1})q'_n\big]\,ds,
\end{aligned}
\tag{3.4.18}
$$

where $\alpha_{n-1} < \sigma_1 < \beta_{n-1}$, $x < \sigma_3 < \alpha_{n-1}$ and $\alpha'_{n-1} < \sigma_2, \sigma_4 < \beta_{n-1}{}'$.

Now, we obtain, from (3.4.18), that

$$
\begin{aligned}
\max_I |q_n(t)| \leq\; & a_1 \max_I |q_{n-1}(t)|^2 + b_1 \max_I \left|q'_{n-1}(t)\right|^2 + c_1 \max_I |p_{n-1}(t)|^2 \\
& +d_1 \max_I \left|p'_{n-1}(t)\right|^2 + C_1 \max_I |q_n(t)| + D_1 \max_I |q'_n(t)|.
\end{aligned}
\tag{3.4.19}
$$

Similarly, from (3.4.13), we find that

$$
\begin{aligned}
\max_I |q'_n(t)| \leq\; & a_{21} \max_I |q_{n-1}(t)|^2 + b_{21} \max_I \left|q'_{n-1}(t)\right|^2 + c_2 \max_I |p_{n-1}(t)|^2 \\
& +d_2 \max_I \left|p'_{n-1}(t)\right|^2 + C_2 \max_I |q_n(t)| + D_2 \max_I |q'_n(t)|.
\end{aligned}
\tag{3.4.20}
$$

Here $a_1 = (L+M)/2$, $a_2 = (L+M)$, $b_1 = (M+N)/2$, $b_2 = (M+N)$, $c_1 = (L+M)/4$, $c_2 = (L+M)/2$, $d_1 = (M+N)/4$, $d_2 = (M+N)/2$ and C_1, C_2, D_1, D_2 are as in (3.4.15) and (3.4.16).

Denote $Y_1 = \begin{pmatrix} a_1 & b_1 \\ a_2 & b_2 \end{pmatrix}$ and $Y_2 = \begin{pmatrix} c_1 & d_1 \\ c_2 & d_2 \end{pmatrix}$.
We can write (3.4.19) and (3.4.20) in the form

$$\begin{pmatrix} \max_I |q_n(t)| \\ \max_I |q_n'(t)| \end{pmatrix} \leq (I^* - X)^{-1} \left\{ Y_1 \begin{pmatrix} \max_I |q_{n-1}(t)|^2 \\ \max_I |q_{n-1}'(t)|^2 \end{pmatrix} + Y \begin{pmatrix} \max_I |q_{n-1}(t)|^2 \\ \max_I |q_{n-1}'(t)|^2 \end{pmatrix} \right\}. \tag{3.4.21}$$

In obtaining (3.4.17) and (3.4.21), we need to suppose that $P + 2Q < 4$ so that $(I^* - X)^{-1}$ exists and is nonnegative.. The estimates (3.4.17) and (3.4.21) prove quadratic convergence and the proof is complete.

Remarks

1. The condition (v) is redundant if we assume

$$(v^*) \quad f_x(t,x,y) \leq 0,\ f_y(t,x,y) \leq 0 \text{ on } \Omega.$$

Then we need to utilize comparison Theorem 3.1.1 instead of comparison Theorem 3.1.2. In this case, we get simpler estimates for the quadratic convergence, which are as follows:

$$\begin{aligned}
\max_I |p_n(t)| &\leq A_1 \max_I |p_{n-1}(t)|^2 + B_1 \max_I |p_{n-1}'(t)|^2, \\
\max_I |p_n'(t)| &\leq A_2 \max_I |p_{n-1}(t)|^2 + B_{21} \max_I |p_{n-1}'(t)|^2, \\
\max_I |q_n(t)| &\leq a_1 \max_I |q_{n-1}(t)|^2 + b_1 \max_I |q_{n-1}'(t)|^2 \\
&\quad + c_1 \max_I |p_{n-1}(t)|^2 + d_1 \max_I |p_{n-1}'(t)|^2
\end{aligned}$$

and

$$\begin{aligned}
\max_I |q_n'(t)| &\leq a_2 \max_I |q_{n-1}(t)|^2 + b_2 \max_I |q_{n-1}'(t)|^2 \\
&\quad + c_2 \max_I |p_{n-1}(t)|^2 + d_2 \max_I |p_{n-1}'(t)|^2.
\end{aligned}$$

2. If our desire is only to obtain a sequence of lower bounds $\{a_n(t)\}$ for the solution of (3.4.1) so that we need to consider only the linear BVP (3.4.10), in which case the condition (v) is redundant and we require the following weaker assumption in Theorem 3.4.3, namely,

   (a) the inequalities in (iii) are replaced by

$$\phi_t(t,x) + \phi_x(t,x)\phi(t,x) + f(t,y,v) + f_x(t,y,v)(x-y)$$

$$+f_{x'}(t,y,v)(\phi(t,x)-v)\leq 0,$$

$$\psi_t(t,x)+\psi_x(t,x)\psi(t,x)+f(t,y,v)+f_x(t,y,v)(x-y)$$
$$+f_{x'}(t,y,v)(\psi(t,x)-v)\geq 0,$$

for every $\alpha_0 \leq y \leq x \leq \beta_0$ and $\phi(t,\alpha_0) \leq v \leq \psi(t,\beta_0)$.

3.5 GENERAL SECOND ORDER BVP (CONTINUED)

In the previous section, we considered the general BVP (3.4.1) under the assumptions that $f(t,x,x')$ is convex in both x and x' and as a result obtained sequences and their derivatives to be monotone. This required a new and interesting mechanism which is not yet popular. In this section, we shall only assume that $f(t,x,x')$ is convex in x and satisfies suitable Nagumo's condition. With some additional condition on f relative to x', we extend generalized quasilinearization technique. We give examples to illustrate the necessity for extra conditions.

Let us consider the general second order BVP

$$-u''=f(t,u,u'),\quad u(0)=a,\quad u(1)=b,\ t\in I=[0,1], \tag{3.5.1}$$

and prove the following main result.

Theorem 3.5.1 *Assume the following hold:*

(i) α_0, β_0 *are lower and upper solutions of* (3.5.1) *such that* $\alpha_0 \leq \beta_0$ *on* I*;*

(ii) There exist functions h_1, $h_2 \in C[R_+,(0,\infty)]$ *such that*

$$|f(t,u,u')|\leq h_1(|u'|) \text{ and } |f_u(t,u,u')|\leq h_2(|u'|) \tag{3.5.2}$$

for $t \in I$ *and* $\alpha_0 \leq u \leq \beta_0$ *on* I*, and*

$$\int^{\infty}\frac{sds}{C_1h_1(s)+C_2h_2(s)+C_0}=\infty, \tag{3.5.3}$$

where C_0, C_1, $C_2 > 0$ *are constants;*

(iii) $f \in C[\Omega\times R, R]$ *is such that* $f_{uu}(t,u,u')$ *exists and is continuous, and* $f_{uu}(t,u,u') \geq 0$ *for* $(t,u,u') \in \Omega\times[-C,C]$*, where* $\Omega = \{(t,u) : \alpha_0(t) \leq u \leq \beta_0(t),\ t\in I\}$ *and* C *is the constant satisfying* $C > \max\{N, \max_I|\alpha_0'|, \max_I|\beta_0'|\}$*,* N *being the Nagumo constant related to* h_1, α_0, β_0*. Moreover,* $f_u(t,u,u') \leq -\lambda < 0$ *for* $(t,u,u') \in \Omega\times[-C,C]$*, where* λ *depends only on* α_0, β_0*.*

(*iv*) *$f(t,u,u')$ is monotone nondecreasing in u' for $u' \leq 0$ and monotone nonincreasing in u' for $u' \geq 0$, where $(t,u) \in \Omega$. Moreover, there exist $h_0 \in [R_+, R_+]$ and $\phi \in [I, R_+]$ such that*

$$|f(t,u,u') - f(t,u,v')| \leq \phi(t) h_0(|u'-v'|), \tag{3.5.4}$$

where h_0 satisfies $h_0(0) = 0$ and ϕ depends on (t,u), $(t,v) \in \Omega$, u' and v'.

Then, there exists a monotone sequence $\{\alpha_n\}$ which converges uniformly to the unique solution of (3.5.1) *and the convergence is quadratic.*

Remark 3.5.1 (*i*) The first inequality in (3.5.2) together with (3.5.3) implies that f satisfies the assumptions in Theorem 3.1.3 for $(t,u) \in \Omega$. Therefore, the Nagumo constant N exists and depends only on h_1, α_0, β_0. The second inequality in (3.5.2) is necessary as can be seen from the example $f(t,u,u') = (u')^2 \exp(-|u'|\, u)$.

(*ii*) In (3.5.4), $h_0(s)$ may take the form $h_0(s) = s^\alpha$, $\alpha > 0$. Also, $\phi(t)$ need not to be uniformly bounded. In the case $0 < \alpha \leq 1$ and $\phi(t) \leq L$ for some constant $L > 0$, (3.5.4) implies the first inequality in (3.5.2) since we may choose $h_1(s) = Lh_0(s) + K$, where K is a constant. When $\alpha = 1$, (3.5.4) is equivalent to a Lipschitz condition of $f(t,u,u')$ in u'.

Proof Let us consider the quasilinear BVP

$$\begin{aligned} -\alpha_1'' &= f(t,\alpha_0,q(\alpha_1')) + f_u(t,\alpha_0,q(\alpha_1'))(\alpha_1 - \alpha_0) \\ &\equiv F(t,\alpha_1,\alpha_1';\alpha_0) \\ \alpha_1(0) &= a \text{ and } \alpha_1(1) = b, \end{aligned} \tag{3.5.5}$$

where

$$q(u') \equiv \max\{-C, \min\{u', C\}\}, \quad \alpha_0 \leq u \leq \beta_0.$$

We first prove that there exists a unique solution α_1 of (3.5.5). Since $|\alpha_0'| < C$ and $|\beta_0'| < C$, we have

$$-\alpha_0'' \leq f(t,\alpha_0,\alpha_0') = F(t,\alpha_0,\alpha_0';\alpha_0)$$

and

$$\begin{aligned} &F(t,\beta_0,\beta_0';\alpha_0) - f(t,\beta_0,\beta_0') \\ &\qquad = f(t,\alpha_0,\beta_0') + f_u(t,\alpha_0,\beta_0')(\beta_0 - \alpha_0) \\ &\qquad\quad - f(t,\beta_0,\beta_0') \\ &\qquad \leq f_u(t,\alpha_0,\beta_0')(\alpha_0 - \beta_0) + f_u(t,\alpha_0,\beta_0')(\beta_0 - \alpha_0) \\ &\qquad = 0, \end{aligned}$$

which implies

$$-\beta_0'' \geq F(t, \beta_0, \beta_0'; \alpha_0).$$

Here, we have used the convexity of $f(t, u, u')$ in u. Therefore, α_0, β_0 are lower and upper solutions of (3.5.5), respectively. To show F satisfies a Nagumo condition, we let $C_2 = \max_I |\beta_0(t) - \alpha_0(t)|$. Then, for $\alpha_0 \leq v \leq \beta_0$,

$$|F(t, v, v'; \alpha_0)| \leq h_1(|q(v')|) + C_2 h_2(|q(v')|).$$

Let $H(|v'|) = h_1(|q(v')|) + C_2 h_2(|q(v')|)$. We have

$$\int^{\infty} \frac{sds}{H(s)} = \int^{C} \frac{sds}{h_1(s) + C_2 h_2(s)} + \int_{C}^{\infty} \frac{sds}{h_1(C) + C_2 h_2(C)} = \infty.$$

Hence, F satisfies the assumptions of Theorem 3.1.3, which guarantees the existence of a solution α_1 of (3.5.5) such that $\alpha_0 \leq \alpha_1 \leq \beta_0$ and $|\alpha_1'| \leq N_0$, where N_0 depends only on α_0, β_0, h_1 and h_2. Since $F(t, \alpha_1, \alpha_1'; \alpha_0)$ is strictly decreasing in α_1, Remark 3.1.1 ensures the uniqueness of solution.

Now, suppose that for $k = n$, there exists a unique solution of

$$\begin{aligned} -\alpha_n'' &= f(t, \alpha_{n-1}, q(\alpha_n')) + f_u(t, \alpha_{n-1}, q(\alpha_n'))(\alpha_n - \alpha_{n-1}) \\ &\equiv F(t, \alpha_n, \alpha_n'; \alpha_{n-1}), \\ \alpha_n(0) &= a \text{ and } \alpha_n(1) = b, \end{aligned}$$

such that $\alpha_{n-1} \leq \alpha_n \leq \beta_{n-1}$ and $|\alpha_n'| \leq N_0$, where N_0 is chosen as before. For $k = n+1$, consider the BVP

$$\begin{aligned} -\alpha_{n+1}'' &= f(t, \alpha_n, q(\alpha_{n+1}')) + f_u(t, \alpha_n, q(\alpha_{n+1}'))(\alpha_{n+1} - \alpha_n) \\ &\equiv F(t, \alpha_{n+1}, \alpha_{n+1}'; \alpha_n), \qquad (3.5.6) \\ \alpha_{n+1}(0) &= a \text{ and } \alpha_{n+1}(1) = b. \end{aligned}$$

Using the convexity of $f(t, u, u')$ in u, we can show that α_n and β_0 are lower and upper solutions of (3.5.6), respectively. Arguing as above, we can assert that there exists a unique solution α_{n+1} of (3.5.6) such that $\alpha_n \leq \alpha_{n+1} \leq \beta_0$ on I and $|\alpha_{n+1}'| \leq N_0$. By induction and employing the standard arguments, we conclude that $\{\alpha_n\}$ is a monotone nondecreasing sequence which converges uniformly to the unique solution of (3.5.1).

To prove the quadratic convergence of $\{\alpha_n\}$, we let $p_n = u - \alpha_n$. Note that $p_n \geq 0$ on I. Then,

$$\begin{aligned} -p_{n+1}'' &= -(u'' - \alpha_{n+1}'') = f(t, u, u') - f(t, \alpha_n, q(\alpha_{n+1}')) - f_u(t, \alpha_n, q(\alpha_{n+1}')) \\ &\quad \times (\alpha_{n+1} - \alpha_n) \\ &= f(t, u, u') - f(t, u, q(\alpha_{n+1}')) + f(t, u, q(\alpha_{n+1}')) - f(t, \alpha_n, q(\alpha_{n+1}')) \\ &\quad + f_u(t, \alpha_n, q(\alpha_{n+1}'))(u - \alpha_{n+1}) - f_u(t, \alpha_n, q(\alpha_{n+1}'))(u - \alpha_n). \end{aligned}$$

In view of the definition of $q(u')$ and using assumption (iv), we have

$$-f(t,u,q(\alpha'_{n+1})) = -f(t,u,\alpha'_{n+1}), \text{ if } |\alpha'_{n+1}| \le C$$

and

$$-f(t,u,q(\alpha'_{n+1})) \le -f(t,u,\alpha'_{n+1}), \text{ if } |\alpha'_{n+1}| > C.$$

Therefore,

$$\begin{aligned} -p''_{n+1} &\le f_u(t,\alpha_n,q(\alpha'_{n+1}))p_{n+1} + \phi(t)h_0(|p'_{n+1}|) + f_{uu}(t,\xi,q(\alpha'_{n+1}))p_n^2 \\ &\le -\lambda p_{n+1} + \phi(t)h_0(|p'_{n+1}|) + K\|p_n\|^2, \end{aligned}$$

with $p_{n+1}(0) = p_{n+1}(1) = 0$, where $\alpha_n < \xi < u$, $K > 0$ is a constant such that $0 \le f_{uu} \le K$ for $(t,u,u') \in \Omega \times [-C,C]$, and $\|p_n\| = \max_I |p_n(t)|$. Let $W = \frac{K}{\lambda}\|p_n\|^2$. It is easy to see that

$$-W'' = -\lambda W + \phi(t)h_0(|W'|) + K\|p_n\|^2, \; W(0) > 0 \text{ and } W(1) > 0.$$

By Remark 3.1.1, we get

$$p_{n+1}(t) \le W = \frac{K}{\lambda}\|p_n\|^2, \text{ for } t \in I,$$

which is the desired inequality. This completes the proof of the theorem.

The following example illustrates the application of Theorem 3.5.1.

Example 3.5.1 Consider

$$-u'' = e^{-u} - 4(u')^2/(1+(u')^2)\sin t, \tag{3.5.7}$$

$$u(0) = u(1) = 0 \text{ on } I. \tag{3.5.8}$$

Let $f(t,u,u')$ denote the right-hand side of equation (3.5.7). Choose $\alpha_0 = 0$ and $\beta_0 = t$. It is easy to check that α_0, β_0 are lower and upper solutions of (3.5.7) and (3.5.8), respectively. Also, all assumptions in Theorem 3.5.1 related to f are satisfied.. For instance, $f_u = -e^{-u} \le -\frac{1}{e} < 0$ for $\alpha_0 \le u \le \beta_0$ on I and $f_{u'} = -4u'/(1+(u')^2)^2$, which implies f is nondecreasing in u' for $u' \le 0$ and nonincreasing in u' for $u' \ge 0$. Hence, Theorem 3.5.1 is applicable to (3.5.7) and (3.5.8).

We observe that, in the case when f is concave in u. we can employ a similar argument to construct a monotone sequence $\{\beta_n\}$ which converges uniformly and quadratically from above to the unique solution of (3.5.1). This result is presented in the next theorem. Since the proof follows along the same lines, we omit the details.

Theorem 3.5.2 *Suppose that assumptions (i) and (ii) of Theorem 3.5.1 hold and in addition,*

$(iii)'$ *$f \in C[\Omega \times R, R]$ is such that $f_{uu}(t,u,u')$ exists and is continuous, and $f_{uu}(t,u,u') \le 0$ for $(t,u,u') \in \Omega \times [-C, C]$, where $\Omega = \{(t,u) : \alpha_0(t) \le u \le \beta_0(t),\ t \in I\}$ and C is the constant satisfying $C > \max\{N, \max_I |\alpha_0'|, \max_I |\beta_0'|\}$, N being the Nagumo constant related to h_1, α_0, β_0. Moreover, $f_u(t,u,u') \le -\lambda < 0$ for $(t,u,u') \in [-C, C]$, where λ depends only on α_0, β_0.*

$(iv)'$ *$f(t,u,u')$ is monotone nonincreasing in u' for $u' \le 0$ and monotone nondecreasing in u' for $u' \ge 0$, where $(t,u) \in \Omega$. Moreover, there exist $h_0 \in [R_+, R_+]$ and $\phi \in [I, R_+]$ such that*

$$|f(t,u,u') - f(t,u,v')| \le \phi(t) h_0(|u' - v'|),$$

where h_0 satisfies $h_0(0) = 0$ and ϕ depends on (t,u), $(t,v) \in \Omega$, u' and v'.

Then, there exists a monotone sequence $\{\beta_n\}$ which converges uniformly from above to the unique solution of (3.5.1) and the convergence is quadratic.

Below, we give an example to demonstrate the application of Theorem 3.5.2.

Example 3.5.2 Consider

$$-u'' = f(t,u,u') \equiv -e^u + |u'|(t^2 - u^2) + t^2, \tag{3.5.9}$$

$$u(0) = u(1) = 0 \text{ on } I. \tag{3.5.10}$$

We observe that $\alpha_0 = 0$ and $\beta_0 = t$ are also the lower and upper solutions of (3.5.9) and (3.5.10). Also,

$$f_{uu} = -e^u - 2|u'| \le 0$$

and

$$f_u = -e^u - 2u|u'| \le -e^0 = -1 < 0,$$

for $0 \le u \le t$. The rest of the hypotheses in Theorem 3.5.2 can be easily verified. Therefore, we can apply the theorem to BVP (3.5.9) and (3.5.10).

3.6 HIGHER ORDER CONVERGENCE

Let us return, in this section, to the simple second order BVP

$$-u'' = f(t,u),\ Bu(\mu) = b_\mu,\ t \in I \equiv [0,1], \tag{3.6.1}$$

where $Bu(\mu) = r_\mu u(\mu) + s_\mu u'(\mu) = b_\mu$, $\mu = 0, 1$, and r_1, $r_2 \geq 0$, s_1, $s_2 > 0$ are real numbers. We shall investigate quasilinearization technique relative to higher order convergence and periodic BVPs. Corresponding results to the general second order BVPs are not yet available. We begin with the following result which provides third order convergence of iterates.

Theorem 3.6.1 *Let the following assumptions hold:*

(*i*) α_0, $\beta_0 \in C^2[I, \mathbb{R}]$ *are lower and upper solutions of* (3.6.1) *such that* $\alpha_0(t) \leq \beta_0(t)$ *on* I;

(*ii*) $f \in C^{0,3}[Q, \mathbb{R}]$ *is* 2*-hyperconvex in* u*, where* $Q = \{(t,u) : \alpha_0(t) \leq u \leq \beta_0(t),\ t \in I\}$;

(*iii*) $f_u(t,u) \leq 0$ *and* $f_{uu}(t,u) \leq 0$ *for* $(t,u) \in Q$.

Then, there exists a monotone nondecreasing sequence $\{\alpha_n\}$ *which converges uniformly to the unique solution* $u(t)$ *of* (3.6.1) *and moreover, the convergence is cubic, that is,there exists a constants* $C > 0$ *such that*

$$\max_I |u(t) - \alpha_{n+1}(t)| \leq C \cdot \max_I |u(t) - \alpha_n(t)|^3. \tag{3.6.2}$$

Proof Let us define

$$F(t,x;y) \equiv \sum_{i=1}^{2} \frac{1}{i!} f^{(i)}(t,y)(x-y)^i,$$

where $f^{(i)}(t,y)$ denotes the ith derivative of f with respect to y for $i \geq 1$, and $f^{(0)}(t,y) \equiv f(t,y)$. For $x \geq y$ and (t,x), $(t,y) \in Q$, the partial derivative of F with respect to x,

$$F_x(t,x;y) = f_x(t,y) + f_{xx}(t,y)(x-y) \leq 0.$$

We consider the following BVP:

$$-\alpha_1''(t) = \sum_{i=0}^{2} \frac{1}{i!} f^{(i)}(t,\alpha_0)(\alpha_1 - \alpha_0)^i \equiv F(t,\alpha_1;\alpha_0), \quad B\alpha_1(\mu) = b_\mu,\ t \in I. \tag{3.6.3}$$

Since

$$F(t,\alpha_0;\alpha_0) = \sum_{i=0}^{2} \frac{1}{i!} f^{(i)}(t,\alpha_0)(\alpha_0-\alpha_0)^i = f(t,\alpha_0),$$

we have

$$-\alpha_0'' \leq F(t,\alpha_0;\alpha_0) \text{ and } B\alpha_0(\mu) \leq b_\mu. \tag{3.6.4}$$

On the other hand,

$$\begin{aligned} F(t,\beta_0;\alpha_0) &= \sum_{i=0}^{2} \frac{1}{i!} f^{(i)}(t,\alpha_0)(\beta_0-\alpha_0)^i \\ &= f(t,\beta_0) - \frac{1}{3!} f^{(3)}(t,\xi)(\beta_0-\alpha_0)^3 \leq f(t,\beta_0), \end{aligned}$$

where $\alpha_0 < \xi < \beta_0$, which implies

$$-\beta_0'' \geq F(t,\beta_0;\beta_0) \text{ and } B\beta_0(\mu) \geq b_\mu. \tag{3.6.5}$$

Inequalities (3.6.4) and (3.6.5) imply that α_0 and β_0 are lower and upper solutions of (3.6.3), respectively. By Theorem 3.1.3, there exists a solution $\alpha_1(t)$ of (3.6.3) such that $\alpha_0 \leq \alpha_1 \leq \beta_0$ on I. Since $F_x \leq 0$, the solution $\alpha_1(t)$ of (3.6.3) is unique.

Assume that α_n is the unique solution of

$$-\alpha_n''(t) = \sum_{i=0}^{2} \frac{1}{i!} f^{(i)}(t,\alpha_{n-1})(\alpha_n-\alpha_{n-1})^i \equiv F(t,\alpha_n;\alpha_{n-1}), \quad B\alpha_n(\mu) = b_\mu,$$

which satisfies $\alpha_{n-1} \leq \alpha_n \leq \beta_0$ on I. It is easy to check that

$$-\alpha_n'' \leq f(t,\alpha_n) \text{ on } I. \tag{3.6.6}$$

We shall show that $\alpha_n \leq \alpha_{n+1} \leq \beta_0$ on I, where $\alpha_{n+1}$is the unique solution of BVP

$$-\alpha_{n+1}''(t) = \sum_{i=0}^{2} \frac{1}{i!} f^{(i)}(t,\alpha_n)(\alpha_{n+1}-\alpha_n)^i \equiv F(t,\alpha_{n+1};\alpha_n), \quad B\alpha_{n+1}(\mu) = b_\mu. \tag{3.6.7}$$

Inequality (3.6.6) implies

$$-\alpha_1''(t) \leq f(t,\alpha_n) \equiv F(t,\alpha_n;\alpha_n), \quad B\alpha_n(\mu) = b_\mu, \, t \in I.$$

We also have

$$\begin{aligned} F(t,\beta_0;\alpha_0) &= \sum_{i=0}^{2} \frac{1}{i!} f^{(i)}(t,\alpha_n)(\beta_0-\alpha_n)^i \\ &= f(t,\beta_0) - \frac{1}{3!} f^{(3)}(t,\xi)(\beta_0-\alpha_n)^3 \leq f(t,\beta_0), \end{aligned}$$

where $\alpha_n < \xi < \beta_0$, which implies

$$-\beta_0'' \geq F(t, \beta_0; \alpha_n) \text{ and } B\beta_0(\mu) \geq b_\mu.$$

Since α_n and β_0 are lower and upper solutions of (3.6.7), respectively and $\alpha_n \leq \beta_0$ on I, there exists a solution α_{n+1} such that $\alpha_n \leq \alpha_{n+1} \leq \beta_0$ on I. Since $F_x(t, \alpha_{n+1}; \alpha_n) \leq 0$, Theorem 3.1.1 guarantees the uniqueness of solution $\alpha_{n+1}(t)$ of (3.6.7). By induction, we easily conclude that there exists a monotone sequence $\{\alpha_n\}$ which satisfies

$$\alpha_0 \leq \alpha_1 \leq \cdots \leq \alpha_n \leq \beta_0 \text{ on } I.$$

Using standard arguments we can show that $\{\alpha_n\}$ converges uniformly to the unique solution $u(t)$ of (3.6.1).

To prove the cubic convergence of $\{\alpha_n\}$, we let $p_n(t) = x(t) - \alpha_n(t)$ for $t \in I$, $p_n \geq 0$ on I. Then, we have

$$p_{n+1}(t) = \int_0^1 G(t,s)\{f(s,u) - F(s, \alpha_{n+1}; \alpha_n)\}\, ds,$$

where

$$G(t,s) = \begin{cases} \frac{1}{c}x(s)y(t), & 0 \leq s < t \leq 1, \\ \frac{1}{c}x(t)y(s), & 0 \leq t < s \leq 1. \end{cases}$$

Here $c = x(t)y'(t) - x'(t)y(t)$ and $x(t) = \frac{r_0}{s_0}t + 1$, $y(t) = \frac{r_1}{s_1}(1-t) + 1$. It is observed that G is nonnegative for t, $s \in I$. Hence,

$$\begin{aligned} p_{n+1}(t) &= \int_0^1 G(t,s)\left\{f(s,u) - f(s,\alpha_{n+1}) + \frac{1}{3!}f^{(3)}(s,\xi)(\alpha_{n+1} - \alpha_n)^3\right\} ds \\ &\leq \int_0^1 G(t,s)\left\{f_u(s,\eta)(u - \alpha_{n+1}) + \frac{1}{3!}f^{(3)}(s,\xi)(u - \alpha_n)^3\right\} ds \\ &= \int_0^1 G(t,s)\left\{f_u(s,\eta)p_{n+1} + \frac{1}{3!}f^{(3)}(s,\xi)\cdot p_n^3\right\} ds, \end{aligned}$$

where $\alpha_n < \xi < \alpha_{n+1}$ and $\alpha_{n+1} < \eta < u$. Let K_1 be a positive constant such that $\frac{1}{3!}f^{(3)}(t,u) \leq K_1$ for $(t,u) \in Q$. Since $f_u(t,u) \leq 0$ for $(t,u) \in Q$, we obtain

$$p_{n+1}(t) \leq K_1 \cdot \max_I |p_n(t)|^3 \int_0^1 G(t,s)ds \leq K_0 K_1 \cdot \max_I |p_n(t)|^3,$$

where $K_0 > 0$ is such that $\int_0^1 G(t,s)ds \leq K_0$ for $t \in I$, from which the desired inequality follows. This completes the proof.

Remark 3.6.1 In Theorem 3.6.1, we have assumed $f_{uu}(t,u) \leq 0$ for $(t,u) \in Q$ and obtained a monotone sequence converging from below to the unique solution of (3.6.1). If instead, one assumes $f_{uu}(t,u) \geq 0$ for $(t,u) \in Q$, then a monotone sequence, converging uniformly from above to the unique solution of (3.6.1), can also be obtained.

The next result offers higher order convergence.

Theorem 3.6.2 *Assume that*

(*i*) $\alpha_0, \beta_0 \in C^2[I, \mathbb{R}]$ *are lower and upper solutions of* (3.6.1) *such that* $\alpha_0(t) \leq \beta_0(t)$ *on* I;

(*ii*) *the function* $f \in C^{0,k+1}[Q, \mathbb{R}]$ *is* k*-hyperconvex with respect to* u;

(*iii*) $f^{(i)}(t,u) \leq 0$, $i = 1, 2, \cdots, k-1$, *for* $(t,u) \in Q$.

Then, there exists a monotone nondecreasing sequence $\{\alpha_n\}$ *which converges uniformly to the unique solution* $u(t)$ *of* (3.6.1). *Moreover, there exists a constant* $C > 0$ *such that*

$$\max_I |u(t) - \alpha_{n+1}(t)| \leq C \cdot \max_I |u(t) - \alpha_n(t)|^{k+1}.$$

Proof The proof is similar to that of Theorem 3.6.1. Let

$$F(t,x;y) = \sum_{i=0}^{k-1} \frac{1}{i!} f^{(i)}(t,y)(x-y)^i.$$

For $x \geq y$, assumption (iii) implies

$$F_u(t,x;y) = \sum_{i=1}^{k-1} \frac{1}{(i-1)!} f^{(i)}(t,y)(x-y)^{i-1} \leq 0.$$

The BVPs under consideration in this case are

$$-\alpha_n'' = \sum_{i=0}^{k-1} \frac{1}{i!} f^{(i)}(t,\alpha_{n-1})(\alpha_n - \alpha_{n-1})^i = F(t,\alpha_n;\alpha_{n-1}), \; B\alpha_n(\mu) = b_\mu,$$

for $n = 1, 2, \cdots$ on I. We omit the rest of the proof.

Remark 3.6.2 If the assumption (2-hyperconvex of f in u) in Theorem 3.6.1 is relaxed to requiring $f + \phi$ to be 2-hyperconvex in u, where $\phi \in C^3[\mathbb{R}, \mathbb{R}]$ satisfies

$\phi^{(3)} \geq 0$ (additional conditions on ϕ will also be needed). The generalized quasilinearization method can be applied to (3.6.1) to obtain cubic convergence by using the following BVPs

$$\begin{aligned} -\alpha_n'' &= \sum_{i=1}^{2} \frac{1}{i!} f^{(i)}(t, \alpha_{n-1})(\alpha_n - \alpha_{n-1})^i + \phi_{uu}(\alpha_{n-1})(\alpha_n - \alpha_{n-1})^2 \\ &\quad - [\phi(\alpha_n) - \phi(\alpha_{n-1})], \qquad \text{and } B\alpha_n(\mu) = b_\mu \text{ on } I, \end{aligned}$$

for $n = 1, 2, \cdots$, to construct monotone sequence $\{\alpha_n\}$.

We consider the periodic boundary value problem (PBVP for short)

$$-u'' = f(t, u),\ u(0) = u(\pi),\ u'(0) = u'(\pi),\ t \in J \equiv [0, \pi]. \tag{3.6.8}$$

Definition 3.6.1 A function $v \in C^2[J, \mathbb{R}]$ is called a lower solution of (3.6.8) if

$$-v'' \leq f(t, v),\ v(0) = v(\pi),\ v'(0) \geq v'(\pi).$$

An upper solution is defined if the above inequalities are reversed.

We need the following known results. For details see [29, 30].

Theorem 3.6.3 *Assume that v, w are lower and upper solutions of* (3.6.8) *such that $v \leq w$ on J. Then there exists a solution $u(t)$ of* (3.6.8) *such that $v \leq u \leq w$ on J.*

Theorem 3.6.4 *Suppose that v, w are lower and upper solutions of* (3.6.8). *If $f(t, u)$ is strictly decreasing in u, then $v \leq w$ on J.*

We are now in a position to prove our main result for PVBP.

Theorem 3.6.5 *Let the following hold:*

(i) $\alpha_0, \beta_0 \in C^2[J, \mathbb{R}]$ are lower and upper solutions of (3.6.8) *such that $\alpha_0(t) \leq \beta_0(t)$ on J;*

(ii) $f \in C^{0,3}[\Omega, \mathbb{R}]$ is a 2-hyperconvex in u, where $\Omega = \{(t, u) : \alpha_0(t) \leq u \leq \beta_0(t),\ t \in$

(iii) $f_u(t,u) < 0$ *and* $f_{uu}(t,u) \le 0$ *for* $(t,u) \in \Omega$.

Then, the conclusions of Theorem 3.6.1 *hold relative to* (3.6.8).

Proof Define

$$F(t,x;y) \equiv \sum_{i=0}^{2} \frac{1}{i!} f^{(i)}(t,y)(x-y)^i.$$

For $x \ge y$ and (t,x), $(t,y) \in \Omega$, we have in this case,

$$F_x(t,x,y) = f_x(t,y) + f_{xx}(t,y)(x-y) < 0.$$

The PBVPs under consideration are:

$$\begin{aligned} -\alpha_n''(t) &= \sum_{i=0}^{2} \frac{1}{i!} f^{(i)}(t,\alpha_{n-1})(\alpha_n - \alpha_{n-1})^i \equiv F(t,\alpha_n;\alpha_{n-1}), \\ & \quad t \in J. \\ \alpha_n(0) &= \alpha_n(\pi) \text{ and } \alpha_n'(0) = \alpha_n'(\pi), \end{aligned} \tag{3.6.9}$$

for $n = 1, 2, \cdots$. The existence of a solution of (3.6.9) for $n \ge 1$ is guaranteed by Theorem 3.6.3. The uniqueness is implied by Theorem 3.6.4 since $F_x(t,x;y) < 0$. Applying Theorem 3.6.4 repeatedly, we can easily obtain, as in the proof of Theorem 3.6.1, a monotone sequence $\{\alpha_n\}$ which converges uniformly from below to the unique solution $u(t)$ of (3.6.8). To show the cubic convergence of $\{\alpha_n\}$ to u, we proceed as before by letting $p_{n+1} = u - \alpha_{n+1}$ and calculating

$$\begin{aligned} -p_{n+1}'' &= f(t,u) - \sum_{i=0}^{2} \frac{1}{i!} f^{(i)}(t,\alpha_n)(\alpha_{n+1} - \alpha_n)^i \\ &\le f_u(t,\eta)(u - \alpha_{n+1}) + \frac{1}{3!} f^{(3)}(t,\xi)(u - \alpha_n)^3 \\ &\le f_u(t,\eta)p_{n+1} + K \cdot \max_I |p_n(t)|^3, \end{aligned}$$

where $\alpha_{n+1} < \eta < u$, $\alpha_n < \xi < \alpha_{n+1}$, and $K > 0$ is a constant such that $\frac{1}{3!} f^{(3)}(t,u) \le K$ for (t,u) in Ω. By assumption (iii), there exists a constant $\lambda > 0$, depending on α_0, β_0 and J, such that $f_u(t,u) \le -\lambda < 0$ for (t,u) in Ω. Let $W = \frac{K}{\lambda} \cdot \max_I |p_n(t)|^3$. It is easy to check that W verifies

$$-W'' \ge f_u(t,\eta)W + K \|p_n\|^3, \; W(0) = W(\pi), \; W'(0) = W'(\pi).$$

By Theorem 3.6.4, $p_{n+1}(t) \le \frac{K}{\lambda} \cdot \max_I |p_n(t)|^3$ on J, which leads to cubic convergence.

Theorem 3.6.6 *Assume that*

(*i*) $\alpha_0, \beta_0 \in C^2[J, I\!R]$ *are lower and upper solutions of* (3.6.8) *such that* $\alpha_0(t) \leq \beta_0(t)$ *on* J;

(*ii*) *the function* $f \in C^{0,k+1}[\Omega, I\!R]$ *is a* k*-hyperconvex with respect to* u;

(*iii*) $f_u(t,u) < 0$ *and* $f^{(i)}(t,u) \leq 0$, $i = 2, 3, \cdots, k-1$, *for* $(t,u) \in \Omega$.

Then, there exists a monotone sequence $\{\alpha_n\}$, *which converges uniformly to the unique solution* $u(t)$ *of* (3.6.8). *Moreover, there exists a constant* $C > 0$ *such that*

$$\max_I |u(t) - \alpha_{n+1}(t)| \leq C \cdot \max_I |u(t) - \alpha_n(t)|^{k+1}.$$

Remark 3.6.3 When $k = 1$, we obtain the quadratic convergence of $\{\alpha_n\}$. In this case, the method of generalized quasilinearization can be employed to relax the convexity assumption of f.

3.7 NOTES

Section 3.1 develops the method of upper, lower solutions for second order boundary value problems. Necessary results that are useful for our purpose are taken from Ladde, Lakshmikantham and Vatsala [21]. See also Bernfeld and Lakshmikantham [4]. The contents of Section 3.2 are due to Shahzad and Vatsala [64], while results of Section 3.3 are taken from Shahzad and Sivasundaram [62]. Section 3.4 consists of the results due to Lakshmikantham and Shahzad [38]. In Section 3.5 and 3.6, we describe the work of Mohapatra, Vajravelu and Yin [47, 48]. For further results in this direction, see Malek and Vatsala [45], Shahzad and Vatsala [65], and Nieto [53, 54].

4 MISCELLANEOUS EXTENSIONS

4.0 INTRODUCTION

In this chapter, we shall introduce several types of dynamic systems and prove typical extensions of the method of generalized quasilinearization to each nonlinear dynamic system. As we shall see that these extensions to a variety of dynamic equations generate new difficulties and therefore await further investigation.

Section 4.1 is devoted to dynamic systems on measure chains (or time scales). This is a new branch of differential equations which permits us to handle both the theory of differential equations and the theory of difference equations simultaneously. Consequently, one gets some insight and a better understanding of the subtle differences of these two different systems. In this section, we develop a method of quasilinearization so that it can be applied to a variety of problems. In Section 4.2, we shall discuss integro-differential equations and in Section 4.3, we shall investigate differential equations with delay. Impulsive differential equations form the contents of Section 4.4, while Stochastic differential equations are studied in Section 4.5. Finally, we discuss differential equations in a Banach space in Section 4.6. In all the cases, we shall prove some typical results bringing out the ideas involved clearly.

4.1 DYNAMIC SYSTEMS ON TIME SCALES

Let us begin by describing necessary concepts. For details see [41]. Let $\boldsymbol{T}$ be a time scale (any subset of R with order and topological structure defined in a canonical way) with $t_0 \geq 0$ as a minimal element and no maximal element. Since a time scale $\boldsymbol{T}$ may or may not be connected, we need the concept of jump operators.

Definition 4.1.1 The mapping $\sigma, \rho : \boldsymbol{T} \to \boldsymbol{T}$ defined as

$$\sigma(t) = \inf\{s \in \boldsymbol{T} : s > t\} \text{ and } \rho(t) = \sup\{s \in \boldsymbol{T} : s < t\}$$

are called jump operators.

Definition 4.1.2 A non-maximal element $t \in \boldsymbol{T}$ is said to be right dense (rd) if $\sigma(t) = t$, right scattered (rs) if $\sigma(t) > t$, left dense (ld) if $\rho(t) = t$, and left scattered (ls) if $\rho(t) < t$.

Definition 4.1.3 The mapping $\mu^* : \boldsymbol{T} \to R^+$ defined by

$$\mu^*(t) = \sigma(t) - t$$

is called graininess.

If $\boldsymbol{T} = R$, $\mu^*(t) = 0$, and when $\boldsymbol{T} = Z$, $\mu^*(t) = 1$.

Definition 4.1.4 The mapping $u : \boldsymbol{T} \to X$, where X is a Banach space is called rd-continuous if at each right dense $t \in \boldsymbol{T}$, it is continuous and at each left-dense point t, the left sided limit $u(t^-)$ exists.

Let $C_{rd}[\boldsymbol{T}, X]$ denote the set of rd-continuous mappings from $\boldsymbol{T}$ to X. It is clear that a continuous mapping is rd-continuous. However, if $\boldsymbol{T}$ contains left-dense and right scattered (ldrs) points, then rd-continuity does not imply continuity. But on a discrete time scale the two notions coincide.

Definition 4.1.5 A mapping $u : \boldsymbol{T} \to X$ is said to be differentiable at $t \in \boldsymbol{T}$, if there exists an $\alpha \in X$ such that for any $\epsilon > 0$ there exists a neighborhood N of t satisfying

$$|u(\sigma(t)) - u(s) - \alpha(\sigma(t) - s)| \leq \epsilon |\sigma(t) - s| \text{ for all } s \in N.$$

Let $u^{\triangle}(t)$ denotes the derivative of u. Note that if $\boldsymbol{T} = R$, $\alpha = u^{\triangle} = \frac{du(t)}{dt}$ and if $\boldsymbol{T} = Z$, $\alpha = u^{\triangle} = u(t+1) - u(t)$. It is easy to see that if u is differentiable at t, then it is continuous at t; if u is continuous at t and t is right-scattered, then u is differentiable and

$$u^{\triangle}(t) = \frac{u(\sigma(t)) - u(t)}{\mu^*(t)}.$$

Definition 4.1.6 The mapping $f : \boldsymbol{T} \times X \to X$ is rd-continuous if it is continuous at each (t, x) with right-dense t, $\lim_{(s,y)\to(t^-,x)} f(s,y) = f(t^-, x)$ exists and at each (t, x) with left dense t, $\lim_{y\to x} f(t, y)$ exists.

Definition 4.1.7 Let h be a mapping from $\boldsymbol{T}$ to X. The mapping $g : \boldsymbol{T} \to X$ is called the antiderivative of h on $\boldsymbol{T}$ if it is differentiable on $\boldsymbol{T}$ and satisfies $g^{\triangle}(t) = h(t)$ for $t \in \boldsymbol{T}$.

The following known properties of the antiderivative are useful.

(a) If $h : \boldsymbol{T} \to X$ is rd-continuous, then h has the antiderivative $g : t \to \int_s^t h(s)ds$, $s, t \in \boldsymbol{T}$.

(b) If the sequence $\{h_n\}_{n\in N}$ of rd-continuous functions $\boldsymbol{T} \to X$ converge uniformly on $[r, s]$ to rd-continuous function h then $\left(\int_r^s h_n(t)dt\right)_{n\in N} \to \int_r^s h(t)dt$, in X.

The basic tool which is employed in the proofs is the following induction principle, well suited for time scales. See [1, 17, 19] for details.

Theorem 4.1.1 *Suppose that for any $t \in \boldsymbol{T}$, there is a statement $A(t)$ such that the following conditions hold:*

(i) $A(t_0)$ is true;

(ii) If t is right-scattered and $A(t)$ is true, then $A(\sigma(t))$ is also true;

(iii) For each right-dense t, there exists a neighborhood N such that whenever $A(t)$ is true, $A(s)$ is also true for all $s \in N$, $s \geq t$;

(iv) For left-dense t, $A(s)$ is true for all $s \in [t_0, t)$ implies $A(t)$ is true.

Then the statement $A(t)$ is true for all $t \in \boldsymbol{T}$.

In case $\boldsymbol{T} = N$, Theorem 4.1.1 reduces to the well known principle of mathematical induction.

Let $\boldsymbol{T}$ be a time scale with $t = 0$ as a minimal element and $\boldsymbol{T} > 0$ as a maximal element which is not left-scattered.

Consider the initial value problem for dynamic system on the time scale

$$u^{\Delta} = f(t,u), \quad u(0) = u_0, \tag{4.1.1}$$

where $f \in C_{rd}[\boldsymbol{T} \times R^n, R^n]$. Relative to (4.1.1) we need the following two known results concerning the upper and lower solutions. See [41].

Theorem 4.1.2 *Let v, $w \in C^1_{rd}[\boldsymbol{T}, R^n]$ satisfying*

$$v^{\Delta}(t) \le f(t, v(t)), \quad w^{\Delta}(t) \ge f(t, w(t)), \quad t \in \boldsymbol{T},$$

where $f \in C_{rd}[\boldsymbol{T} \times R^n, R^n]$, $f(t,x)$ is quasimonotone nondecreasing in x, and for each i, $1 \le i \le n$, $f_i(t,x)\mu^*(t) + x_i$ is nondecreasing in x_i for each $t \in \boldsymbol{T}$. Then $v(t_0) \le w(t_0)$ implies $v(t) \le w(t)$, $t \in \boldsymbol{T}$, provided f satisfies

$$f_i(t,x) - f_i(t,y) \le L \sum_{i=1}^{n} (x_i - y_i), \quad x \ge y.$$

Theorem 4.1.3 *Let v, $w \in C^1_{rd}[\boldsymbol{T}, R^n]$ be lower and upper solutions of (4.1.1) such that $v(t) \le w(t)$ on $\boldsymbol{T}$ and $f \in C_{rd}[\Omega, R^n]$ where $\Omega = \{(t,u) : v(t) \le u \le w(t),\ t \in \boldsymbol{T}\}$, f is bounded on Ω, f is quasimonotone nondecreasing in x, $f_i(t,x)\mu^*(t) + x_i$ is nondecreasing in x_i for each $t \in \boldsymbol{T}$. Then there exists a solution $u(t)$ of (4.1.1) such that $v(t) \le u(t) \le w(t)$ on $\boldsymbol{T}$, provided that $v(0) \le u_0 \le w(0)$.*

We are now in a position to prove the main result.

Consider the dynamic system on time scale

$$x^{\Delta} = f(t,x), \quad x(0) = x_0 \tag{4.1.2}$$

where $f \in C_{rd}[\Omega_0, R^n]$, $\Omega_0 = [(t,x) : \alpha_0(t) \le x \le \beta_0(t),\ t \in \boldsymbol{T}]$ and α_0, $\beta_0 \in C_{rd}[\boldsymbol{T}, R^n]$ such that $\alpha_0(t) \le \beta_0(t)$ on $\boldsymbol{T}$. Let $\Omega = [(t,x,y,z) : \alpha_0(t) \le x,y,z \le \beta_0(t),\ t \in \boldsymbol{T}]$.

Relative to (4.1.2), we now prove the following main result.

Theorem 4.1.4 *Suppose that there exist a functions α_0, β_0, S_j, $j = 1,2$ such that the following assumptions are satisfied:*

(H_0) $S_j \in C_{rd}[\Omega, R^n]$, $\Omega = [(t,x,y,z) : t \in \boldsymbol{T}, \alpha_0(t) \le x,y,z \le \beta_0(t)]$, S_j bounded on Ω, $S_j(t,x,y,z)$ is quasimonotone nondecreasing in x and

$S_{ji}(t,x,y,z)\mu^*(t)+x_i$ is nondecreasing in x_i for each i, $1\le i\le n$ and $\alpha_0(t)\le y,z\le\beta_0(t)$;

(H_1) α_0, $\beta_0\in C^1_{rd}[\boldsymbol{T},R^n]$, $\alpha_0^\Delta\le S(t,\alpha_0,\alpha_0,\beta_0)$, $\beta_0^\Delta\ge S(t,\beta_0,\alpha_0,\beta_0)$ and $\alpha_0(t)\le\beta_0(t)$ on $\boldsymbol{T}$;

(H_2) $S_1(t,y,y,z)=f(t,y)$, $S_2(t,z,y,z)=f(t,z)$ and $S_j(t,x,x,x)=f(t,x)$;

(H_3) $S_1(t,x,y,z)\le S_1(t,x,x,z)$ if $y\le x$, for each $\alpha_0(t)\le z\le\beta_0(t)$ and $S_2(t,x,y,z)\ge S_2(t,x,y,x)$ if $x\le z$, for each $\alpha_0(t)\le y\le\beta_0(t)$;

(H_4) $S_{ji}(t,x_1,y,z)-S_{ji}(t,x_2,y,z)\le L\sum_{i=1}^n(x_{1i}-x_{2i})$ whenever $\alpha_0(t)\le x_2\le x_1\le\beta_0(t)$ for some $L\ge 0$ and for $t\in\boldsymbol{T}$, $x,u,v,w\in\Omega$

$$|S_j(t,x,x,x)-S_j(t,u,v,w)|\le M|x-u|+N\left[|x-v|^{1+\eta}+|x-w|^{1+\eta}\right],$$

$0<\eta\le 1$, M, N being nonnegative constant matrices.

Then there exist monotone sequences $\{\alpha_n(t)\}$, $\{\beta_n(t)\}$ which converge locally, uniformly to the unique solution of (4.1.2) on $\boldsymbol{T}$ and the convergence is superlinear.

Proof We consider the initial value problems

$$\alpha_1^\Delta = S_1(t,\alpha_1,\alpha_0,\beta_0),\quad \alpha_1(0)=x_0, \tag{4.1.3}$$
$$\beta_1^\Delta = S_2(t,\beta_1,\alpha_0,\beta_0),\quad \beta_1(0)=x_0, \tag{4.1.4}$$

where $\alpha_0(0)\le x_0\le\beta_0(0)$. Because of (H$_1$) and (H$_2$), we have

$$\begin{aligned}\alpha_0^\Delta &\le f(t,\alpha_0)=S_1(t,\alpha_0,\alpha_0,\beta_0),\\ \beta_0^\Delta &\ge f(t,\beta_0)=S_1(t,\beta_0,\alpha_0,\beta_0).\end{aligned}$$

Using Theorems 4.1.2 and 4.1.3, we get the existence of a unique solution $\alpha_1(t)$ of (4.1.3) on $\boldsymbol{T}$ satisfying $\alpha_0(t)\le\alpha_1(t)\le\beta_0(t)$ on $\boldsymbol{T}$.

Similarly, in view of (H$_1$) and (H$_2$), we also have

$$\begin{aligned}\alpha_0^\Delta &\le f(t,\alpha_0)\le S_2(t,\alpha_0,\alpha_0,\beta_0),\\ \beta_0^\Delta &\ge f(t,\beta_0)=S_2(t,\beta_0,\alpha_0,\beta_0),\end{aligned}$$

which by Theorems 4.1.2 and 4.1.3 imply the existence of a unique solution $\beta_1(t)$ of (4.1.4) such that $\alpha_0(t)\le\beta_1(t)\le\beta_0(t)$ on $\boldsymbol{T}$.

Now, since $\alpha_0 \leq \alpha_1$ and $\beta_1 \leq \beta_0$, we see using (H_3),

$$\begin{aligned} \alpha_1^{\Delta} &= S_1(t, \alpha_1, \alpha_0, \beta_0) \leq S_1(t, \alpha_1, \alpha_1, \beta_0), \\ \beta_1^{\Delta} &= S_2(t, \beta_1, \alpha_0, \beta_0) \geq S_2(t, \beta_1, \alpha_0, \beta_1). \end{aligned}$$

Hence by Theorem 4.1.2, $\alpha_1(t) \leq \beta_1(t)$ on $\boldsymbol{T}$ and thus,

$$\alpha_0 \leq \alpha_1 \leq \beta_1 \leq \beta_0, \text{ on } \boldsymbol{T}. \tag{4.1.5}$$

Assume now that for some $k > 1$ and $t \in \boldsymbol{T}$,

$$\alpha_k^{\Delta} \leq f(t, \alpha_k), \ \beta_k^{\Delta} \geq f(t, \beta_k), \ \ \alpha_k(t) \leq \beta_k(t),$$

and consider the initial value problems

$$\alpha_{k+1}^{\Delta} = S_1(t, \alpha_{k+1}, \alpha_k, \beta_k), \quad \alpha_{k+1}(0) = x_0, \tag{4.1.6}$$

$$\beta_{k+1}^{\Delta} = S_2(t, \beta_{k+1}, \alpha_k, \beta_k), \quad \beta_{k+1}(0) = x_0. \tag{4.1.7}$$

Clearly, $\alpha_k^{\Delta} \leq f(t, \alpha_k) = S_1(t, \alpha_k, \alpha_k, \beta_k)$, and $\beta_k^{\Delta} \geq f(t, \beta_k) = S_1(t, \beta_k, \alpha_k, \beta_k)$.

One obtains therefore by Theorems 4.1.2 and 4.1.3, the existence of a unique solution $\alpha_{k+1}(t)$ of (4.1.6) satisfying

$$\alpha_k(t) \leq \alpha_{k+1}(t) \leq \beta_k(t) \text{ on } \boldsymbol{T}.$$

Similar reasoning shows that there exists a unique solution $\beta_{k+1}(t)$ of (4.1.7) such that $\alpha_k(t) \leq \beta_{k+1}(t) \leq \beta_k(t)$, on $\boldsymbol{T}$. Using (H_3), together with the facts $\alpha_k \leq \alpha_{k+1}$, $\beta_{k+1} \leq \beta_k$, we see that

$$\begin{aligned} \alpha_{k+1}^{\Delta} &= S_1(t, \alpha_{k+1}, \alpha_k, \beta_k) \leq S_1(t, \alpha_{k+1}, \alpha_{k+1}, \beta_k), \\ \text{and } \beta_{k+1}^{\Delta} &= S_2(t, \beta_{k+1}, \alpha_k, \beta_k) \geq S_2(t, \beta_{k+1}, \alpha_k, \beta_{k+1}). \end{aligned}$$

Hence, Theorem 4.1.2 yields $\alpha_{k+1}(t) \leq \beta_{k+1}(t)$ on $\boldsymbol{T}$. We have for $t \in \boldsymbol{T}$,

$$\alpha_k \leq \alpha_{k+1} \leq \beta_{k+1} \leq \beta_k. \tag{4.1.8}$$

By induction, we then have for all n,

$$\alpha_0 \leq \alpha_1 \leq \alpha_2 \leq \cdots \leq \alpha_n \leq \beta_n \leq \cdots \leq \beta_1 \leq \beta_0, \ t \in \boldsymbol{T},$$

with $\alpha_{n+1}^{\Delta} = S_1(t, \alpha_{n+1}, \alpha_n, \beta_n)$, $\beta_{n+1}^{\Delta} = S_2(t, \beta_{n+1}, \alpha_n, \beta_n)$, $\alpha_{n+1}(0) = x_0, \beta_{n+1}(0) = x_0$.

Employing standard arguments, it is now easy to conclude that

$$\lim_{n\to\infty} \alpha_n(t) = \rho(t), \ \lim_{n\to\infty} \beta_n(t) = r(t)$$

locally, uniformly and monotonically on $\boldsymbol{T}$ and ρ, r are solutions of (4.1.2). An application of Theorem 4.1.2 implies that $x(t) = \rho(t) = r(t)$ is the unique solution of (4.1.2) on $\boldsymbol{T}$.

To prove superlinear convergence of α_n β_n to x, we set $p_{n+1}(t) = x(t) - \alpha_{n+1}(t)$, $q_{n+1}(t) = \beta_{n+1}(t) - x(t)$, so that $p_{n+1}(t) \geq 0$, $q_{n+1}(t) \geq 0$, $p_{n+1}(0) = 0 = q_{n+1}(0)$. Using ($H_4$), we then obtain

$$p_{n+1}^{\Delta}(t) \leq M|p_{n+1}(t)| + N\left[|p_n(t)|^{1+\eta} + |q_n(t)|^{1+\eta}\right], \ t \in \boldsymbol{T}.$$

Using the variation of parameters formula [41], we now get

$$p_{n+1}(t) \leq \int_0^t \Phi_M(t, \sigma(s)) N\left[|p_n(t)|^{1+\eta} + |q_n(s)|^{1+\eta}\right] \Delta s$$

where $\Phi_M(t,s)$ is the fundamental matrix solution of the matrix differential equation

$$X^{\Delta} = MX, \ X(0) = I(\text{Identity matrix}).$$

It then follows that

$$\max_T |p_{n+1}(t)| \leq K\left[\max_T \left(|p_n(t)|^{1+\eta} + |q_n(t)|^{1+\eta}\right)\right].$$

Similarly one can obtain the estimate

$$\max_T |q_{n+1}(t)| \leq K\left[\max_T \left(|p_n(t)|^{1+\eta} + |q_n(t)|^{1+\eta}\right)\right].$$

The proof is complete.

As an application of Theorem 4.1.4, we shall give below conditions that guarantee the existence of S_j satisfying the assumptions of Theorem 4.1.4.

Assume that

(A_0) $\alpha_0, \beta_0 \in C^1_{rd}[\boldsymbol{T}, R^n]$, $\alpha_0^{\Delta} \leq f(t, \alpha_0)$, $\beta_0^{\Delta} \geq f(t, \beta_0)$ and $\alpha_0 \leq \beta_0$, on $\boldsymbol{T}$.

(A_1) $f \in C_{rd}[\boldsymbol{T} \times R^n, R^n]$ and on $\Omega_0 = [(t,x) : t \in \boldsymbol{T}, \ \alpha_0(t) \leq x \leq \beta_0(t)]$, $f(t,x)$ is quasimonotone nondecreasing in x. Further $f_x(t,x)$ exists such that whenever $x \geq y$ on Ω_0, we have

$$0 \leq f_x(t,x) - f_x(t,y) \leq L|x-y|^{\alpha} \text{ for } 0 < \alpha \leq 1$$

where L is an $n \times n$ matrix and $|x| = (|x_1|, \cdots, |x_n|)$.

We now set

$$\begin{aligned} S_1(t,x,y,z) &= f(t,y)+f_x(t,y)(x-y), \\ S_2(t,x,y,z) &= f(t,z)+f_x(t,y)(x-z) \end{aligned}$$

for (t,x,y,z) on Ω with $y \le x$. Now it is easy to verify S_1, S_2 satisfies all the hypothesis of Theorem 4.1.4. If $\alpha = 1$, we have quadratic convergence. In this case we can as well assume that $f(t,x)$ is uniformly convex in x. We note that if $f(t,x)$ is not convex in x on Ω_0, then we can still construct S_1 and S_2 and obtain quadratic convergence, provided there exists a uniform function $\phi(t,x)$ on Ω_0 such that $f(t,x)+\phi(t,x)$ is uniformly convex on Ω_0. In addition the matrix given by

$$A(t,\alpha_0,\beta_0) = a_{ij}(t,\alpha_0,\beta_0)$$

should be such that $a_{ij}(t,\alpha_0,\beta_0) \ge 0$ for $i \ne j$. Here the matrix $A(t,\alpha_0,\beta_0)$ is given by

$$A(t,\alpha_0,\beta_0) = f_x(t,\alpha_0)+\phi_x(t,\alpha_0)-\phi_x(t,\beta_0) = F_x(t,\alpha_0)-\phi_x(t,\beta_0).$$

Now one can define S_1 and S_2 as

$$\begin{aligned} S_1(t,x,y,z) &= f(t,x)+A(t,x,y)(x-y), \\ S_2(t,x,y,z) &= f(t,z)+A(t,x,y)(x-z). \end{aligned}$$

Note that these assumptions imply the inequality

$$f(t,x) \ge f(t,y)+[f_x(t,y)+\phi_x(t,y)](x-y)-[\phi(t,x)-\phi(t,y)]$$

on Ω_0 whenever $x \ge y$. It is now easy to see that this new S_1 and S_2 satisfy all the hypothesis of Theorem 4.1.4. In view of the following Lemma, it is easy to verify that the assumptions (H_0) to (H_4) of Theorem 4.1.4 are satisfied by $S_1(t,x,y,z)$ and $S_2(t,x,y,z)$.

Lemma 4.1.1 *Let $f \in C_{rd}[\boldsymbol{T} \times R^n, R^n]$ and $f_x(t,x)$ exists and continuous. If $f(t,x)$ is uniformly convex in x, then*

$$f(t,x+h)-f(t,x) \le f_x(t,x)h,$$

where the inequalities are componentwise.

Proof Since $f_x(t,x)$ exists and continuous, we have

$$f(t,x+h) = f(t,x)+f_x(t,x)h+|h|\eta(t,x,h), \; h,x \in R^n \tag{4.1.9}$$

where $|\eta(t,x,h)| \to 0$ as $|h| \to 0$. Since $f(t,x)$ is convex,

$$f(t, \lambda x + (1-\lambda)z) \leq \lambda f(t,x) + (1-\lambda)f(t,z), \ 0 < \lambda < 1. \tag{4.1.10}$$

Set $z = x + h/(1-\lambda)$, $0 < \lambda < 1$, so that $x + h = \lambda x + (1-\lambda)z$. Consequently, in view of (4.1.9) and (4.1.10) we get

$$\begin{aligned} \lambda f(t,x) + (1-\lambda)f(t,z) &\geq f(t, \lambda x + (1-\lambda)z) \\ &= \lambda f(t,x) + (1-\lambda)f(t,x) + f_x(t,x)h + |h|\eta(t,x,h), \end{aligned}$$

which implies

$$\begin{aligned} |h|\eta(t,x,h) &\leq (1-\lambda)\left[f(t,z) - f(t,x) - f_x(t,x)\frac{h}{1-\lambda}\right] \\ &= (1-\lambda)\eta\left(t,x,\frac{h}{1-\lambda}\right)\left|\frac{h}{1-\lambda}\right| \\ &= \eta\left(t,x,\frac{h}{1-\lambda}\right)|h|. \end{aligned}$$

Thus, we have $\eta(t,x,h) \leq \eta\left(t,x,\frac{h}{1-\lambda}\right)$ for $\lambda \in (0,1)$. Now taking $\eta = y/n$, $y \in R^n$ and $\lambda = 1 - 1/n$, we obtain

$$\eta\left(t,x,\frac{y}{n}\right) \leq \eta(t,x,y),$$

which yields $\eta(t,x,y) \geq 0$, because $\lim_{n\to\infty} \eta\left(t,x,\frac{y}{n}\right) \to 0$. The proof is complete in view of (4.1.9).

This lemma is a special case of Lemma 4.5.1 in [28]. We have given here for completeness.

4.2 INTEGRO-DIFFERENTIAL EQUATIONS

Integro-differential equations arise quite frequently as mathematical models in diverse disciplines. The theory and applications of Volterra integro-differential equations have emerged as a new area of investigation. For a basic theory and qualitative properties of solutions of Volterra integro-differential equations see the recent monograph [36]. In this section, we shall discuss the method of quasilinearization relative to Volterra integro-differential equations.

We consider the IVP

$$\begin{aligned} x'(t) &= f(t,x(t)) + \int_{t_0}^{t} K(t,s,x(s))ds, \ x(t_0) = x_0, \\ &\text{with } t \in J = [t_0, t_0+T], t_0 \geq 0, T > 0, \end{aligned} \tag{4.2.1}$$

and assume suitable conditions on f and K.

We need the following known Lemma [36].

Lemma 4.2.1 *Consider the IVP* (4.2.1) *and that*

(A_1) $f \in C[J \times R, R]$, $K \in C[J \times J \times R, R]$ *and* $K(t,s,u)$ *is monotone nondecreasing in* u *for each fixed* $(t,s) \in J \times J$;

(A_2) $v,\ w \in C^1[J,R]$,

$$\begin{aligned} v' &\leq f(t,v) + \int_{t_0}^{t} K(t,s,v(s))ds, \\ w' &\geq f(t,w) + \int_{t_0}^{t} K(t,s,w(s))ds, \quad t \in J; \end{aligned}$$

(A_3) *for* $(t,s) \in J \times J$, $x \geq y$, $L \geq 0$, $f(t,x) - f(t,y) \leq L(x-y)$, *and* $K(t,s,x) - K(t,s,y) \leq N(x-y)$, $N > 0$, $t \in J$.

Then we have $v(t) \leq w(t)$ *for* $t \in J$, *provided* $v(t_0) \leq w(t_0)$.

Now we are in a position to prove our first result.

Theorem 4.2.1 *Assume that:*

(i) *condition* (A_1) *in Lemma* 4.2.1 *holds together with* $f \in C^2[J \times R, R]$, $K \in C^2[J \times J \times R, R]$;

(ii) α_0, $\beta_0 \in C^1[J \times R, R]$ *such that* $\alpha_0(t) \leq \beta_0(t)$ *and*

$$\begin{aligned} \alpha_0' &\leq f(t,\alpha_0) + \int_{t_0}^{t} K(t,s,\alpha_0(s))ds, \\ \beta_0' &\geq f(t,\beta_0) + \int_{t_0}^{t} K(t,s,\beta_0(s))ds, \end{aligned}$$

where $\alpha_0(t_0) \leq x_0 \leq \beta_0(t_0)$;

(iii) $f_{xx}(t,x) \geq 0$ *for each* $t \in J$ *and* $K_{xx}(t,s,x) \geq 0$ *for each* $(t,s) \in J \times J$.

Then there exist monotone sequences $\{\alpha_n(t)\}$, $\{\beta_n(t)\}$ *which converge uniformly and quadratically to the unique solution of* (4.2.1).

Proof In view of assumption (iii), we obtain for $u \geq v$:

$$f(t,u) \geq f(t,v) + f_x(t,v)(u-v), \tag{4.2.2}$$

$$K(t,s,u) \geq K(t,s,v) + K_x(t,s,v)(u-v),$$
$$f(t,u) - f(t,v) \leq L(u-v)$$

and

$$K(t,s,u) - K(t,s,v) \leq N(u-v),$$

for $t \in J$, $(t,s) \in J \times J$, $L > 0$, $N > 0$.

Since f and K satisfy the one-sided Lipschitz condition, it follows by the repeated use of Lemma 4.2.1 that IVP (4.2.1) possesses the unique solution.

Let α_1, β_1 be the solutions of the IVPs

$$\begin{aligned} \alpha_1' &= f(t,\alpha_0) + f_x(t,\alpha_0)(\alpha_1 - \alpha_0) \\ &\quad + \int_{t_0}^{t} [K(t,s,\alpha_0(s)) + K_x(t,s,\alpha_0(s))(\alpha_1(s) - \alpha_0(s))]\, ds, \\ &\quad \alpha_1(t_0) = x_0, \end{aligned} \tag{4.2.3}$$

and

$$\begin{aligned} \beta_1' &= f(t,\beta_0) + f_x(t,\alpha_0)(\beta_1 - \beta_0) \\ &\quad + \int_{t_0}^{t} [K(t,s,\beta_0(s)) + K_x(t,s,\alpha_0(s))(\beta_1(s) - \beta_0(s))]\, ds, \\ &\quad \beta_1(t_0) = x_0. \end{aligned} \tag{4.2.4}$$

We shall prove the inequalities

$$\alpha_0(t) \leq \alpha_1(t) \leq \beta_1(t) \leq \beta_0(t), \ t \in J. \tag{4.2.5}$$

(ia) Set $p = \alpha_0 - \alpha_1$. Clearly $p(t_0) \leq 0$. Then we obtain

$$\begin{aligned} p' &= \alpha_0' - \alpha_1' \\ &\leq f(t,\alpha_0) + \int_{t_0}^{t} K(t,s,\alpha_0(s))ds - f(t,\alpha_0) - f_x(t,\alpha_0)(\alpha_1 - \alpha_0) \\ &\quad - \int_{t_0}^{t} [K(t,s,\alpha_0(s)) + K_x(t,s,\alpha_0(s))(\alpha_1(s) - \alpha_0(s))]\, ds \\ &\leq f_x(t,\alpha_0)p + \int_{t_0}^{t} K_x(t,s,\alpha_0(s))p(s)ds. \end{aligned}$$

Using the special case of Lemma 4.2.1 for the linear problem with $p(t) = \alpha(t)$ and $0 = \beta(t)$ yields $p(t) \leq 0$, $t \geq t_0$, which in turn proves $\alpha_0 \leq \alpha_1$, $t \in J$.

(*ib*) By a similar process, we can prove that $\beta_1(t) \leq \beta_0(t)$, $t \in J$.

(*ic*) It remains to show that $\alpha_1(t) \leq \beta_1(t)$. With $\alpha_0 \leq \alpha_1$, $t \in J$, using the inequalities (4.2.2), we get

$$\alpha_1' = f(t, \alpha_0) + f_x(t, \alpha_0)(\alpha_1 - \alpha_0)$$
$$+ \int_{t_0}^{t} [K(t, s, \alpha_0(s)) + K_x(t, s, \alpha_0(s))(\alpha_1(s) - \alpha_0(s))]\, ds$$
$$\leq f(t, \alpha_1) + \int_{t_0}^{t} K(t, s, \alpha_1(s)) ds,$$

and similarly

$$\beta_1' \geq f(t, \beta_1) + \int_{t_0}^{t} K(t, s, \beta_1(s)) ds.$$

Also note that $\alpha_1(t_0) = \beta_1(t_0) = x_0$. The application of Lemma 4.2.1 yields $\alpha_1(t) \leq \beta_1(t)$, $t \in J$.

From (ia) through (ic), the inequalities (4.2.5) are established.
Assume that for some $n > 1$,

$$\alpha_n' \leq f(t, \alpha_n) + \int_{t_0}^{t} K(t, s, \alpha_n(s)) ds,$$
$$\beta_n' \geq f(t, \beta_n) + \int_{t_0}^{t} K(t, s, \beta_n(s)) ds,$$

and $\alpha_0(t) \leq \alpha_n(t) \leq \beta_n(t) \leq \beta_0(t)$ on J and α_{n+1} and β_{n+1} are the solutions of the IVP

$$\begin{aligned} \alpha_{n+1}' &= f(t, \alpha_n) + f_x(t, \alpha_n)(\alpha_{n+1} - \alpha_n) \\ &\quad + \int_{t_0}^{t} [K(t, s, \alpha_n(s)) + K_x(t, s, \alpha_n(s))(\alpha_{n+1}(s) - \alpha_n(s))]\, ds, \\ &\quad \alpha_{n+1}(t_0) = x_0, \end{aligned}$$

and

$$\begin{aligned} \beta_{n+1}' &= f(t, \beta_n) + f_x(t, \alpha_n)(\beta_{n+1} - \beta_n) \\ &\quad + \int_{t_0}^{t} [K(t, s, \beta_n(s)) + K_x(t, s, \alpha_n(s))(\beta_{n+1}(s) - \beta_n(s))]\, ds, \\ &\quad \beta_{n+1}(t_0) = x_0. \end{aligned}$$

Set $p = \alpha_n - \alpha_{n+1}$ so that

$$\begin{aligned} p' &= \alpha_n' - \alpha_{n+1}' \\ &\le f(t, \alpha_n) + \int_{t_0}^t K(t, s, \alpha_n(s))ds - f(t, \alpha_n) - f_x(t, \alpha_n)(\alpha_{n+1} - \alpha_n) \\ &\quad - \int_{t_0}^t \left[K(t, s, \alpha_n(s)) + K_x(t, s, \alpha_n(s))(\alpha_{n+1}(s) - \alpha_n(s))\right] ds \\ &= f_x(t, \alpha_n)(\alpha_n - \alpha_{n+1}) + \int_{t_0}^t K_x(t, s, \alpha_n(s))(\alpha_n(s) - \alpha_{n+1}(s))ds \\ &= f_x(t, \alpha_n)p + \int_{t_0}^t K_x(t, s, \alpha_n(s))p(s)ds. \end{aligned}$$

Now, since $p(t_0) = 0$, by Lemma 4.2.1, we get $p(t) \le 0$ and this implies $\alpha_n(t) \le \alpha_{n+1}(t)$, $t \in J$.

Similar arguments yield $\beta_{n+1}(t) \le \beta_n(t)$ and $\alpha_{n+1}(t) \le \beta_{n+1}(t)$ which together yield $\alpha_n(t) \le \alpha_{n+1}(t) \le \beta_{n+1}(t) \le \beta_n(t)$, $t \in J$.

Hence, by induction, we have, for all n,

$$\alpha_0(t) \le \alpha_1(t) \le \cdots \le \alpha_n(t) \le \beta_n(t) \le \cdots \le \beta_1(t) \le \beta_0(t), \ t \in J.$$

By standard arguments, it can be shown that the sequences $\{\alpha_n(t)\}$ and $\{\beta_n(t)\}$ converge uniformly and monotonically to $x(t)$, the unique solution of (4.2.1) on J.

Now we shall show that the convergence is quadratic. For this purpose

(IA) let $p_{n+1} = x - \alpha_{n+1}$. Clearly, $p_{n+1}(t_0) = 0$ and $p_{n+1} \ge 0$, $t \in J$. Then

$$\begin{aligned} p_{n+1}' &= x' - \alpha_{n+1}' \\ &= f(t, x) + \int_{t_0}^t K(t, s, x(s))ds - f(t, \alpha_n) - f_x(t, \alpha_n)(\alpha_{n+1} - \alpha_n) \\ &\quad - \int_{t_0}^t \left[K(t, s, \alpha_n(s)) + K_x(t, s, \alpha_n(s))(\alpha_{n+1}(s) - \alpha_n(s))\right] ds \\ &= f_x(t, \xi_1)(x - \alpha_n) - f_x(t, \alpha_n)(\alpha_{n+1} - x + x - \alpha_n) \\ &\quad + \int_{t_0}^t K_x(t, s, \xi_2(s))(x(s) - \alpha_n(s))ds \\ &\quad - \int_{t_0}^t K_x(t, s, \alpha_n(s))(\alpha_{n+1} - x(s) + x(s) - \alpha_n(s))ds \end{aligned}$$

by the mean value theorem with $\alpha_n \le \xi_1, \xi_2 \le x$. Thus

$$p_{n+1}' = f_x(t, \xi_1)p_n + f_x(t, \alpha_n)p_{n+1} - f_x(t, \alpha_n)p_n$$

$$
\begin{aligned}
&+\int_{t_0}^{t} K_x(t,s,\xi_2(s))p_n(s)ds + \int_{t_0}^{t} K_x(t,s,\alpha_n(s))p_{n+1}(s)ds \\
&-\int_{t_0}^{t} K_x(t,s,\alpha_n(s))p_n(s)ds \\
\leq\ & f_{xx}(t,\xi_3)p_n^2 + f_x(t,\alpha_n)p_{n+1} \\
&+\int_{t_0}^{t} K_{xx}(t,s,\xi_4(s))p_n^2(s)ds + \int_{t_0}^{t} K_x(t,s,\alpha_n(s))p_{n+1}ds \\
\leq\ & Mp_n^2 + Lp_{n+1} + M_1\int_{t_0}^{t} p_n^2(s)ds + L_1\int_{t_0}^{t} p_{n+1}(s)ds
\end{aligned}
$$

where $\alpha_n \leq \xi_3 \leq \xi_1$, $\alpha_n(s) \leq \xi_4(s) \leq \xi_2(s)$, $f_x(t,x) \leq L$, $f_{xx}(t,x) \leq M$, $K_x(t,s,x) \leq L_1$, $K_{xx}(t,s,x) \leq M_1$. By Lemma 4.2.1, we now have that $p_{n+1}(t) \leq r(t)$, $t \in J$ where $r(t)$ is the solution of the related linear integro-differential equation

$$
r'(t) = Lr(t) + L_1\int_{t_0}^{t} r(s)ds + F(t),\ p_{n+1}(t_0) = r(t_0) = 0, \tag{4.2.6}
$$

where $F(t) = Mp_n^2(t) + M_1\int_{t_0}^{t} p_n^2(s)ds$. We note that

$$
F(t) \leq (M + M_1T)\max_{t\in J} p_n^2(t),\ t \in J. \tag{4.2.7}
$$

We now find an estimate for $r(t)$ given by (4.2.6). Let

$$
z(t) = \int_{t_0}^{t} r(s)ds. \tag{4.2.8}
$$

Then the equation (4.2.6) becomes $z'' = Lz' + L_1 z + F(t)$, $z(t_0) = 0$, $z'(t_0) = 0$. This is a second order, linear nonhomogeneous differential equation with constant coefficients. Using the variation of parameters method, in view of (4.2.7) and (4.2.8), it follows that

$$
p_{n+1}(t) \leq r(t) \leq \frac{2e^{LT}}{\sqrt{L^2 + 4L_1}}F(t).
$$

Hence,

$$
\max_{t\in J} p_{n+1}(t) \leq \frac{2e^{LT}}{\sqrt{L^2 + 4L_1}}(M + M_1T)\max_{t\in J} p_n^2(t).
$$

Thus $\{\alpha_n(t)\}$ converges quadratically to $x(t)$, the unique solution of (4.2.1).

It remains to show that $\{\beta_n(t)\}$ converges quadratically to $x(t)$.

Let $q_{n+1} = \beta_{n+1} - x$. Clearly, $q_{n+1}(t_0) = 0$, $q_{n+1} \geq 0$, $t \in J$. By a process similar to (IA), we obtain

$$q_{n+1}(t) \leq r(t) \leq \frac{2e^{LT}}{\sqrt{L^2 + 4L_1}} F(t),$$

$$\max_{t \in J} q_{n+1}(t) \leq \frac{2e^{LT}}{\sqrt{L^2 + 4L_1}} (M + M_1 T) \max_{t \in J} q_n^2(t).$$

The proof is complete.

Observe that the conclusion of Theorem 4.2.1 is based on the fact that $K(t, s, x)$ is monotone nondecreasing in x for each fixed $(t, s) \in J \times J$, which is a natural assumption in treating integro-differential equations. We now consider a nonincreasing property of K and extend the method of quasilinearization. For this purpose, we need the following comparison result:

Lemma 4.2.2 *Let $a \in C[J, R]$, $b \in C[J \times J, R_+]$. Consider the integro-differential inequality*

$$p'(t) \leq a(t)p(t) - \int_{t_0}^{t} b(t, s)p(s)ds, \; t \in J, \; \text{where } p(t_0) \leq 0. \tag{4.2.9}$$

Assume that

$$\max_{t \in J} \int_{t_0}^{t} \left[\exp\left(-\int_{s}^{t} a(\tau)d\tau \right) b(t, s) \right] ds \leq \frac{1}{T}, \tag{4.2.10}$$

then $p(t) \leq 0$, $t \in J$.

Proof Let $m(t) = p(t) \exp\left(-\int_{t_0}^{t} a(s)ds \right)$. Then it follows from (4.2.9) that

$$m'(t) \leq -\int_{t_0}^{t} \exp\left(-\int_{s}^{t} a(\tau)d\tau \right) b(t, s)m(s)ds, \; t \in J. \tag{4.2.11}$$

It is enough to show that $m(t) \leq 0$, $t \in J$.

Suppose that this claim is not true. Then we consider two possibilities: (a) $m(t_0) = 0$, (b) $m(t_0) < 0$. In the case of (a), suppose that there exists $t_1 \in (t_0, T]$ such that $m(t_1) > 0$ and that $m(t) \geq 0$ for $t \in [t_0, t_1]$. Then from (4.2.9), we have $m'(t) \leq 0$ for $t \in [t_0, t_1]$, which means that $m(t)$ is nonincreasing in $[t_0, t_1]$. Then $m(t_1) \leq m(t_0) = 0$. This is a contradiction.

Alternately it follows that when $m(t_0) < 0$, there exists a $t_2 \in [t_0, t_1]$ such that

$$m(t_2) = \min_{t \in [t_0, t_1]} m(t) = -\mu, \; \mu > 0.$$

By the mean value theorem applied over the interval $[t_2, t_1]$, we have

$$m'(t^*) = \frac{m(t_1) - m(t_2)}{t_1 - t_2} = \frac{m(t_1) + \mu}{t_1 - t_2} > \frac{\mu}{T} \tag{4.2.12}$$

where $t_2 \leq t^* \leq t_1$.

On the other hand, from (4.2.10) and (4.2.11) it follows that

$$\begin{aligned} m'(t^*) &\leq -\int_{t_0}^{t^*} \exp\left(-\int_s^{t^*} a(\tau)d\tau\right) b(t^*, s)m(s)ds \\ &\leq \mu \int_{t_0}^{t^*} \exp\left(-\int_s^{t^*} a(\tau)d\tau\right) b(t^*, s)ds \\ &\leq \frac{\mu}{T}. \end{aligned} \tag{4.2.13}$$

Inequalities (4.2.12) and (4.2.13) together lead us to a contradiction. The proof of Lemma 4.2.2 is complete.

We shall now consider the following theorem which extends the scope of Theorem 4.2.1.

Theorem 4.2.2 *Consider the equation* (4.2.1) *where* $f \in C^2[J \times R, R]$, $K \in C^2[J \times J \times R, R]$. *Assume that:*

(a) $\alpha_0, \beta_0 \in C[J \times R, R]$, *such that*

$$\alpha_0' \leq f(t, \alpha_0) + \int_{t_0}^{t} K(t, s, \alpha_0(s))ds$$

$$\text{and } \beta_0' \geq f(t, \beta_0(s)) + \int_{t_0}^{t} K(t, s, \beta_0(s))ds$$

$$\text{and } \alpha_0(t_0) \leq x_0 \leq \beta_0(t_0), \; t \in J;$$

(b) $f_{xx}(t, x) \geq 0$ *for each* $t \in J$, $K_{xx}(t, s, x) \geq 0$ *for each* $(t, s) \in J \times J$ *and* $K_x(t, s, x) \leq 0$;

(c) $1 + T \max_{t \in J} \int_{t_0}^{t} \exp\left(-\int_s^t f_x(\tau, v(\tau))d\tau\right) K_x(t, s, v(s))ds \geq 0$, $\alpha_0(t) \leq v(t) \leq \beta_0(t)$, $t \in J$.

Then there exist monotone sequences $\{\alpha_n(t)\}$, $\{\beta_n(t)\}$ *which converge uniformly and quadratically to the unique solution of* (4.2.1).

Proof Define α_1' and β_1' as in (4.2.3) and (4.2.4). We now prove the inequalities $\alpha_0(t) \le \alpha_1(t) \le \beta_1(t) \le \beta_0(t)$, $t \in J$. As before, set $p = \alpha_0 - \alpha_1$. Clearly $p(t_0) \le 0$. Then we obtain

$$\begin{aligned} p' &= \alpha_0' - \alpha_1' \\ &\le f(t,\alpha_0) + \int_{t_0}^{t} K(t,s,\alpha_0(s))ds - f(t,\alpha_0) - f_x(t,\alpha_0)(\alpha_1 - \alpha_0) \\ &\quad - \int_{t_0}^{t} [K(t,s,\alpha_0(s))ds + K_x(t,s,\alpha_0(s))(\alpha_1(s) - \alpha_0(s))]\, ds \\ &\le f_x(t,\alpha_0)p + \int_{t_0}^{t} K_x(t,s,\alpha_0(s))p(s)ds. \end{aligned}$$

We now recall Lemma 4.2.1 and note that $a(t) = f_x(t,\alpha_0)$, $b(t,s) = -K_x(t,s,\alpha_0(s))$ and $v(t) = \alpha_0(t)$. This fact together with assumptions (a) and (b) yields $p(t) \le 0$, $t \ge t_0$, which in turn proves $\alpha_0 \le \alpha_1$.

The remaining inequalities are proven using similar techniques shown in the proof of Theorem 4.2.1, but employing Lemma 4.2.1 and condition (c). Additionally, by using the arguments employed in Theorem 4.2.2 with Lemma 4.2.1, it can be shown that $\alpha_n(t) \le \alpha_{n+1}(t) \le \beta_{n+1}(t) \le \beta_n(t)$ for $n > 1$, $t \in J$. Thus, we obtain the sequences $\{\alpha_n(t)\}$ and $\{\beta_n(t)\}$ as previously. The proof that these sequences converge quadratically is again similar. We omit the details. This completes the proof.

4.3 FUNCTIONAL DIFFERENTIAL EQUATIONS

The future state of a physical system depends, in many circumstances, not only on the present state but also on its past history. Functional differential equations provide a mathematical model for such physical systems in which the rate of change of the system may depend on the influence of its hereditarily effects. The simplest type of such a system is a differential-difference equation of the form

$$x' = f(t, x(t), x(t-\tau)),$$

where $\tau > 0$ is a constant. More general systems may be described by

$$x'(t) = f(t, x_t)$$

where f is a suitable functional. If x is a function defined on some interval $[t_0 - \tau, t_0 + a]$, $a > 0$, the symbol x_t is the graph of x on $[t - \tau, t]$ shifted to the interval $[-\tau, 0]$. In this section, we consider such functional differential equations and establish a typical result concerning the method of generalized quasilinearization.

Consider the IVP for FDE

$$x' = f(t, x_t),\ x_0 = \phi_0 \in C,\ t \in [0, T] = J, \tag{4.3.1}$$

where $f \in C[J \times C, R]$, $C = C[[-\tau, 0], R]$, $\tau > 0$ and for any $\tau \in J$, $x_t = x(t+s)$, $-\tau \leq s \leq 0$. Let $\overline{J} = [-\tau, T]$.

We need the following lemma.

Lemma 4.3.1 *Suppose that*

(*i*) $u,\ v \in C[\overline{J}, R] \cap C^1[J, R]$, $f \in C[J \times C, R]$, *and*

$$v' \geq f(t, v_t),\quad u' \leq f(t, u_t),\quad t \in J;$$

(*ii*) $f(t, \phi) - f(t, \psi) \leq L \int_{\tau}^{0} [\phi(s) - \psi(s)] ds$, $t \in J$, *whenever* $\phi(s) \geq \psi(s)$, $-\tau \leq s \leq 0$, $L > 0$ *is a constant, and* $f(t, \phi)$ *is quasinondecreasing in* ϕ *for each* $t \in J$, *that is, whenever* $\psi(s) \leq \phi(s)$, $-\tau \leq s < 0$, *and* $\psi(0) = \phi(0)$, $f(t, \psi) \leq f(t, \phi)$, $t \in J$;

Then $u(s) \leq v(s)$, $-\tau \leq s \leq 0$ *implies* $u(t) \leq v(t)$, $t \in J$, *provided that* $L + e^{-L\tau} > 1$.

Proof Let us first prove the lemma for strict inequalities. Suppose that

$$v' > f(t, v_t),\ u' \leq f(t, u_t),\ t \in J,\ u_0(s) < v_0(s), -\tau \leq s \leq 0. \tag{4.3.2}$$

We show that $u(t) < v(t)$, $t \in J$.

Suppose that this is not true. Then there exists $t_1 \in (0, T]$ such that

$$u(t_1) = v(t_1),\ u(t) < v(t),\ -\tau \leq t \leq t_1.$$

Hence $u_t < v_t$, $0 \leq t < t_1$. For each $h > 0$ sufficiently small, we get

$$u(t_1 - h) - u(t_1) \leq v(t_1 - h) - v(t_1)$$

and hence $u'(t_1) \geq v'(t_1)$. In view of (4.3.2) and the fact that $f(t, \phi)$ is quasinondecreasing in ϕ,

$$f(t_1, u_{t_1}) \geq u'(t_1) \geq v'(t_1) > f(t, v_{t_1}) \geq f(t_1, u_{t_1}).$$

This is a contradiction. Hence $u(t) < v(t)$, $t \in J$.

Now, set for $\epsilon > 0$,

$$\bar{v}(t) = v(t) + \epsilon e^{Lt}, \ t \in \overline{J}. \tag{4.3.3}$$

Clearly $\bar{v}_t > v_t$, $t \in J$. Further because of (ii)

$$f(t, \bar{v}_t) - f(t, v_t) \leq L\epsilon \int_{-\tau}^{0} e^{L(t+s)} ds = \epsilon e^{Lt}\left(1 - e^{-L\tau}\right). \tag{4.3.4}$$

Employing (4.3.3) and (4.3.4), we then obtain

$$\begin{aligned} \bar{v}' &= v'(t) + L\epsilon e^{Lt} \geq f(t, v_t) + L\epsilon e^{Lt} \\ &\geq f(t, \bar{v}_t) - \epsilon e^{Lt}(1 - e^{-L\tau}) + L\epsilon e^{Lt} \\ &> f(t, \bar{v}_t) \end{aligned}$$

since we have assumed that $L + e^{-L\tau} > 1$.

Now we have

$$\bar{v}' > f(t, \bar{v}_t), \ u'(t) \leq f(t, u_t), \ u_0(s) < \bar{v}_0(s), \ -\tau \leq s \leq 0 \ \text{ and } t \in J.$$

Applying foregoing part of the proof, we conclude that $u(t) < \bar{v}(t)$, $t \in J$. Letting $\epsilon \to 0$ in (4.3.3), we obtain $u(t) \leq v(t)$ and $t \in J$. The proof is complete.

We list the following hypotheses:

(H_1) $f \in C[J \times C, R]$;

(H_2) u_0, $v_0 \in C[\overline{J}, R] \cap C^1[J, R]$ and $v_0'(t) \geq f(t, v_{0,t})$, $u_0' \leq f(t, u_{0,t})$, $t \in J$, $u_0(s) \leq v_0(s)$, $-\tau \leq s \leq 0$;

(H_3) The Frechet derivative $f_\phi(t, \phi)$ exists, is a continuous linear operator satisfying

(*i*) $f_\phi(t, \phi)\psi \leq L \int_{-\tau}^{0} \psi(s) ds$, ϕ, $\psi \in C$, $u_{0,t} \leq \phi, \psi \leq v_{0,t}$ and $L + e^{-L\tau} > 1$;

(*ii*) $f(t, \phi) \geq f(t, \psi) + f_\phi(t, \psi)(\phi - \psi)$ whenever $u_{0,t} \leq \psi \leq \phi \leq v_{0,t}$, $t \in J$; and

(*iii*) $\psi_1(s) \leq \psi_2(s)$, $-\tau \leq s \leq 0$, $\psi_1(0) = \psi_2(0)$ implies that $f_\phi(t, \psi)\psi_1 \leq f_\phi(t, \psi)\psi_2$, ψ_1, $\psi_2 \in C$, $t \in J$ and $u_{0,t} \leq \psi_1 \leq \psi_2 \leq v_{0,t}$.

(H_4) $|f_\phi(t, \psi_1) - f_\phi(t, \psi_2)| \leq L_2 |\psi_1 - \psi_2|_0^\beta$ where $t \in J$, ψ_1, $\psi_2 \in C$, $L_2 > 0$, $|\phi|_0 = \max_{-\tau \leq s \leq 0} |\phi(s)|$ and $0 \leq \beta \leq 1$.

We are now in position to prove the following existence result.

Theorem 4.3.1 *Let the hypotheses (H_1) to (H_4) hold. Then there exists monotone sequences $\{u_n(t)\}$, $\{v_n(t)\}$ which converge uniformly to the unique solution of (4.3.1) on J and that the convergence is superlinear.*

Proof In view of (H_3)(ii), it is clear that $f(t,\phi)$ satisfies a one-sided Lipschitz condition in ϕ for $(t,\phi) \in J \times C$.

Let u_1 and v_1 be the solutions of the IVPs for FDEs

$$u_1' = f(t, u_{0,t}) + f_\phi(t, u_{0,t})(u_{1,t} - u_{0,t}), \ t \in J, \ u_{1,0} = \phi_0$$

and

$$v_1' = f(t, v_{0,t}) + f_\phi(t, u_{0,t})(v_{1,t} - v_{0,t}), \ t \in J, \ v_{1,0} = \phi_0$$

respectively, where $u_{0,0} \le \phi_0 \le v_{0,0}$. We shall show that

$$u_0(t) \le u_1(t) \le v_1(t) \le v_0(t), \ t \in J.$$

To do this, let $p(t) = u_0(t) - u_1(t)$, $t \in \overline{J}$, so that on $-\tau \le s \le 0$,

$$p(s) = u_0(s) - u_1(s) = u_0(s) - \phi_0(s) \le 0.$$

Further for $t \in J$, we have

$$\begin{aligned} p'(t) &= u_0' - u_1'(t) \\ &\le f(t, u_{0,t}) - [f(t, u_{0,t}) + f_\phi(t, u_{0,t})(u_{1,t} - u_{0,t})] \\ &= f_\phi(t, u_{0,t})p(t). \end{aligned}$$

Application of Lemma 4.3.1, yields that $p(t) \le 0$ proving that $u_0(t) \le u_1(t)$, $t \in J$.

Set $p(t) = u_1(t) - v_0(t)$, $t \in \overline{J}$, so that for $-\tau \le s \le 0$, $p(s) = u_1(s) - v_0(s) = \phi_0(s) - v_0(s) \le 0$. Also for $t \in J$

$$\begin{aligned} p'(t) &= u_1'(t) - v_0'(t) \\ &\le f(t, v_{0,t}) - f_\phi(t, u_{0,t})(v_{0,t} - u_{0,t}) \\ &\quad + f_\phi(t, u_{0,t})(u_{1,t} - u_{0,t}) - f(t, v_{0,t}) \\ &= f_\phi(t, u_{0,t})(u_{1,t} - v_{0,t}) = f_\phi(t, u_{0,t})p_t. \end{aligned}$$

By Lemma 4.3.1, we get $u_1(t) \le v_0(t)$, $t \in J$.

Similarly, we can obtain that $u_0(t) \le v_1(t) \le v_0(t)$, $t \in J$. We now prove $u_1(t) \le v_1(t)$, $t \in J$. We already have $u_0(t) \le u_1(t) \le v_0(t)$, $t \in J$. Employing (H_3), we obtain, since $u_{0,t} \le u_{1,t}$,

$$u_1'(t) = f(t, u_{0,t}) + f_\phi(t, u_{0,t})(u_{1,t} - u_{0,t}) \le f(t, u_{1,t}).$$

Similarly, we get $v_1'(t) \geq f(t, v_{1,t})$. Observe that $u_1(s) \leq v_1(s)$, $-\tau \leq s \leq 0$.

Recalling Lemma 4.3.1, we claim that $u_1(t) \leq v_1(t)$, $t \in J$. Thus we have so far obtained

$$u_0(t) \leq u_1(t) \leq v_1(t) \leq v_0(t), \ t \in J.$$

Now assume that for some $n > 1$

$$u_n'(t) \leq f(t, u_{n,t}), \ v_n' \geq f(t, v_{n,t}), \ t \in J$$

and $u_n(s) \leq v_n(s)$, $-\tau \leq s \leq 0$. We now aim to show that

$$u_n(t) \leq u_{n+1}(t) \leq v_{n+1}(t) \leq v_n(t), \ t \in J \tag{4.3.5}$$

where $u_{n+1}(t)$ and $v_{n+1}(t)$ are the solutions of the IVPs

$$u_{n+1}' = f(t, u_{n,t}) + f_\phi(t, u_{n,t})(u_{n+1,t} - u_{n,t}), \ t \in J, \ u_{n+1,0} = \phi_0$$

and

$$v_{n+1}' = f(t, v_{n,t}) + f_\phi(t, v_{n+1,t} - v_{n,t}), \ t \in J, \ v_{n+1,0} = \phi_0$$

respectively.

Set $p(t) = u_n(t) - u_{n+1}(t)$, $t \in \overline{J}$. As before, we get $p'(t) \leq f_\phi(t, u_{n,t})p_t$, $t \in J$, and $p_0 = 0$ which implies that $p(t) \leq 0$, $t \in J$.

Similarly, let $p(t) = u_{n+1}(t) - v_n(t)$, $t \in \overline{J}$, we arrive at

$$p'(t) \leq f(t, u_{n,t}) + f(t, u_{n,t})(u_{n+1,t} - u_{n,t}) - f(t, v_{n,t}).$$

Since $u_n(t) \leq v_n(t)$, $t \in J$, (H_3) yields

$$p'(t) \leq f_\phi(t, u_{n,t})p_t, \ t \in J, \ p_0 = 0,$$

which shows that $p(t) \leq 0$, $t \in J$. Thus we have

$$u_n(t) \leq u_{n+1}(t) \leq v_n(t), \ t \in J.$$

Similarly, we can prove that $u_n(t) \leq v_{n+1}(t) \leq v_n(t)$, $t \in J$.

By (H_3), we see that

$$\begin{aligned} u_{n+1}' &= f(t, u_{n,t}) + f_\phi(t, u_{n,t})(u_{n+1,t} - u_{n,t}) \\ &\leq f(t, u_{n+1,t}) - f_\phi(t, u_{n,t})(u_{n+1,t} - u_{n,t}) \\ &\quad + f_\phi(t, u_{n,t})(u_{n+1,t} - u_{n,t}) \\ &= f(t, u_{n+1,t}). \end{aligned}$$

Similarly, we get $v_{n+1}' \geq f(t, v_{n+1,t})$.

Using Lemma 4.3.1, we conclude that $u_{n+1}(t) \le v_{n+1}(t)$, $t \in J$. The inequalities (4.3.5) are thus established. By induction, we therefore, get

$$u_0(t) \le u_1(t) \le \cdots \le u_n(t) \le v_n(t) \le \cdots \le v_1(t) \le v_0(t),\ t \in J.$$

Following the usual argument, one can show that the sequences $\{u_n(t)\}$, $\{v_n(t)\}$ converge uniformly and monotonically to the unique solution $x(t)$ of (4.3.1) on J.

To prove superlinear convergence, let

$$p_n(t) = x(t) - u_n(t),\ q_n(t) = v_n(t) - x(t),\ t \in [-\tau, T].$$

Hence $p_n(s) = q_n(s) = 0$, $-\tau \le s \le 0$. Further, note that $p_n(t) \ge 0$, $q_n(t) \ge 0$. Also

$$\begin{aligned} p_n'(t) &= x'(t) - u_n'(t) \\ &= f(t, x_t) - [f(t, u_{n-1,t}) + f_\phi(t, u_{n-1,t})(u_{n,t} - u_{n-1,t})] \\ &= \int_0^1 f_\phi(t, sx_t + (1-s)u_{n-1,t})p_{n-1,t}ds \\ &\quad - f_\phi(t, u_{n-1,t})(p_{n-1,t} - p_{n,t}) \\ &= \int_0^1 [f_\phi(t, sx_t + (1-s)u_{n-1,t}) - f_\phi(t, u_{n-1,t})]p_{n-1,t}ds \\ &\quad + f_\phi(t, u_{n-1,t})p_{n,t}. \end{aligned}$$

From (H$_3$) and (H$_4$), it follows that

$$\begin{aligned} p_n'(t) &\le \int_0^1 L_2|sx_t + (1-s)u_{n-1,t} - u_{n-1,t}|_0^\beta |p_{n-1,t}|_0 ds + L\int_{-\tau}^0 p_{n,t}(s)ds \\ &\le L_2\int_0^1 s^\beta |p_{n-1,t}|_0^{\beta+1} ds + L\int_{-\tau}^0 p_{n,t}(s)ds \\ &\le L_2|p_{n-1,t}|_0^{\beta+1} + L\int_{-\tau}^0 p_{n,t}(s)ds = w'(t),\ \text{say}. \end{aligned}$$

Clearly, $w'(t) \ge 0$. Since $p_n(t) \le w(t)$, and $w(t)$ is nondecreasing in t, we get

$$w(t) \le L\tau \int_0^t w(s)ds + L_2\int_0^t |p_{n-1,s}|_0^{\beta+1} ds$$

which implies, noting that $w(0) = 0$, by Gronwall's inequality

$$\begin{aligned} p_n(t) \le w(t) &\le L_2\int_0^t e^{L\tau(t-s)}|p_{n-1,s}|_0^{\beta+1} ds \\ &\le \frac{L_2 e^{L\tau T}}{L\tau} \max_J |p_{n-1,t}|_0^{\beta+1}. \end{aligned}$$

Consequently,

$$\max_J p_n(t) \le \frac{L_2 e^{L\tau T}}{L\tau} \max_J |p_{n-1,t}|_0^{\beta+1}.$$

Similarly,

$$\begin{aligned}
q_n'(t) &= v_n'(t) - x'(t) \\
&= f(t, v_{n-1,t}) + f_\phi(t, u_{n-1,t})(v_{n,t} - v_{n-1,t}) - f(t, x_t) \\
&= \int_0^1 f_\phi(t, sv_{n-1,t} + (1-s)x_t)(v_{n-1,t} - x_t)ds \\
&\quad + f_\phi(t, u_{n-1,t})(v_{n,t} - v_{n-1,t}) \\
&= \int_0^1 f_\phi(t, sv_{n-1,t} + (1-s)x_t)q_{n-1,t}ds \\
&\quad + f_\phi(t, u_{n-1,t})(q_{n,t} - q_{n-1,t}) \\
&= \int_0^1 [f_\phi(t, sv_{n-1,t} + (1-s)x_t) - f_\phi(t, x_t)]q_{n-1,t}ds \\
&\quad + [f_\phi(t, x_t) - f_\phi(t, u_{n-1,t})]q_{n-1,t} + f_\phi(t, u_{n-1,t})q_{n,t}
\end{aligned}$$

Employing (H_3) and (H_4), we get

$$\begin{aligned}
q_n'(t) &\le \int_0^1 L_2|sv_{n-1,t} + (1-s)x_t - x_t|_0^\beta |q_{n-1,t}|_0 ds \\
&\quad + L_2|x_t - u_{n-1,t}|_0^\beta |q_{n-1,t}|_0 + L\int_{-\tau}^0 q_{n,t}(s)ds \\
&\le L_2 \int_0^1 |s|^\beta |v_{n-1,t} - x_t|_0^\beta |q_{n-1,t}|_0 ds \\
&\quad + L_2|p_{n-1,t}|_0^\beta |q_{n-1,t}|_0 + L\int_{-\tau}^0 q_{n,t}(s)ds \\
&\le L_2|q_{n-1,t}|_0^{\beta+1} + L_2|p_{n-1,t}|_0^\beta |q_{n-1,t}|_0 + L\int_{-\tau}^0 q_{n,t}(s)ds \\
&= L_2\left[|q_{n-1,t}|_0^{\beta+1} + |p_{n-1,t}|_0^\beta |q_{n-1,t}|_0\right] + L\int_{-\tau}^0 q_{n,t}(s)ds.
\end{aligned}$$

Let $z'(t) = L_2\left[|q_{n-1,t}|_0^{\beta+1} + |p_{n-1,t}|_0^\beta |q_{n-1,t}|_0\right] + L\int_{-\tau}^0 q_{n,t}(s)ds$. Then $z'(t) \ge 0$. Since $q_n(t) \le z(t)$, and $z(t)$ is nondecreasing in t, we get

$$z(t) \le L\tau \int_0^t z(s)ds + L_2 \int_0^t \left[|q_{n-1,s}|_0^{\beta+1} + |p_{n-1,s}|_0^\beta |q_{n-1,s}|_0\right] ds$$

which implies, noting that $z(0) = 0$, by Gronwall's inequality

$$\begin{aligned} q_n(t) \le z(t) &\le L_2 \int_0^t e^{L_T(t-s)} \left[|q_{n-1,s}|_0^{\beta+1} + |p_{n-1,s}|_0^{\beta} |q_{n-1,s}|_0 \right] ds \\ &\le \frac{L_2 e^{L_T T}}{L_T} \max_J \left[|q_{n-1,t}|_0^{\beta+1} + |p_{n-1,t}|_0^{\beta} |q_{n-1,t}|_0 \right]. \end{aligned}$$

Consequently

$$\max_J q_n(t) \le \frac{L_2 e^{L_T T}}{L_T} \max_J \left[|q_{n-1,t}|_0^{\beta+1} + |p_{n-1,t}|_0^{\beta+1} |q_{n-1,t}|_0 \right].$$

The proof is complete.

Remark 4.3.1 (i) If $\beta = 1$, then the convergence of sequences $\{u_n(t)\}$, $\{v_n(t)\}$, $t \in J$, is quadratic.

(ii) The foregoing discussion heavily depends on the condition $(H_3)(ii)$ which is equivalent to the fact that $f(t, \phi)$is convex with respect to ϕ in the space of continuous functions.

4.4 IMPULSIVE DIFFERENTIAL EQUATIONS

Many evolution processes are characterized by the fact that at certain moments of time they experience a change of state abruptly. These processes are subject to short term perturbations whose derviation is negligible in comparison with the derivation of the process. Consequently, it is natural to assume that these perturbations act instantaneously, that is, in the form of impulses. Thus impulsive differential equations are a natural description of evolution phenomena of several real world problems. Moreover, the theory of impulsive differential equations is much richer than the corresponding theory of differential equations without impulsive effects. Recently, the theory of impulsive differential equations has attracted much attention. In this section, we shall extend the method of generalized quasilinearization to impulsive differential equations and exhibit the interplay of impulses in the development.

We start with the following definitions.

Definition 4.4.1 Let $0 \le t_0 < t_1 < t_2 \cdots < t_k < \cdots$ and $t_k \to \infty$ as $k \to \infty$. Then we say that $h \in PC[\mathbb{R}_+ \times \mathbb{R}, \mathbb{R}]$, if $h : (t_{k-1}, t_k] \times \mathbb{R} \to \mathbb{R}$ is continuous in $(t_{k-1}, t_k] \times \mathbb{R}$ and for every $x \in \mathbb{R}$

$$\lim_{(t,y)\to(t_k^+,x)} h(t, y) = h(t_k^+, x) \text{ exists, } k = 1, 2, \cdots$$

Definition 4.4.2 We say that $h \in PC^j[\mathbb{R}_+ \times \mathbb{R}, \mathbb{R}]$ if h is piecewise continuous and is differentiable j times in each interval $(t_k, t_{k+1}]$.

Consider the IVP

$$\begin{aligned} u' &= f(t,u),\ t \neq t_k,\ t \in J = [0,T] \\ u(t_k^+) &= I_k(u(t_k)) \\ u(0) &= u_0, \end{aligned} \tag{4.4.1}$$

where $f \in PC[\Omega, \mathbb{R}]$ and $I_k : \mathbb{R} \to R$ with $\Omega = J \times \mathbb{R}$, where $0 < t_1 < t_2 < \cdot\cdot < t_N < T$.

A function $v \in PC^1[J, \mathbb{R}]$ is said to be an upper solution of (4.4.1) if

$$\begin{aligned} v' &\geq f(t,v),\ t \neq t_k \\ v(t_k^+) &\geq I_k(v(t_k)) \\ v(0) &\geq v_0, \end{aligned}$$

and a lower solution of (4.4.1) if the reverse inequalities hold.

We need the following two theorems to prove our main result.

Theorem 4.4.1 *Let α, $\beta \in PC^1[\Omega, \mathbb{R}]$ be lower and upper solutions of (4.4.1) on J. Further, assume that*

$$f(t,u) - f(t,v) \leq L(u-v),\ L \geq 0$$

whenever $v \leq u$, and $I_k(u)$ is nondecreasing in u, for each k, $k = 1, 2, \cdots, n, \cdots$ then $\alpha(0) \leq \beta(0)$ implies $\alpha(t) \leq \beta(t)$ on J.

Proof Using the basic comparison results in the theory of ordinary differential equations [27] and Theorem 1.4.3 in [25], the proof follows easily.

The following result gives the existence of a solution in a sector.

Theorem 4.4.2 *Let $f \in PC[\Omega, \mathbb{R}]$, $I_k(u)$ be nondecreasing in u, for each k, $k = 1, 2, \cdots, n, \cdots$, and α, $\beta \in PC^1[J, \mathbb{R}]$ be lower and upper solutions of (4.4.1) such that $\alpha(t) \leq \beta(t)$ on J. Then there exists a solution $u(t)$ of (4.4.1) satisfying $\alpha(t) \leq u(t) \leq \beta(t)$ on J.*

Proof The proof follows from Theorem 1.1.4 in [21] and the comparison results in the theory of impulsive differential equations in [25].

The next theorem is needed to prove the quadratic convergence part of the main result. See [25] for the proof.

Theorem 4.4.3 *Assume that*
(A_0) *the sequence* $\{t_k\}$ *satisfies* $0 \leq t_0 < t_1 < t_2 < \cdots t_k$, *with* $\lim_{k\to\infty} t_k = \infty$;
(A_1) $m \in PC^1[\mathbb{R}_+, \mathbb{R}]$ *and* $m(t)$ *is left-continuous at* t_k, $k = 1, 2, \cdots$;
(A_3) *for* $k = 1, 2, \cdots$, $t \geq t_0$,

$$\begin{aligned} m'(t) &\leq p(t)m(t) + q(t), \ t \neq t_k \\ m(t_k^+) &\leq d_k m(t_k) + b_k \end{aligned}$$

where $q, p \in C[\mathbb{R}_+, \mathbb{R}]$, $d_k \geq 0$ *and* b_k *are constants.*
Then

$$\begin{aligned} m(t) \leq\ & m(t_0) \prod_{t_0 < t_k < t} d_k \exp\left(\int_{t_0}^{t} p(s)ds\right) \\ &+ \sum_{t_0 < t_k < t} \left(\prod_{t_k < t_j < t} d_j \exp\left(\int_{t_k}^{t} p(s)ds\right)\right) b_k \\ &+ \int_{t_0}^{t} \prod_{s < t_k < t} d_k \exp\left(\int_{s}^{t} p(\sigma)d\sigma\right) q(s)ds, \ t \geq t_0. \end{aligned}$$

We are now in a position to prove our main result. We construct two monotone sequences which acts as lower and upper bounds for the solution and they converge uniformly to unique solution $u(t)$ of (4.4.1) and the convergence is quadratic.

Theorem 4.4.4 *Assume that*

(*i*) $v_0, w_0 \in PC^1[J, \mathbb{R}]$ *are lower and upper solutions of (4.4.1) such that* $v_0(t) \leq w_0(t)$ *on* J;

(*ii*) $F(t, u) = f(t, u) + \varphi(t, u)$ *such that* $F_u(t, u)$, $F_{uu}(t, u)$ *exist and are continuous such that* $F_{uu}(t, u) \geq 0$ *for some convex function* $\varphi(t, u)$. *(i.e.* $\varphi(t, u)$ *is such that* φ_u, φ_{uu} *exist such that* $\varphi_{uu}(t, u) \geq 0$ *on* $\Omega = \{(t, u) : v_0(t) \leq u \leq w_0(t)\}$*)*

(*iii*) $\tilde{I}_k(u) = I_k(u) + \psi_k(u)$ *be such that* $\tilde{I}_{ku}(u)$, $\tilde{I}_{kuu}$ *exist and are continuous over* J *with*

$$I_{ku}(v_0(t_k)) \geq \psi_{ku}(w_0(t_k)) - \psi_{ku}(v_0(t_k))$$

and $\tilde{I}_{kuu} \geq 0$, *for some convex function* $\psi_k(u)$ *for* $k = 1, \cdots, N$.

Then there exists monotone sequences $\{v_n(t)\}$, $\{w_n(t)\}$ *which converges uniformly to the unique solution of* (4.4.1) *and the convergence is quadratic.*

Proof It is easy to observe that, from assumption (ii)

$$f(t,u) \geq f(t,v) + F_u(t,v)(u-v) - \varphi(t,u) + \varphi(t,v), \text{ for } u \geq v \tag{4.4.2}$$

and

$$f(t,v) \leq f(t,u) + F_u(t,v)(v-u) - \varphi(t,v) + \varphi(t,u). \tag{4.4.3}$$

Similarly from assumption (iii)

$$\begin{aligned} I_k(u(t_k)) \geq\ & I_k(v(t_k)) + \tilde{I}_{ku}(v(t_k))(u(t_k) - v(t_k)) \\ & -\psi_k(u(t_k)) + \psi_k(v(t_k)), \text{ for } u(t_k) \geq v(t_k) \end{aligned} \tag{4.4.4}$$

and also

$$I_k(v(t_k)) \leq I_k(u(t_k)) + \tilde{I}_{kv}(v(t_k))(v(t_k) - u(t_k)) + \psi_k(u(t_k)) - \psi_k(v(t_k)). \tag{4.4.5}$$

Further for any $v(t_k)$, $w(t_k)$ such that $v_0(t_k) \leq v(t_k) \leq w(t_k) \leq w_0(t_k)$, we have from (iii)

$$\tilde{I}_{ku}(v(t_k)) - \psi_{ku}(w(t_k)) \geq \tilde{I}_{ku}(v_0(t_k)) - \psi_{ku}(w_0(t_k)) \geq 0. \tag{4.4.6}$$

Now consider the following two linear impulsive differential equations

$$\begin{aligned} v_1' &= f(t,v_0) + (F_u(t,v_0) - \varphi_u(t,w_0))(v_1 - v_0), \ t \neq t_k \\ v_1(t_k^+) &= I_k(u_0(t_k)) + (\tilde{I}_{ku}(v_0(t_k)) - \psi_{ku}(w_0(t_k)))(v_1(t_k) - v_0(t_k)) \\ v_1(0) &= u_0 \end{aligned} \tag{4.4.7}$$

and

$$\begin{aligned} w_1' &= f(t,w_0) + (F_u(t,v_0) - \varphi_u(t,w_0))(w_1 - w_0), \ t \neq t_k \\ w_1(t_k^+) &= I_k(w_0(t_k)) + (I_{ku}(v_0(t_k)) - \psi_{ku}(w_0(t_k)))(w_1(t_k) - w_0(t_k)) \\ w_1(0) &= u_0. \end{aligned} \tag{4.4.8}$$

It is easy to see that $v_1(t)$ and $w_1(t)$ exist and are unique and PC^1 on J. We prove

$$v_0(t) \leq v_1(t) \leq w_1(t) \leq w_0(t) \text{ on } J. \tag{4.4.9}$$

For that purpose set $p(t) = v_0 - v_1$. Then

$$\begin{aligned} p' = v_0' - v_1' &\leq f(t,v_0) - [f(t,v_0) + (F_u(t,v_0) - \varphi_u(t,w_0))(v_1 - v_0)], \ t \neq t_k \\ &\leq (F_u(t,v_0) - \varphi_u(t,w_0))p(t). \end{aligned}$$

That is $p'(t) \le (F_u(t,v_0) - \varphi_u(t,w_0))p(t)$, $t \ne t_k$ and

$$\begin{aligned} p(t_k^+) &= v_0(t_k^+) - v_1(t_k^+) \\ &\le I_k(v_0(t_k)) - [I_k(v_0(t_k)) + (\tilde{I}_{ku}(v_0(t_k)) - \psi_{ku}(w_0(t_k)))(v_1(t_k) - v_0(t_k))] \\ &\le (\tilde{I}_{ku}(v_0(t_k)) - \psi_{ku}(w_0(t_k)))p(t_k^+) \end{aligned}$$

and $p(0) \le 0$. Now using assumption (iii) and Theorem 4.4.1 it follows that $p(t) \le 0$ on J, i.e. $v_0(t) \le v_1(t)$ on J.

Similarly if we set $q(t) = w_1 - w_0$, then

$$\begin{aligned} q'(t) &= w_1' - w_0' \\ &\le f(t,w_0) + (F_u(t,w_0) - \varphi_u(t,w_0))(w_1 - w_0) - f(t,w_0) \\ &= (F_u(t,v_0) - \varphi_u(t,w_0))q(t),\ t \ne t_k. \end{aligned}$$

$$\begin{aligned} q(t_k^+) &= w_1(t_k^+) - w_0(t_k^+) \\ &\le I_k(w_0(t_k)) + (\tilde{I}_{ku}(v_0(t_k)) - \psi_{ku}(w_0(t_k)))(w_1(t_k) - w_0(t_k)) - I_k(w_0(t_k)) \\ &\le (\tilde{I}_{ku}(v_0(t_k)) - \psi_{ku} w_0(t_k))q(t_k) \end{aligned}$$

and $q(0) \le 0$. As before using the special case of Theorem 4.4.1, it follows that $q(t) \le 0$ on J. That is $w_1(t) \le w_0(t)$ on J. Next we prove $v_1(t)$ is a lower solution of (4.4.1) and $w_1(t)$ is an upper solution of (4.4.1). For that consider

$$\begin{aligned} v_1'(t) &= f(t,v_0) + (F_u(t,v_0) - \varphi_u(t,w_0))(v_1 - v_0),\ t \ne t_k \\ &\le f(t,v_1) \end{aligned}$$

from (4.4.2) since $v_1 \ge v_0$. Also from (4.4.4) and the increasing nature of $\psi_k(u)$ in u it follows

$$\begin{aligned} v_1(t_k^+) &= I_k(v_0(t_k)) + (\tilde{I}_{ku}(v_0(t_k)) - \psi_{ku}(w_0(t_k)))(v_1(t_k) - v_0(t_k)) \\ &\le I_k(v_1(t_k)), \\ v(0) &= u_0. \end{aligned}$$

Similarly we can prove that $w_1(t)$ is an upper solution of (4.4.1). Since $f(t,u)$ is Lipschitzian and $I_k(u)$ is nondecreasing in u, it follows from Theorem 4.4.1 that $v_1(t) \le w_1(t)$ on J. This proves (4.4.9).

Now assume that for $l > 1$, that

$$v_0(t) \le v_1(t) \le \cdots \le v_l(t) \le w_l(t) \le \cdots \le w_1(t) \le w_0(t) \tag{4.4.10}$$

where $v_l(t)$ and $w_l(t)$ are the unique solution of the linear impulsive differential equations

$$\begin{aligned} v_l' &= f(t, v_{l-1}) + (F(t, v_{l-1}) - \varphi(t, w_{l-1}))(v_l - v_{l-1}),\ t \neq t_k \\ v_l(t_k^+) &= I_k(v_{l-1}(t_k)) \\ &\quad + (\tilde{I}_{ku}(v_{l-1}(t_k)) - \psi_{ku}(w_{l-1}(t_k)))(v_l(t_k) - v_{l-1}(t_k)) \\ v_l(0) &= u_0 \end{aligned} \tag{4.4.11}$$

and

$$\begin{aligned} w_l' &= f(t, w_{l-1}) + (F(t, v_{l-1}) - \varphi(t, w_{l-1}))(w_l - w_{l-1}),\ t \neq t_k \\ w_l(t_k^+) &= I_k(w_{l-1}(t_k)) \\ &\quad + (\tilde{I}_{ku}(w_{l-1}(t_k)) - \psi_{ku}(w_{l-1}(t_k)))(w_l(t_k) - w_{l-1}(t_k)) \\ w_l(0) &= u_0, \end{aligned} \tag{4.4.12}$$

respectively. Further let v_l and w_l are lower and upper solutions of (4.4.1). Now let v_{l+1}, w_{l+1} be the solutions of (4.4.11) and (4.4.12) with l replaced by $l+1$. It is easy to see that v_{l+1}, w_{l+1} exist and are unique on J. We claim $v_l \leq v_{l+1} \leq w_{l+1} \leq w_l$ and further v_{l+1} and w_{l+1} are lower and upper solutions of (4.4.1) respectively. Set $p(t) = v_l(t) - v_{l+1}(t)$. Then

$$\begin{aligned} p'(t) &= v_l'(t) - v_{l+1}'(t) \\ &\leq f(t, v_l) - [f(t, v_l) + (F_u(t, v_l) - \varphi_u(t, w_l))(v_{l+1} - v_l)],\ t \neq t_k \\ &= (F(t, v_l) - \varphi(t, w_l))p(t), \\ p(t_k^+) &= v_l(t_k^+) - v_{l+1}(t_k^+) \\ &\leq I_k(v_l(t_k)) - [I_k(v_l(t_k)) - (\tilde{I}_{ku}(v_l(t_k)) \\ &\quad - \psi_{ku}(w_l(t_k)))(v_{l+1}(t_k) - v_l(t_k))] \\ &\leq (I_{ku}(v_l(t_k) - \psi_{ku}(w_l(t_k)))p(t_k) \\ p(0) &= 0. \end{aligned}$$

Also if we denote $\mathcal{L}_k(u) = (\tilde{I}_{ku}(v_l(t_k)) - \psi_{ku}(w_l(t_k)))u$, then we have to show that $\mathcal{L}_k(u)$ is nondecreasing in u. It is enough to show $\tilde{I}_k(v_l(t_k)) - \psi_k(w_l(t_k)) \geq 0$. Since $v_l(t_k) \geq v_0(t_k)$ and $w_l(t_k) \leq w_0(t_k)$ and $\tilde{I}_{ku}(u)$ is nondecreasing in u and $\psi_{ku}(u)$ is nondecreasing in u, we get

$$\begin{aligned} \tilde{I}_k(v_l(t_k)) - \psi_k(w_l(t_k)) &\geq \tilde{I}_{ku}(v_0(t_k)) - \psi_{ku}(w_0(t_k)) \\ &= I_{ku}(v_0(t_k)) + \psi_{ku}(v_0(t_k)) - \psi_{ku}(w_0(t_k)) \\ &\geq 0 \end{aligned}$$

from (iii). This prove $\mathcal{L}_k(u)$ is nondecreasing in u. From comparison theorem 4.4.1 it follows that $p(t) \leq 0$ on J. This proves $v_l(t) \leq v_{l+1}(t)$ on J. Similarly we can prove that $w_{l+1}(t) \leq w_l(t)$. Further using (4.4.1), (4.4.2) one can easily prove that v_{l+1} and w_{l+1} are lower and upper solutions of (4.4.1). It follows using mathematical induction, (4.4.10) is true for all l. Now employing the standard arguments to each subinterval $(t_k, t_{k+1}] \subseteq J$, we can conclude that $\{v_n(t)\}$, and $\{w_n(t)\}$ converges uniformly and monotonically to the unique solution of (4.4.1) on J. We now claim the convergence of the sequences $\{v_n(t)\}$ and $\{w_n(t)\}$ to $u(t)$ the unique solution of (4.4.1) is quadratic. For that purpose we set $p_n(t) = u(t) - v_n(t) \geq 0$ and $q_n(t) = w_n(t) - u(t)$, so that $p_n(0) = 0 = q_n(0)$ for each $n > 1$.

Now, for $t \neq t_k$, using (ii) and The mean value theorem, it follows that

$$\begin{aligned}
p_n'(t) &= f(t,u) - [f(t,v_{n-1}) + (F_u(t,v_{n-1}) - \varphi_u(t,w_{n-1}))] \\
&\quad \times (v_n(t) - v_{n-1}(t)) \\
&= F(t,u) - [F(t,v_{n-1}) + F_u(t,v_{n-1})(v_n(t) - v_{n-1}(t))] \\
&\quad -\varphi(t,u) + \varphi(t,v_{n-1}) + \varphi_u(t,w_{n-1})(v_n(t) - v_{n-1}(t)) \\
&\leq F(t,v_{n-1}) + F_u(t,v_{n-1})(u - v_{n-1}) + F_{uu}(t,\xi)\frac{(u-v_{n-1})^2}{2!} \\
&\quad -F(t,v_{n-1}) - F_u(t,v_{n-1})(u - v_{n-1}) + F_u(t,v_{n-1})(u - v_n) \\
&\quad -\varphi_u(t,v_{n-1})(u - v_{n-1}) + \varphi_u(t,w_{n-1})(u - v_{n-1}) \\
&\quad -\varphi_u(t,w_{n-1})(u - v_n) \\
&= (F_u(t,v_{n-1}) - \varphi_u(t,w_{n-1}))p_n(t) + \frac{F_{uu}(t,\xi)}{2!}p_{n-1}^2 \\
&\quad +\varphi_{uu}(t,\eta)(u - v_{n-1})(w_{n-1} - v_{n-1}) \\
&\leq (F_u(t,v_{n-1}) - \varphi_u(t,w_{n-1}))p_n(t) + (\frac{F_{uu}(t,\xi)}{2!} + \frac{3}{2}\varphi_{uu}(t,\xi))p_{n-1}^2(t) \\
&\quad +\frac{1}{2}\varphi_{uu}(t,\xi)q_{n-1}^2(t) \\
&\leq A_1 p_n(t) + B_1 p_{n-1}^2 + C_1 q_{n-1}^2
\end{aligned}$$

where $|F_u(t,\cdot) - \varphi_u(t,\cdot)| < A_1$, $\left|\frac{F_{uu}(t,\xi)}{2!} + \frac{3}{2}\varphi_{uu}(t,\xi)\right| < B_1$ and $\frac{1}{2}\varphi_{uu}(t,\xi) \leq C_1$.

At $t = t_k$, we have

$$\begin{aligned}
p_n(t_k^+) &= u(t_k^+) - v_n(t_k^+) \\
&= I_k(u(t_k) - (I_k(v_{n-1}(t_k)) \\
&\quad +(\tilde{I}_{ku}(v_{n-1}(t_k)) - \psi_{ku}(w_{n-1}(t_k)))(v_n(t_k) - v_{n-1}(t_k))
\end{aligned}$$

$$
\begin{aligned}
&= \tilde{I}_k(u(t_k)) - \tilde{I}_k(v_{n-1}(t_k)) - \tilde{I}_{ku}(v_{n-1}(t_k))(u - v_{n-1}(t_k)) \\
&\quad -\psi_{ku}(w_{n-1}(t_k))(u(t_k) - v_{n-1}(t_k)) + (\tilde{I}_{ku}(v_{n-1}(t_k)) \\
&\quad -\psi_{ku}(w_{n-1}(t_k)))(u(t_k) - v_n(t_k)) - \psi_k(u(t_k)) + \psi_k(v_{n-1}(t_k)) \\
&\leq I_{kuu}(\xi)\frac{(u(t_k) - v_{n-1}(t_k))^2}{2!} + \psi_{ku}(w_{n-1}(t_k))(u(t_k) - v_{n-1}(t_k)) \\
&\quad -\psi_{ku}(v_{n-1}(t_k))p_{n-1}(t_k) + (\tilde{I}_{ku}(v_{n-1}(t_k)) \\
&\quad -\psi_{ku}(w_{n-1}(t_k)))p_n(t_k) \\
&\leq I_{kuu}(\xi)p_{n-1}^2(t_k) + \psi_{kuu}(\eta)p_{n-1}(t_k)(p_{n-1}(t_k) + q_{n-1}(t_k)) \\
&\quad +(\tilde{I}_{ku}(v_{n-1}(t_k)) - \psi_{ku}(w_{n-1}(t_k))p_n(t_k) \\
&\leq (I_{kuu}(\xi) + \frac{3}{2}\psi_{kuu}(\eta))p_{n-1}^2(t_k) + \frac{\psi_{kuu}(\eta)}{2!}q_{n-1}^2(t_k) \\
&\quad +(\tilde{I}_{ku}(v_{n-1}(t_k)) - \psi_{ku}(w_{n-1}(t_k)))p_n(t_k) \\
&\leq L_{1k}p_{n-1}^2(t_k) + M_{1k}q_{n-1}^2(t_k) + N_{1k}p_n(t_k),
\end{aligned}
$$

using the nature of $\tilde{I}_k(u)$ and $\psi_k(u)$ and the mean value theorem and by letting $\left|I_{kuu}() + \frac{3}{2}\psi_{kuu}()\right| < L_{1k}$, $\left|\frac{\psi_{kuu}}{2!}\right| < M_{1k}$ and $\left|\tilde{I}_{ku}(v_{n-1}(t_k)) - \psi_{ku}(w_{n-1}(t_k))\right| < N_{1k}$. Now using Theorem 4.4.3

$$
\begin{aligned}
p_n'(t) &\leq A_1 p_n(t) + B_1 p_{n-1}^2(t) + C_1 q_{n-1}^2(t) \\
p_n(t_k^+) &\leq N_{1k}p_n(t_k) + L_{1k}p_{n-1}^2(t_k) + M_{1k}q_{n-1}(t_k) \\
p_n(0) &= 0.
\end{aligned}
$$

We get

$$
\begin{aligned}
0 &< p_n(t) \leq \sum_{0<t_k<t}\left(\prod_{t_k<t_j<t} N_{1j}e^{A_1(t-t_k)}\right)(L_{1k}p_{n-1}^2(t_k) + M_{1k}q_{n-1}^2(t_k)) \\
&\quad + \int_0^T \prod_k N_{1k}e^{A_1(T-s)}(B_1 p_{n-1}^2(s) + C_1 q_{n-1}^2(s))ds \\
&\leq e^{A_1 T}\Omega_k \sum_{0<t_k<T} e^{-A_1 t_k}(L_{1k}p_{n-1}^2(t_k) + M_{1k}q_{n-1}^2(t_k)) \\
&\quad + \int_0^T \Omega_k e^{A_1 T}(B_1 p_{n-1}^2(s) + C_1 q_{n-1}^2(s))e^{-A_1 s}ds
\end{aligned}
$$

where $\Omega_k = N_{11}N_{12}\cdots N_{1k}$.

This in turn yields

$$
\begin{aligned}
|p_n|_0 &\leq e^{A_1 T}\Omega_k((\textstyle\sum^k L_{1i})\left|p_{n-1}^2\right|_0 + (\textstyle\sum^k M_{1i})\left|q_{n-1}^2\right|_0) \\
&\quad +\Omega_k\frac{e^{A_1 T}}{A_1}(B_1\left|p_{n-1}^2\right|_0 + C_1\left|q_{n-1}^2\right|_0)
\end{aligned}
$$

where $|\cdot|_0 = \sup_J \|\cdot\|$.

Similarly, using the mean value theorem and the increasing nature of the function $F_u(t,u)$ and $\varphi_u(t,u)$, for $t \neq t_k$, we obtain

$$\begin{aligned}
q_n'(t) &= w_n'(t) - u'(t) \\
&= f(t, w_{n-1}) + (F_u(t, v_{n-1}) - \varphi_u(t, w_{n-1}))(w_n - w_{n-1}) - f(t, u) \\
&= F(t, w_{n-1}) + F(t, v_{n-1})q_n - F_u(t, v_{n-1})q_{n-1} - \varphi_u(t, w_{n-1})q_n \\
&\quad + \varphi_u(t, w_{n-1})q_{n-1} + \varphi(t, u) - \varphi(t, w_{n-1}) - F(t, u) \\
&\leq F_u(t, w_{n-1})q_{n-1} - F_u(t, v_{n-1})q_{n-1} + (F_u(t, v_{n-1}) - \varphi_u(t, w_{n-1}))q_n \\
&\quad + \varphi_u(t, w_{n-1})q_{n-1} - \varphi_u(t, u)q_{n-1} \\
&\leq (F_u(t, v_{n-1}) - \varphi_u(t, w_{n-1}))q_n + F_{uu}(t, \sigma)(q_{n-1} + p_{n-1})q_{n-1} \\
&\quad + \varphi_{uu}(t, \delta)q_{n-1}^2.
\end{aligned}$$

Further if we use the estimate

$$|F_u(t, v_{n-1}) - \varphi_u(t, w_{n-1})| < A_2, \left|\frac{3}{2}F_{uu} + \frac{\varphi_{uu}(t,\delta)}{2}\right| < C_2, \text{ and } |\varphi_{uu}(t,\delta)| < B_2,$$

we get

$$q_n'(t) \leq A_2 q_n(t) + B_2 p_{n-1}^2(t) + C_2 q_{n-1}^2(t).$$

For $t = t_k$, we get

$$\begin{aligned}
q_n(t_k^+) &= w_n(t_k^+) - I_k(u) \\
&= I_k(w_{n-1}(t_k)) + (\tilde{I}_{ku}(v_{n-1}(t_k)) - \psi_{ku}(w_n(t_k)))(w_{n-1}(t_k) - w_{n-1}(t_k)) \\
&\quad - I_k(u(t_k)) \\
&= \tilde{I}_k(w_{n-1}(t_k)) - \tilde{I}_k(u(t_k)) + (\tilde{I}_{ku}v_{n-1}(t_k) - \psi_{ku}(w_{n-1}(t_k)))q_n(t_k) \\
&\quad - (\tilde{I}_{ku}(v_{n-1}(t_k)) - \psi_{ku}(w_n(t_k)))q_{n-1}(t_k) + \psi_k(u(t_k)) - \psi_k(w_{n-1}(t_k)) \\
&\leq \tilde{I}_{ku}(w_{n-1}(t_k))q_{n-1}(t_k) - \tilde{I}_{ku}(v_{n-1}(t_k))q_{n-1}(t_k) + \psi_{ku}(w_n(t_k))q_{n-1}(t_k) \\
&\quad - \psi_{ku}(u(t_k))q_{n-1}(t_k) + \tilde{I}_{ku}(v_{n-1}(t_k) - \psi_{ku}(w_{n-1}(t_k)))q_n(t_k) \\
&\leq I_{kuu}(\gamma_1)(w_{n-1}(t_k) - v_{n-1}(t_k))q_{n-1}(t_k) + \psi_{kuu}(\gamma_2)q_{n-1}^2(t_k) \\
&\quad + \tilde{I}_{ku}(v_{n-1}(t_k) - \psi_{ku}(w_{n-1}(t_k)))q_n(t_k) \\
&\leq N_{2k}q_n(t_k) + L_{1k}p_{n-1}^2(t_k) + M_{1k}q_{n-1}^2(t_k)
\end{aligned}$$

where

$$\begin{aligned}
\left|\tilde{I}_{ku}(v_{n-1}(t_k)) - \psi_{ku}(w_{n-1}(t_k))\right| &\leq N_{2k} \\
\left|\frac{3}{2}I_{kuu}(\gamma_1) + \psi_{kuu}(\gamma_2)\right| &< M_{1k} \\
|\psi_{kuu}| &\leq L_{1k}.
\end{aligned}$$

Now using Theorem 4.4.3 on

$$\begin{aligned} q_n'(t) &\leq A_2 q_n(t) + B_2 p_{n-1}^2(t) + C_2 q_{n-1}^2(t) \\ q_n'(t_k^+) &\leq N_{2k} q_n(t_k) + L_{1k} p_{n-1}^2(t_k) + N_{1k} q_{n-1}^2(t_k) \\ q_n(0) &= 0, \end{aligned}$$

as before, we get

$$\begin{aligned} 0 &< q_n(t) \leq \left(\sum_{0<t_k<t} \prod_{t_k<t_j<t} N_{2j} e^{A_2(t-t_k)} \right) (L_{2k} p_{n-1}^2(t_k) + M_{2k} q_{n-1}^2(t_k)) \\ &+ \int_0^T \prod N_{2k} e^{A_2(T-s)} (B_2 p_{n-1}^2(s) + C_2 q_{n-1}^2(s)) ds \end{aligned}$$

where $\bar{\Omega}_k = N_{21} \times N_{22} \times .....N_{2k}$. This in turn yields

$$\begin{aligned} |q_n|_0 &\leq e^{A_2 T} \bar{\Omega}_k ((\sum^k L_{2i}) \left| p_{n-1}^2 \right|_0 + (\sum^k M_{2i}) \left| q_{n-1}^2 \right|_0) \\ &+ \bar{\Omega}_k \frac{e^{A_2 T}}{A_2} (B_2 \left| p_{n-1}^2 \right|_0 + C_2 \left| q_{n-1}^2 \right|_0) \end{aligned}$$

where $|\cdot|_0 = \sup_J \|\cdot\|$ .This proves the quadratic convergence.

4.5 STOCHASTIC DIFFERENTIAL EQUATIONS

The mathematical modeling of several real world problems leads to differential equations that involve some inherent randomness due to ignorance or uncertainties. Consequently, the study of nonlinear random differential equations is a very important area in applied mathematics.

In this section, we shall concentrate in developing the method of generalized quasilinearization for stochastic differential equations. For necessary preliminary concepts we refer to [20].

Throughout this section, (Ω, A, P) denotes a probability measurable space. Let $x_0 : \Omega \to R$ be a given measurable function.

We consider

$$\begin{aligned} x'(t,\omega) &= F(t, x(t,\omega), \omega) \text{ a.e. on } J = [0,T] \qquad (4.5.1) \\ & x(0,\omega) = x_0(\omega) \end{aligned}$$

where $F : J \times R \times \Omega \to R$.

Suppose

(1) F admits a decomposition $F = f + g$, where $f(t,u,\cdot)$, $g(t,u,\cdot)$ are measurable for all (t,u), $f(\cdot,u,\cdot)$, $g(\cdot,u,\cdot)$ are product measurable for every u and $f(\cdot,u,\cdot)$, $g(\cdot,u,\cdot)$ are continuous in u for each (t,ω).

(2) $|F(t,u,\omega)| \leq K(t,\omega)$ on $J \times R \times \Omega$, where $K : J \times \Omega \to R_+$ is measurable in t and $\int_0^T K(s,\omega)ds < \infty$ on Ω.

A stochastic process $x : J \times \Omega \to R$ is called a sample solution of (4.5.1) if $x(0,\omega) = x_0(\omega)$ and is absolutely continuous (*a.c.*) on J and satisfies

$$x'(t,\omega) = F(t,x(t),\omega) \ a.e. \ \text{on } J.$$

A stochastic process $\alpha : J \times \Omega \to R$ is said to be a sample lower solution of (4.5.1) if for P-almost all $\omega \in \Omega$, $\alpha(\cdot,\omega)$ is *a.c.* and $\alpha'(t,\omega) \leq F(t,\alpha(t,\omega),\omega)$ *a.e.* on J. The definition of sample upper solution is obtained by reversing the inequality above.

The following known results [20, 42] are required in our discussion.

Theorem 4.5.1 *Let u, v be sample lower and upper solutions of* (4.5.1) *respectively. Suppose further that F satisfies* (2) *and that*

(3) $F(t,u_1,\omega) - F(t,u_2,\omega) \leq L(t,\omega)(u_1 - u_2)$ *a.e. on J,*

whenever $u_1 \geq u_2$, where $L(t,\omega) \geq 0$ is measurable for every t and $\int_0^T L(t,\omega)dt < \infty$ on Ω. Then $u(0,\omega) \leq v(0,\omega)$ on Ω implies that $u(t,\omega) \leq v(t,\omega)$ on $J \times \Omega$.

Proof Suppose that the conclusion $u \leq v$ on $J \times \Omega$ is false. Then there exists an $\omega \in \Omega$ and a $t_1 \geq 0$ such that

$$u(t_1,\omega) = v(t_1,\omega) \text{ and } u(t,\omega) > v(t,\omega) \text{ for } t_1 < t < t_1 + h,$$

where $h > 0$ is sufficiently small. This implies, using (3) and the fact α, β are sample lower and upper solutions,

$$\begin{aligned} u'(t,\omega) - v'(t,\omega) &\leq F(t,u(t,\omega),\omega) - F(t,v(t,\omega),\omega) \\ &\leq L(t,\omega)[u(t,\omega) - v(t,\omega)], \ t_1 < t < t_1 + h. \end{aligned}$$

Hence it follows that

$$u(t,\omega) - v(t,\omega) \leq [u(t_1,\omega) - v(t_1,\omega)] \exp\left[\int_{t_1}^t L(s,\omega)ds\right] = 0$$

for $t_1 < t < t_1 + h$. Consequently we have a contradiction and the proof is complete.

When we know the existence of sample upper and lower solutions u, v such that $u \leq v$ on $J \times \Omega$, we can prove the existence of a sample solution in the sector $[u,v] = [x : J\times\Omega \to R$ such that $x(t,\cdot)$ is measurable, $x(\cdot,\omega)$ is absolutely continuous and $u(t,\omega) \leq x(t,\omega) \leq v(t,\omega)$ on $J \times \Omega]$.

Theorem 4.5.2 *Let u, v be sample lower and upper solutions of* (4.5.1) *such that $u \leq v$ on $J\times\Omega$. If F satisfies* (2), *then there exists a sample solution $x(t,\omega)$ of* (4.5.1) *satisfying $u \leq x \leq v$ on $J \times \Omega$ provided $u(0,\omega) \leq x_0(\omega) \leq v(0,\omega)$ on Ω.*

Proof Define $\tilde{F} : J \times R \times \Omega \to R$ by

$$\tilde{F}(t,u,\omega) = F(t,p(t,u,\omega),\omega) + r(t,u,\omega)$$

where

$$p(t,x,\omega) = \max[u(t,\omega), \min(x, v(t,\omega))],$$

$$r(t,u,\omega) = \begin{cases} \frac{v(t,\omega)-x}{1+x^2} & \text{if} \quad x > v(t,\omega), \\ 0 & \text{if} \quad u \leq x \leq v, \\ \frac{u(t,\omega)-x}{1+x^2} & \text{if} \quad x < u(t,\omega). \end{cases}$$

It is easy to see that $\tilde{F}$ satisfies condition (2) and therefore the IVP

$$x'(t,\omega) = F(t,x(t,\omega),\omega), \; x(0,\omega) = x_0(\omega),$$

has a sample solution $x(t,\omega)$ by Theorem 1 in [12]. If we can show that this $x(t,\omega)$ satisfies the relation $u \leq x \leq v$ on $J \times \Omega$, then we are done in view of the definition of $\tilde{F}$.

We shall only prove $u \leq x$ on $J \times \Omega$, since the inequality $x \leq v$ on $J \times \Omega$ follows similarly. If $u \leq x$ on $J \times \Omega$ is false, then there exists an $\omega \in \Omega$ and a $t_1 \in (0,T)$ such that

$$u(t_1,\omega) = x(t_1,\omega) \text{ and } u(t,\omega) > x(t,\omega) \text{ for } t_1 < t < t_1 + h,$$

$h > 0$ being sufficiently small. Then $p(t,u,\omega) = u(t,\omega)$ and

$$r(t,u,\omega) = \frac{u(t,\omega) - x(t,\omega)}{1 + x^2(t,\omega)} > 0 \text{ for } t_1 < t < t_1 + h.$$

Hence it follows that $u'(t,\omega) - x'(t,\omega) \leq 0$ *a.e.* on $t_1 < t < t_1 + h$ and consequently we arrive at a contradiction $u(t,\omega) - x(t,\omega) \leq 0$ for $t_1 < t < t_1 + h$. The proof is therefore complete.

We are now in the position to prove our main result.

Theorem 4.5.3 *Assume that*

(*i*) F *satisfies* (1);

(*ii*) α_0, β_0 *are lower and upper sample solutions of* (4.5.1) *such that* $\alpha_0 \leq \beta_0$ *on* $J \times \Omega$;

(*iii*) f *satisfies* (3) *and* $g_{xx}(t,x,\omega) + \phi_{xx}(t,x,\omega) \geq 0$ *where* $\phi : J \times R \times \Omega \to R$, $\phi(t,x,\omega)$ *is measurable in* ω, *is product measurable in* (t,x), *continuous in* x *and* $g_x(t,x,\omega)$, $\phi_x(t,x,\omega)$, $g_{xx}(t,x,\omega)$, $\phi_{xx}(t,x,\omega)$ *exist, are measurable in* ω, *continuous in* x, *product measurable in* (t,ω) *and* $\phi_{xx}(t,x,\omega) > 0$;

(*iv*) f, g, ϕ, g_x, ϕ_x, g_{xx}, ϕ_{xx} *satisfy condition* (2) *with different bounds.*

Then there exist monotone sequences $\{\alpha_n(t,\omega)\}$, $\{\beta_n(t,\omega)\}$ *of sample solutions which converge for p-almost all* $\omega \in \Omega$, *to the unique sample solution* $x(t,\omega)$ *of* (4.5.1) *and the convergence is quadratic.*

Proof In view of assumption $g_{xx}(t,x,\omega) + \phi_{xx}(t,x,\omega) \geq 0$ we obtain

$$\begin{aligned} g(t,x,\omega) \geq\; & g(t,y,\omega) + [g_x(t,y,\omega) + \phi_x(t,y,\omega)](x-y) \\ & -[\phi(t,x,\omega) - \phi(t,y,\omega)] \end{aligned} \tag{4.5.2}$$

for $x \geq y$.

Now, consider the IVPS's

$$\begin{aligned} \text{(a) } u' &= f(t,u,\omega) + g(t,\alpha_0,\omega) + [g_x(t,\alpha_0,\omega) + \phi_x(t,\alpha_0,\omega)](u-\alpha_0) \\ &\quad -[\phi(t,u,\omega) - \phi(t,\alpha_0,\omega)], \\ \text{(b) } v' &= f(t,v,\omega) + g(t,\beta_0,\omega) + [g_x(t,\alpha_0,\omega) + \phi_x(t,\alpha_0,\omega)](v-\beta_0) \\ &\quad -[\phi(t,v,\omega) - \phi(t,\beta_0,\omega)], \end{aligned} \tag{4.5.3}$$

which we shall denote for convenience in the following form,

$$\begin{cases} u' = g^*(t,u,\omega;\alpha_0) \equiv H(t,u,\omega) + M(t,\alpha_0,\omega) + M_x(t,\alpha_0,\omega)(u-\alpha_0), \\ v' = G(t,v,\omega;\alpha_0,\beta_0) \equiv H(t,v,\omega) + M(t,\beta_0,\omega) + M_x(t,\alpha_0,\omega)(v-\beta_0), \end{cases}$$

a.e. on J where $H(t,x,\omega) = f(t,x,\omega) - \phi(t,x,\omega)$, $M(t,x,\omega) = g(t,x,\omega) + \phi(t,x,\omega)$, $u(0,\omega) = x_0(\omega) = v(0,\omega)$ and $\alpha_0(0,\omega) \leq x_0(\omega) \leq \beta_0(0,\omega)$.

It is clear that H, g^* and G satisfy, for any $\alpha_0(t,\omega) \leq u_2 \leq u_1 \leq \beta_0(0,\omega)$.

$$\begin{cases} H(t,u_1,\omega) - H(t,u_2,\omega) \leq L_1(t,\omega)(u_1-u_2) \\ g^*(t,u_1,\omega;\alpha_0) - g^*(t,u_2,\omega;\alpha_0) \leq L_2(t,\omega)(u_1-u_2) \\ G(t,u_1,\omega;\alpha_0,\beta_0) - G(t,u_2,\omega;\alpha_0,\beta_0) \leq L_3(t,\omega)(u_1-u_2) \end{cases} \tag{4.5.4}$$

a.e. on J, where $L_1(t,\omega)$, $L_2(t,\omega)$, $L_3(t,\omega) > 0$, are measurable in ω, product measurable in (t,ω) and $\int_J L_i(s,\omega)ds < \infty$ for $i = 1,2,3$ and every $\omega \in \Omega$. The assumption (ii) and inequality (4.5.2) imply

$$\alpha_0' \leq F(t,\alpha_0,\omega) = f(t,\alpha_0,\omega) + g(t,\alpha_0,\omega) \equiv g^*(t,\alpha_0,\omega;\alpha_0) \tag{4.5.5}$$

$$\begin{aligned}
\beta_0' &\geq F(t,\beta_0,\omega) = f(t,\beta_0,\omega) + g(t,\beta_0,\omega) \\
&\geq f(t,\beta_0,\omega) + g(t,\alpha_0,\omega) + [g_x(t,\alpha_0,\omega) + \phi_x(t,\alpha_0,\omega)](\beta_0 - \alpha_0) \\
&\quad -[\phi(t,\beta_0,\omega) - \phi(t,\alpha_0,\omega)] \\
&= H(t,\beta_0,\omega) + M(t,\alpha_0,\omega) + M_x(t,\alpha_0,\omega)(\beta_0 - \alpha_0) \\
&\equiv g^*(t,\beta_0,\omega;\alpha_0).
\end{aligned} \tag{4.5.6}$$

Hence by Theorem 4.5.2 it follows that there exists a unique sample solution $\alpha_1(t,\omega)$ of (4.5.3)(a) such that

$$\alpha_0(t,\omega) \leq \alpha_1(t,\omega) \leq \beta_0(t,\omega) \text{ on } J \times \Omega.$$

Similarly, using (4.5.2) we get

$$\begin{aligned}
\alpha_0' &\leq F(t,\alpha_0,\omega) \\
&= f(t,\alpha_0,\omega) + g(t,\alpha_0,\omega) \\
&\leq f(t,\alpha_0,\omega) + g(t,\beta_0,\omega) - [g_x(t,\alpha_0,\omega) - \phi_x(t,\alpha_0,\omega)](\beta_0 - \alpha_0) \\
&\quad +[\phi(t,\beta_0,\omega) - \phi(t,\alpha_0,\omega)] \\
&= H(t,\alpha_0,\omega) + M(t,\beta_0,\omega) + M_x(t,\alpha_0,\omega)(\alpha_0 - \beta_0) \\
&\equiv G(t,\alpha_0,\omega;\alpha_0,\beta_0)
\end{aligned} \tag{4.5.7}$$

and

$$\beta_0' \geq F(t,\beta_0,\omega) = f(t,\beta_0,\omega) + g(t,\beta_0,\omega) \equiv G(t,\beta_0,\omega;\alpha_0,\beta_0). \tag{4.5.8}$$

Hence by Theorem 4.5.2, there exists a unique sample solution $\beta_1(t,\omega)$ of (4.5.3)(b) such that

$$\alpha_0(t,\omega) \leq \beta_1(t,\omega) \leq \beta_0(t,\omega) \text{ on } J \times \Omega.$$

Now, since $\alpha_0' = g^*(t,\alpha_1,\omega;\alpha_0)$, using (4.5.2) we get

$$\begin{aligned}
\alpha_1' &= H(t,\alpha_1,\omega) + M(t,\alpha_0,\omega) + M_x(t,\alpha_0,\omega)(\alpha_1 - \alpha_1) \\
&= f(t,\alpha_1,\omega) + g(t,\alpha_0,\omega) + [g_x(t,\alpha_0,\omega) + \phi_x(t,\alpha_0,\omega)](\alpha_1 - \alpha_0) \\
&\quad -[\phi(t,\alpha_1,\omega) - \phi(t,\alpha_0,\omega)] \\
&\leq f(t,\alpha_1,\omega) + g(t,\alpha_1,\omega) = F(t,\alpha_1,\omega),
\end{aligned}$$

and similarly, since $\beta_1' = G(t, \beta_1, \omega; \alpha_0, \beta_0)$, using (4.5.2) again,

$$\begin{aligned}
\beta_1' &= H(t,\beta_1,\omega) + M(t,\beta_0,\omega) + M_x(t,\alpha_0,\omega)(\beta_1 - \beta_0) \\
&= f(t,\beta_1,\omega) + g(t,\beta_0,\omega) - [g_x(t,\alpha_0,\omega) + \phi_x(t,\alpha_0,\omega)](\beta_0 - \beta_1) \\
&\quad +[\phi(t,\beta_0,\omega) - \phi(t,\beta_1,\omega)] \\
&\geq f(t,\beta_1,\omega) + g(t,\beta_1,\omega) = F(t,\beta_1,\omega).
\end{aligned}$$

Then by Theorem 4.5.1 and the fact F is one-sided Lipschitzian, it follows that $\alpha_1(t,\omega) \leq \beta_1(t,\omega)$ on $J \times \Omega$ and consequently, we have

$$\alpha_0(t,\omega) \leq \alpha_1(t,\omega) \leq \beta_1(t,\omega) \leq \beta_0(t,\omega) \text{ on } J \times \Omega. \tag{4.5.9}$$

Next, consider the IVP's

$$\begin{cases} \text{(a) } u' = g^*(t,u,\omega;\alpha_1), \ u(0,\omega) = x_0(\omega) \\ \text{(b) } v' = G(t,v,\omega;\alpha_1,\beta_1), \ v(0,\omega) = x_0(\omega). \end{cases} \tag{4.5.10}$$

Observe that

$$\begin{aligned}
\alpha_1 &\leq F(t,\alpha_1,\omega) = f(t,\alpha_1,\omega) + g(t,\alpha_1,\omega) \\
&\equiv g^*(t,\alpha_1,\omega;\alpha_1),
\end{aligned} \tag{4.5.11}$$

and using (4.5.2), it follows that

$$\begin{aligned}
\beta_1' &\geq F(t,\beta_1,\omega) = f(t,\beta_1,\omega) + g(t,\beta_1,\omega) \\
&\geq f(t,\beta_1,\omega) + g(t,\alpha_1,\omega) + [g_x(t,\alpha_1,\omega) + \phi_x(t,\alpha_1,\omega)](\beta_1 - \alpha_1) \\
&\quad -[\phi(t,\beta_1,\omega) - \phi(t,\alpha_1,\omega)] \\
&= H(t,\beta_1,\omega) + M(t,\alpha_1,\omega) + M_x(t,\alpha_1,\omega)(\beta_1 - \alpha_1) \\
&\equiv g^*(t,\beta_1,\omega;\alpha_1).
\end{aligned} \tag{4.5.12}$$

Thus, by Theorem 4.5.2, there exists a unique sample solution of (4.5.10)(a) such that $\alpha_1(t,\omega) \leq \alpha_2(t,\omega) \leq \beta_1(t,\omega)$ on $J \times \Omega$.

Similarly, using (4.5.2) again, we have

$$\begin{aligned}
\alpha_1' &\leq F(t,\alpha_1,\omega) = f(t,\alpha_1,\omega) + g(t,\alpha_1,\omega) \\
&\leq f(t,\alpha_1,\omega) + g(t,\beta_1,\omega) - [g_x(t,\alpha_1,\omega) + \phi_x(t,\alpha_1,\omega)](\beta_1 - \alpha_1) \\
&\quad +[\phi(t,\beta_1,\omega) - \phi(t,\alpha_1,\omega)] \\
&= H(t,\alpha_1,\omega) + M(t,\beta_1,\omega) + M_x(t,\alpha_1,\omega)(\alpha_1 - \beta_1) \\
&\equiv G(t,\alpha_1,\omega;\alpha_1,\beta_1),
\end{aligned} \tag{4.5.13}$$

and

$$\begin{aligned}\beta_1' &\geq F(t,\beta_1,\omega)\\ &= f(t,\beta_1,\omega)+g(t,\beta_1,\omega)\equiv G(t,\beta_1,\omega;\alpha_1,\beta_1). \qquad (4.5.14)\end{aligned}$$

Hence, in view of Theorem 4.5.2, there exists $\beta_2(t,\omega)$ which is the unique sample solution of (4.5.10)(b) such that

$$\alpha_1(t,\omega)\leq\beta_2(t,\omega)\leq\beta_1(t,\omega) \text{ on } J\times\Omega.$$

Furthermore, using (4.5.2), we obtain

$$\begin{aligned}\alpha_2' &= g^*(t,\alpha_2,\omega;\alpha_1)\\ &\leq F(t,\alpha_2,\omega)\end{aligned}$$

and, using (4.5.2)

$$\begin{aligned}\beta_2' &= G(t,\beta_2,\omega;\alpha_1,\beta_1)\\ &= H(t,\beta_2,\omega)+M(t,\beta_1,\omega)+M_x(t,\alpha_1,\omega)(\beta_2-\beta_1)\\ &= f(t,\beta_2,\omega)+g(t,\beta_1,\omega)-[g_x(t,\alpha_1,\omega)+\phi_x(t,\alpha_1,\omega)](\beta_1-\beta_2)\\ &\quad -[\phi(t,\beta_2,\omega)-\phi(t,\beta_1,\omega)]\\ &\geq f(t,\beta_2,\omega)+g(t,\beta_2,\omega)\\ &= F(t,\beta_2,\omega).\end{aligned}$$

Therefore, by Theorem 4.5.1 and the fact that F is one-sided Lipschitzian, we have $\alpha_2(t,\omega)\leq\beta_2(t,\omega)$ on $J\times\Omega$ and consequently, $\alpha_0(t,\omega)\leq\alpha_1(t,\omega)\leq\alpha_2(t,\omega)\leq\beta_2(t,\omega)\leq\beta_1(t,\omega)$ on $J\times\Omega$. This process can be continued successively to obtain

$$\alpha_0(t,\omega)\leq\alpha_1(t,\omega)\leq\cdots\leq\alpha_n(t,\omega)\leq\beta_n(t,\omega)\leq\cdots\leq\beta_1(t,\omega)\leq\beta_0(t,\omega)$$

on $J\times\Omega$, where $\{\alpha_n(t,\omega)\}$, $\{\beta_n(t,\omega)\}$ are the unique solutions of the following initial value problems,

$$\begin{cases}\text{(a) } \alpha_{n+1}' = g^*(t,\alpha_{n+1},\omega;\alpha_n),\ \alpha_{n+1}(0,\omega)=x_0(\omega)\\ \text{(b) } \beta_{n+1}' = G(t,\beta_{n+1},\omega;\alpha_n,\beta_n),\ \beta_{n+1}(0,\omega)=x_0(\omega).\end{cases} \qquad (4.5.15)$$

Let us note that for each fixed $t\in J$,

$$\begin{aligned}a(t,\omega) &= \sup_{n\geq0}\alpha_n(t,\omega)=\lim_{n\to\infty}\alpha_n(t,\omega),\\ b(t,\omega) &= \inf_{n\geq0}\beta_n(t,\omega)=\lim_{n\to\infty}\beta_n(t,\omega)\end{aligned}$$

exists w.p.l. and $a(t,\omega)$, $b(t,\omega)$ are measurable functions in ω for each $t \in J$. To show that there is unique sample solution $a = b = x$ on $J \times \Omega$ of (4.5.1), we see from (4.5.15)

$$\begin{aligned} \alpha_{n+1}(t,\omega) &= x_0(\omega) + \int_0^t g^*(s, \alpha_{n+1}(s,\omega), \omega; \alpha_n(s,\omega))ds, \\ \beta_{n+1}(t,\omega) &= x_0(\omega) + \int_0^t G(s, \beta_{n+1}(s,\omega), \omega; \alpha_n(s,\omega), \beta_n(s,\omega))ds. \end{aligned}$$

By standard arguments [12, 46, 57, 59] it is easily seen that $\{\alpha_{n+1}(t,\omega)\}$ and $\{\beta_{n+1}(t,\omega)\}$ are sample bounded and equicontinuous for P-almost all $\omega \in \Omega$, and Lebesgue dominated convergence yields that

$$\begin{aligned} a(t,\omega) &= x_0(\omega) + \int_0^t F(s, a(s,\omega), \omega)ds, \\ b(t,\omega) &= x_0(\omega) + \int_0^t F(s, b(s,\omega), \omega)ds. \end{aligned}$$

Thus $a(t,\omega) = b(t,\omega) = x(t,\omega)$ on $J \times \Omega$ is the unique sample solution of (4.5.1) in view of the fact that F is one-sided Lipschitzian.

Next we shall show that the convergence of the sequences $\{\alpha_n(t,\omega)\}$, $\{\beta_n(t,\omega)\}$ to the unique sample solution $x(t,\omega)$ of (4.5.1) is quadratic. For this purpose, consider

$$\begin{aligned} p_{n+1}(t,\omega) &= x(t,\omega) - \alpha_{n+1}(t,\omega) \geq 0, \\ q_{n+1}(t,\omega) &= \beta_{n+1}(t,\omega) - x(t,\omega) \geq 0, \end{aligned}$$

and note that $p_{n+1}(0,\omega) = q_{n+1}(0,\omega) = 0$. Also,

$$\begin{aligned} p'_{n+1} &= x' - \alpha'_{n+1} \\ &= f(t,x,\omega) + g(t,x,\omega) - [H(t,\alpha_{n+1},\omega) + M(t,\alpha_n,\omega) \\ &\quad + M_x(t,\alpha_n,\omega)(\alpha_{n+1} - \alpha_n)] \\ &= f(t,x,\omega) - \phi(t,x,\omega) + g(t,x,\omega) + \phi(t,x,\omega) \\ &\quad - [H(t,\alpha_{n+1},\omega) + M(t,\alpha_n,\omega) + M_x(t,\alpha_n,\omega)(\alpha_{n+1} - \alpha_n)] \\ &= [H(t,x,\omega) - H(t,\alpha_{n+1},\omega)] + \{M(t,x,\omega) - M(t,\alpha_n,\omega)\} \\ &\quad - M_x(t,\alpha_n,\omega)(\alpha_{n+1} - \alpha_n) \\ &= \{H(t,x,\omega) - H(t,\alpha_{n+1},\omega)\} + M_x(t,\xi,\omega)(x - \alpha_n) \\ &\quad + M_x(t,\alpha_n,\omega)(\alpha_n - \alpha_{n+1}) \\ &\leq \{H(t,x,\omega) - H(t,\alpha_{n+1},\omega)\} + M_x(t,x,\omega)(x - \alpha_n) \end{aligned}$$

$$
\begin{aligned}
& +M_x(t,\alpha_n,\omega)(\alpha_n-\alpha_{n+1}) \\
= \; & \{H(t,x,\omega)-H(t,\alpha_{n+1},\omega)\}+\{M_x(t,x,\omega)-M_x(t,\alpha_n,\omega)\}(x-\alpha_n) \\
& +M_x(t,\alpha_n,\omega)(x-\alpha_{n+1}) \\
= \; & \{H(t,x,\omega)-H(t,\alpha_{n+1},\omega)\}+M_{xx}(t,\sigma,\omega)(x-\alpha_n)^2 \\
& +M_x(t,\alpha_n,\omega)(x-\alpha_{n+1}) \\
\leq \; & L_1(t,\omega)(x-\alpha_{n+1})+M_{xx}(t,\sigma,\omega)(x-\alpha_n)^2+M_x(t,\alpha_n,\omega)(x-\alpha_{n+1})
\end{aligned}
$$

where $\alpha_n < \xi$, $\sigma < x$. Hence

$$p'_{n+1} \leq K(t,\omega)p_{n+1}+[N(t,\omega)+P(t,\omega)]p_n^2$$

where $K(t,\omega)=L_1(t,\omega)+Q(t,\omega)$, $|g_{xx}(t,x,\omega)| \leq N(t,\omega)$,

$$|\phi_{xx}(t,x,\omega)| \leq P(t,\omega),\ |M_x(t,x,\omega)| \leq Q(t,\omega),$$

$$\int_J K(t,\omega)dt = X(\omega) < \infty \text{ and } \int_J [N(t,\omega)+P(t,\omega)]dt = Y(\omega) < \infty. \tag{4.5.16}$$

Thus, by Gronwall's lemma, we get

$$
\begin{aligned}
0 \leq p_{n+1}(t,\omega) \; & \leq \int_0^t \left[\exp \int_s^t K(s,\omega)ds\right]\{N(s,\omega)+P(s,\omega)\}p_n^2(s,\omega)ds \text{ on } J, \\
& \leq \int_0^T \left[\exp \int_0^T K(s,\omega)ds\right]\{N(s,\omega)+P(s,\omega)\}p_n^2(s,\omega)ds.
\end{aligned}
$$

It therefore follows that

$$\max_J |x(t,\omega)-\alpha_{n+1}(t,\omega)| \leq e^{X(\omega)}Y(\omega)\max_J |x(t,\omega)-\alpha_n(t,\omega)|^2$$

P-almost all $\omega \in \Omega$.

Similarly,

$$
\begin{aligned}
q'_{n+1} \; & = \beta'_{n+1}-x' \\
& = [H(t,\beta_{n+1},\omega)+M(t,\beta_n,\omega)+M_x(t,\alpha_n,\omega)(\beta_{n+1}-\beta_n) \\
& \quad -f(t,x,\omega)-g(t,x,\omega) \\
& = \{H(t,\beta_{n+1},\omega)-H(t,x,\omega)\}+\{M(t,\beta_n,\omega)-M(t,x,\omega)\} \\
& \quad +M_x(t,\alpha_n,\omega)(\beta_{n+1}-\beta_n) \\
& = \{H(t,\beta_{n+1},\omega)-H(t,x,\omega)\}+M_x(t,\xi,\omega)(\beta_n-x) \\
& \quad +M_x(t,\alpha_n,\omega)(\beta_{n+1}-\beta_n)
\end{aligned}
$$

$$\begin{aligned}
&\le \{H(t,\beta_{n+1},\omega) - H(t,x,\omega)\} + \{M_x(t,\beta_n,\omega)(\beta_n - x) \\
&\quad + M_x(t,\alpha_n,\omega)\}(\beta_{n+1} - \beta_n) \\
&= \{H(t,\beta_{n+1},\omega) - H(t,x,\omega)\} + \{M_x(t,\beta_n,\omega) \\
&\quad - M_x(t,\alpha_n,\omega)\}(\beta_n - x) \\
&\quad + M_x(t,\alpha_n,\omega)(\beta_{n+1} - x) \\
&= \{H(t,\beta_{n+1},\omega) - H(t,x,\omega)\} + M_{xx}(t,\sigma,\omega)(\beta_n - \alpha_n)(\beta_n - x) \\
&\quad + M_x(t,\alpha_n,\omega)(\beta_{n+1} - x) \\
&\le L_1(t,\omega)(\beta_{n+1} - x) + M_{xx}(t,\sigma,\omega)(\beta_n - x)^2 \\
&\quad + M_{xx}(t,\sigma,\omega)(x - \alpha_n)(\beta_n - x) \\
&\quad + M_x(t,\alpha_n,\omega)(\beta_{n+1} - x)
\end{aligned}$$

where $x < \xi < \beta_n$ and $\alpha_n < \sigma < \beta_n$.

Consequently, we have

$$\begin{aligned}
q'_{n+1} &\le K(t,\omega)q_{n+1} + [N(t,\omega) + P(t,\omega)](q_n^2 + p_n q_n) \\
&\le K(t,\omega)q_{n+1} + [N(t,\omega) + P(t,\omega)](2q_n^2 + p_n^2)
\end{aligned}$$

where K, N ,P are as in (4.5.16).

Thus, by Gronwall's lemma, we get

$$\begin{aligned}
0 \le q_{n+1}(t,\omega) &\le \int_0^t \left[\exp \int_s^t K(s,\omega)ds\right] \{N(s,\omega) + P(s,\omega)\} \\
&\quad \times (2q_n^2(s,\omega) + p_n^2(s,\omega))ds \\
&\le \int_0^T \left[\exp \int_0^T K(s,\omega)ds\right] \{N(s,\omega) + P(s,\omega)\} \\
&\quad (2q_n^2(s,\omega) + p_n^2(s,\omega))ds.
\end{aligned}$$

It therefore follows that

$$\begin{aligned}
&\max_J |\beta_{n+1}(t,\omega) - x(t,\omega)| \\
\le\ & e^{X(\omega)}Y(\omega) \left[2\max_J |\beta_n(t,\omega) - x(t,\omega)|^2 + \max_J |x(t,\omega) - \alpha_n(t,\omega)|^2\right]
\end{aligned}$$

P-almost all $\omega \in \Omega$.

4.6 DIFFERENTIAL EQUATIONS IN A BANACH SPACE

Let K be a cone in E, a Banach space, that is, closed, convex subset of E such that $K \cap \{-K\} = o$, where o denotes the null element of E. The cone

K induces the order relation in E defined by $x \le y$, $x, y \in E$ if and only if $y - x \in K$.

Let us consider the IVP

$$x' = f(t, x), \ t \in J = [0, T], \ x(0) = x_0, \tag{4.6.1}$$

where $f \in C[J \times E, E]$.

A function $v \in C^1[J, E]$ is said to be an upper solution of (4.6.1) if $f(t, v) \le v'$, $x_0 \le v(0)$ and a lower solution of (4.6.1) if the reverse inequalities hold.

We need the following result in our discussion which is a special case of Theorem 4.4.3 in [28].

Theorem 4.6.1 *Let K be a cone in E. Assume that*

(i) u, $v \in C^1[J, E]$, $f \in C[J \times E, E]$ and $f(t, x)$ is quasimonotone nondecreasing in x relative to K, for each $t \in J$;

(ii) $u'(t) \le f(t, u(t))$, $f(t, v(t)) \le v'(t)$, $t \in J$;

(iii) $\|f(t, x) - f(t, y)\| \le L\|x - y\|$, $L > 0$, $x \in E - K$, $y \in \delta K$, where δK denotes the boundary of K;

(iv) K is a distance set.

Then $u(0) \le v(0)$ implies $u(t) \le v(t)$, $t \in J$.

We can now extend the method of quasilinearization to the present set up.

Theorem 4.6.2 *Assume that*

(i) $f \in C[J \times E, E]$ and there exists constant M such that $\|f(t, x)\| \le M$ on $J \times \Omega$, where $\Omega = \{x \in E : u_0(t) \le x \le v_0(t), t \in J\}$;

(ii) $f(t, x)$ is quasimonotone nondecreasing in x relative to K, for each $t \in J$, where K is a cone in E and is a distance set;

(iii) u_0, $v_0 \in C^1[J, E]$ such that $u_0' \le f(t, u_0)$, $f(t, v_0) \le v_0'$ and $u_0(t) \le v_0(t)$, $t \in J$;

(iv) the Frechet derivative $f_x(t, x)$ exists and is continuous and $\|f_x(t, x)\| \le L_1$ for $(t, x) \in J \times \Omega$, for some $L_1 > 0$ and $f(t, y) \le f(t, x) - f_x(t, y)(x - y)$ where $u_0(t) \le y \le x \le v_0(t)$, $t \in J$;

(v) $\|f_x(t, x) - f_x(t, y)\| \le L_2\|x - y\|^\gamma$, $t \in J$, where L_2 is a positive constant and $0 \le \gamma < 1$.

Then there exist monotone sequences $\{u_n(t)\}$, $\{v_n(t)\}$ *which converge uniformly to the unique solution of* (4.6.1) *and the convergence is superlinear.*

Proof In view of assumption (iv), it is clear that $f(t,x)$ satisfies a Lipschitz condition in x for $(t,x) \in J \times \Omega$. Furthermore we have

$$f(t,y) \leq f(t,x) - f_x(t,y)(x-y), \quad \text{whenever } u_0(t) \leq y \leq x \leq v_0(t) \text{ on } J. \tag{4.6.2}$$

Let u_1, v_1 be the solutions of linear IVPs

$$u_1' = f(t,u_0) + f_x(t,u_0)(u_1 - u_0), \ u_1(0) = x_0 \tag{4.6.3}$$

and

$$v_1' = f(t,v_0) + f_x(t,u_0)(v_1 - v_0), \ v_1(0) = x_0, \tag{4.6.4}$$

where $u_0(0) \leq x_0 \leq v_0(0)$. We shall show that $u_0 \leq u_1$, on J. To do this, let $p = u_0 - u_1$, so that $p(0) \leq 0$. Then

$$\begin{aligned} p' &= u_0' - u_1' \\ &\leq f(t,u_0) - [f(t,u_0) + f_x(t,u_0)(u_1 - u_0)] \\ &= f_x(t,u_0)p. \end{aligned}$$

Theorem 4.6.1 gives $p(t) \leq 0$ on J proving that $u_0(t) \leq u_1(t)$, on J. Now set $p = u_1 - v_0$ and note that $p(0) \leq 0$. Also, using (4.6.2)

$$\begin{aligned} p' &= u_1' - v_0' \\ &\leq f(t,u_0) + f_x(t,u_0)(u_1 - u_0) - f(t,v_0) \\ &\leq f(t,v_0) - f_x(t,u_0)(v_0 - u_0) + f_x(t,u_0)(u_1 - u_0) - f(t,v_0) \\ &\leq f_x(t,u_0)p, \end{aligned}$$

which again implies $u_1(t) \leq v_0(t)$ on J.

Similarly, we can obtain that $u_0(t) \leq v_1(t) \leq v_0(t)$, on J. In order to prove $u_1(t) \leq v_1(t)$, on J, we proceed as follows, since $u_0 \leq u_1 \leq v_0$, using (4.6.2), we see that

$$u_1'(t) = f(t,u_0) + f_x(t,u_0)(u_1 - u_0) \leq f(t,u_1).$$

Similarly, $f(t,v_1) \leq v_1'(t)$ and therefore by Theorem 4.6.1 it follows that $u_1(t) \leq v_1(t)$ on J which shows that

$$u_0(t) \leq u_1(t) \leq v_1(t) \leq v_0(t) \ \text{ on } J.$$

Assume now that for some $n > 1$, $u_n' \le f(t, u_n)$, $f(t, v_n) \le v_n'$ and $u_n \le v_n$, on J. We shall show that

$$u_n(t) \le u_{n+1}(t) \le v_{n+1}(t) \le v_n(t) \text{ on } J, \tag{4.6.5}$$

where $u_{n+1}(t)$ and $v_{n+1}(t)$ are the solutions of linear IVPs

$$u_{n+1}' = f(t, u_n) + f_x(t, u_n)(u_{n+1} - u_n), \; u_{n+1}(0) = x_0, \tag{4.6.6}$$

and

$$v_{n+1}' = f(t, v_n) + f_x(t, u_n)(v_{n+1} - v_n), \; v_{n+1}(0) = x_0. \tag{4.6.7}$$

Hence setting $p = u_n - u_{n+1}$, it follows as before $p' \le f_x(t, u_n)p$ on J and $p(0) = 0$ which again implies that $p(t) \le 0$ on J. On the other hand, letting $p = u_{n+1} - v_n$ yields

$$p' \le f(t, u_n) + f_x(t, u_n)(u_{n+1} - u_n) - f(t, v_n).$$

Since $u_n \le v_n$, (4.6.2) gives $p' \le f_x(t, u_n)p$. This proves that $p(t) \le 0$ on J and hence $u_n \le u_{n+1} \le v_n$ on J. In a similar manner, we can prove that $u_n \le v_{n+1} \le v_n$ on J.

Using (4.6.2), we obtain

$$\begin{aligned} u_{n+1}' &= f(t, u_n) + f_x(t, u_n)(u_{n+1} - u_n) \\ &\le f(t, u_{n+1}) - f_x(t, u_n)(u_{n+1} - u_n) + f_x(t, u_n)(u_{n+1} - u_n) \\ &= f(t, u_{n+1}). \end{aligned}$$

Similar arguments yield $f(t, v_{n+1}) \le v_{n+1}'$ and hence Theorem 4.6.1 shows that $u_{n+1} \le v_{n+1}$ on J which proves (4.6.5) is true. So by induction we obtain

$$u_0 \le u_1 \le \cdots \le u_n \le v_n \le \cdots \le v_1 \le v_0 \quad \text{on} \quad J.$$

Now using standard arguments (see [28]), it can be shown that the sequences $\{u_n(t)\}$, $\{v_n(t)\}$ converge uniformly and monotonically to the unique solution $x(t)$ of (4.6.1) on J. Finally, to prove superlinear convergence, we let $p_n = x - u_n$, $q_n = v_n - x$ so that $p_n(0) = 0$, $q_n(0) = 0$. Also

$$\begin{aligned} p_n' &= x' - u_n' \\ &= f(t, x) - [f(t, u_{n-1}) + f_x(t, u_{n-1})(u_n - u_{n-1})] \\ &= \int_0^1 f_x(t, sx + (1-s)u_{n-1})(x - u_{n-1})ds - f_x(t, u_{n-1})(u_n - u_{n-1}) \\ &= \int_0^1 f_x(t, sx + (1-s)u_{n-1})p_{n-1}ds - f_x(t, u_{n-1})(p_{n-1} - p_n) \\ &= \int_0^1 [f_x(t, sx + (1-s)u_{n-1}) - f_x(t, u_{n-1})]p_{n-1}ds + f_x(t, u_{n-1})p_n. \end{aligned}$$

From (iv) and (v), it follows that

$$\begin{aligned}
\|p_n'\| &\le \int_0^1 L_2\|sx+(1-s)u_{n-1}-u_{n-1}\|^\gamma\|p_{n-1}\|ds + L_1\|p_n\| \\
&\le \int_0^1 L_2\|sx-su_{n-1}\|^\gamma\|p_{n-1}\|ds + L_1\|p_n\| \\
&\le \int_0^1 L_2\|sp_{n-1}\|^\gamma\|p_{n-1}\|ds + L_1\|p_n\| \\
&\le \int_0^1 L_2\|p_{n-1}\|^{\gamma+1}ds + L_1\|p_n\| \\
&= L_2\|p_{n-1}\|^{\gamma+1} + L_1\|p_n\|.
\end{aligned}$$

Then setting $a_n = \|p_n\|$, we find,

$$a_n' \le \|p_n'\| \le L_2(a_{n-1})^{\gamma+1} + L_1 a_n.$$

Now Gronwall's inequality implies,

$$0 \le a_n(t) \le L_2\int_0^t \exp[L_1(t-s)](a_{n-1}(s))^{\gamma+1}ds \text{ on } J,$$

which yields the estimate

$$\max_J\|p_n(t)\| \le L_2\frac{\exp(L_1T)}{L_1}\max_J\|p_{n-1}(t)\|^{\gamma+1}.$$

Similarly,

$$\begin{aligned}
q_n' &= v_n' - x' \\
&= f(t,v_{n-1}) + f_x(t,u_{n-1})(v_n - v_{n-1}) - f(t,x) \\
&= \int_0^1 f_x(t, sv_{n-1}+(1-s)x)(v_{n-1}-x)ds + f_x(t,u_{n-1})(v_n - v_{n-1}) \\
&= \int_0^1 f_x(t, sv_{n-1}+(1-s)x)q_{n-1}ds + f_x(t,u_{n-1})(q_n - q_{n-1}) \\
&= \int_0^1 [f_x(t, sv_{n-1}+(1-s)x) - f_x(t,x)]q_{n-1}ds \\
&\quad + [f_x(t,x) - f_x(t,u_{n-1})]q_{n-1} + f_x(t,u_{n-1})q_n.
\end{aligned}$$

We find, using (iv) and (v), that

$$\begin{aligned}
\|q_n'\| &\le \int_0^1 L_2\|sv_{n-1}+(1-s)x-x\|^\gamma\|q_{n-1}\|ds \\
&\quad + L_2\|x-u_{n-1}\|^\gamma\|q_{n-1}\| + L_1\|q_n\| \\
&\le L_2\|q_{n-1}\|^{\gamma+1} + L_2\|p_{n-1}\|^\gamma\|q_{n-1}\| + L_1\|q_n\|.
\end{aligned}$$

Setting $b_n = \|q_n\|$ and $a_{n-1} = \|p_{n-1}\|$, it is easily follows that

$$b'_n \leq L_2(b_{n-1})^{\gamma+1} + L_2(a_{n-1})^{\gamma} b_{n-1} + L_1 b_n.$$

An application of Gronwall's inequality yields

$$\|q_n(t)\| \leq L_2 \int_0^t \exp[L_1(t-s)] \left[\|q_{n-1}(s)\|^{\gamma+1} + \|p_{n-1}(s)\|^{\gamma}\|q_{n-1}(s)\|\right] ds \text{ on } J$$

and hence

$$\max_J \|q_n(t)\| \leq L_2 \frac{\exp(L_1 T)}{L_1} \left[\max_J \|q_{n-1}(t)\|^{\gamma+1} + \max_J \left(\|p_{n-1}(t)\|^{\gamma}\|q_{n-1}(t)\|\right)\right].$$

The proof is therefore complete.

Remark 4.6.1 Instead of assumption (v) of Theorem 4.6.2 if we assume that

$$\|f_x(t,x) - f_x(t,y)\| \leq L_2\|x-y\|, \; t \in J, \text{ where } L_2 \text{ is a positive constant,}$$

then we can see that the convergence is quadratic. Also, instead of condition

$$f(t,x) \geq f(t,y) + f_x(t,y)(x-y), \; x \geq y,$$

if we suppose that $f(t,x)$ is uniformly convex in x, then by Lemma 4.5.1 in [28], the assumed inequality results. Moreover, the quasimonotonicity of $f(t,x)$ in x implies by Lemma 4.5.2 in [28] that $f_x(t,\alpha(t))x$ is also quasimonotone in x.

4.7 NOTES

The results of Section 4.1 are taken from Leela and Vatsala [44]. For details on Dynamic Systems on Measure Chains, see Lakshmikantham, Sivasundaram, and Kaymakcalan [41]. For the contents of Section 4.2, refer to the work of Deo and Knoll [8, 9]. For the necessary comparison results used in Section 4.2, see Lakshmikantham and RamMohana Rao [36]. Extension of the method of quasilinearization to functional differential equations which is reported in Section 4.3, is adapted from the work of Deo and Sivasundaram [11]. The details on functional differential equations may be seen in Lakshmikantham and Leela Volume II [27]. In Section 4.4, we discuss impulsive differential equations, see for details Lakshmikantham, Bainov and Simenov [25]. The contents of Section 4.4 are taken from Vasundara Devi, Chandrakala and Vatsala [67]. Stochastic differential equations are considered in Section 4.5. For necessary preliminary concepts, see Ladde and Lakshmikantham [20]. For necessary existence and comparison results we refer to Deimling [12], and Lakshmikantham and Zhang

[42]. The main results on quasilinearization are taken from Shahzad and McRae [61]. For further results in this direction, see McRae [46] and Shahzad and McRae [60, 59]. The extension of the method of quasilinearization considered in Section 4.6 is due to Shahzad [57]. The necessary abstract results are taken from Lakshmikantham and Leela [28]. For earlier work in this direction, see Lakshmikantham, Mitchell, and Sety [34], and Drici [15].

References

[1] Aulbach, B., and Hilger, S., *Linear Dynamic Process with Inhomogeneous Time Scales, Nonlinear Dynamic Systems*, Akademic-Verlag, Berlin, 1990.

[2] Bellman, R., *Methods of Nonlinear Analysis*, Vol. II, Academic Press, New York 1973.

[3] Bellman, R., and Kalaba, R., *Quasilinearization and Nonlinear Boundary Value Problems*, American Elsevier, New York 1965.

[4] Bernfeld, S.R., and Lakshmikantham, V., *An Introduction to Nonlinear Boundary Value Problems*, Academic Press, 1974.

[5] Cabada, A., and Nieto, Juan, J., 'Rapid convergence of the Iterative Technique for First Order Initial Value Problems', *Applied & Math. Computation*, to appear.

[6] Cesari, L., and Kannan, R., 'An abstract theorem at resonance', *Proc. Amer. Math. Soc.* **63** (1977), 221-225.

[7] Chong, Jose, and Vatsala, A.S., 'First Order Systems and Faster Convergence Methods', to appear.

[8] Deo, S.G., and McGloin Knoll, C., 'Further Extension of the Method of Quasilinearization to Integro-Differential Equations', *J. of Nonlinear Differential Equations*, TMAA, to appear.

[9] Deo, S.G., and McGloin Knoll, C., 'Extension of the Method of Quasilinearization to Integro-Differential Equations', *Nonlinear World*, to appear.

[10] Deo, S.G., and McGloin Knoll, C., 'K^{th} Order Convergence of an Iterative Method for Integro Differential Equations'.

[11] Deo, S.G., and Sivasundaram, S., 'Extension of the Method of Quasilinearization for Functional Differential Equations', to appear.

[12] Deimling, K., 'Sample solutions of Stochastic Ordinary Differential Equations', *Stochastic Analysis and Appl.* **3** (1985), 15-21.

[13] Doddaballapur, V. and Eloe, P., 'Monotone and Quadratic Convergence of Approximate Solutions of Ordinary Differential Equations with Impulse', Comm. Appl. Analysis, to appear.

[14] Doddaballapur, V., Eloe, P. and Zhang, Y., 'Quadratic Convergence of Approximate Solutions of Two-Point Boundary Value Problems with Impulse', submitted to the Proceedings of the Conference on Differential Equations and Computer Simulation.

[15] Drici, Z., 'Generalized quasilinearization in a Banach space', to appear.

[16] Eloe, P.W. and Zhang, Y., 'A Quadratic Monotone Iteration Scheme for Two-Point Boundary Value Problems for Ordinary Differential Equations', Nonlinear Analysis, to appear.

[17] Hilger, S., 'Analysis on Measure Chains: A unified approach to Continuous and Discrete Calculus', *Res. in Mathematics* **18** (1990), 18-58.

[18] Kavoch, T., Dissertation, Florida Institute of Technology, 1996.

[19] Kaymakcalan, B., 'Existence and Comparison result for Dynamic Systems on Time Scale', *J.M.A.A.* **172** (1993), 243-255.

[20] Ladde, G.S., and Lakshmikantham, V., *Random Differential Inequalities*, Academic Press, New York 1980.

[21] Ladde, G.S., Lakshmikantham, V., and Vatsala, A.S., *Monotone Iterative Techniques for Nonlinear Differential Equations*, Pitman, Boston 1985.

[22] Lakshmikantham, V., 'Periodic Boundary Value Problems of First and Second Order Differential Equations', *J. Appl. Math. and Simulation* **2** (1989), 131-138.

[23] Lakshmikantham, V., 'An Extension of the Method of Quasilinearization', *JOTA* Vol. 82, **2** (1994), pp 315-321.

[24] Lakshmikantham, V., 'Further Improvement of Generalized Quasilinearization', *Nonlinear Analysis*, **27** (1996), 223-227.

[25] Lakshmikantham, V., Bainov, D.D., and Simenov, P.S., *Theory of Impulsive Differential Equations*, World Scientific, Singapore 1989.

[26] Lakshmikantham, V., and Köksal, S., 'Another Extension of the Method of Quasilinearization', *Proc. of Dynamic Systems and Applications* (1994), 205-211.

[27] Lakshmikantham, V., and Leela, S., *Differential and Integral Inequalities* **I**, **II**, Academic Press, 1969.

[28] Lakshmikantham, V., and Leela, S., *Nonlinear Differential Equations in Abstract Space*, Pergamon Press, Oxford 1981.

[29] Lakshmikantham, V., and Leela, S., 'Existence and monotone method for periodic solutions for first order differential equations', *J. Math. Anal. and Appl.* **91** (1983), 237-243.

[30] Lakshmikantham, V., and Leela, S., 'Remark on first and second order Periodic Boundary Value Problems', *J. Appl. Math. and Simulation* **2** (1989), 131-138.

[31] Lakshmikantham, V., Leela, S., and McRae, F.A., 'Improved Generalized Quasilinearization Method', *Nonlinear Analysis* **24** (1995), 1627-1637.

[32] Lakshmikantham, V., Leela, S., and Sivasundaram, S., 'Extension of the Method of Quasilinearization', *JOTA* Vol. 87, **2** (1995), pp 379-401.

[33] Lakshmikantham, V., and Malek, S., 'Generalized Quasilinearization', *Nonlinear World* 1 (1994), 59-63.

[34] Lakshmikantham, V., Mitchell, A.R., and Sety, D., 'Method of quasilinearization and positivity of solutions in abstract cones', *JOTA* **22** (1977), 353-372.

[35] Lakshmikantham, V., and Nieto, J.J., 'Generalized Quasilinearization for Nonlinear First Order Ordinary Differential Equations', *Nonlinear Times and Digest*, Vol. II, No. 1 (1995), 1-9.

[36] Lakshmikantham, V., and M. Rama Mohana Rao, *Theory of Integro-Differential Equations*, Gordan & Breach, London, 1995.

[37] Lakshmikantham, V., and Shahzad, N., 'Further Generalization of Generalized Quasilinearization Method', *J. Appl. Math. and Stoch. Anal.* **7** (1994), 545-552.

[38] Lakshmikantham, V., and Shahzad, N., 'Method of Quasilinearization for General Second Order Boundary Value Problems', *Nonlinear World* **2** No 2(1995), 233-246.

[39] Lakshmikantham, V., and Shahzad, N., 'On Monotone Sequences, Fixed Points and Quadratic Convergence', *Nonlinear Math. Anal. and its Applications*, (ed. by T.M. Rassias) (1995), to appear.

[40] Lakshmikantham, V., Shahzad, N., and Nieto, J.J., 'Methods of Generalized Quasilinearization for Periodic Boundary Value Problems', *Nonlinear Analysis* **27** (1996), 143-151.

[41] Lakshmikantham, V., Sivasundaram, S., and Kaymakcalan, B., *Dynamic Systems on Measure Chains*, Kluwer Academic publishers 1996.

[42] Lakshmikantham, V., and Zhang, B.G., 'On the method of sample upper and lower solutions for Stochastic Differential Equations', *Stochastic Analysis and Applications* **3** (1985), 341-347.

[43] Lee, E. Stanley, *Quasilinearization and Invariant Imbedding*, Academic Press, 1968.

[44] Leela, S., and Vatsala, A.S., 'Dynamic systems on time scales and generalized quasilinearization', *Nonlinear Studies*, Vol. 3, No. 2 (1996), 179-186.

[45] Malek, S., and Vatsala, A.S., 'Method of Generalized Quasilinearization for Second Order Boundary Value Problem', *Inequalities and Applications*, WSSIAA **3** (1994), 449-457.

[46] McRae, F.A., 'Generalized Quasilinearization of Stochastic Initial Value Problems', *Stoch. Anal. and Appl.* **13**, 3 (1995), 210-224.

[47] Mohapatra, R.N., Vajravelu, K., and Yin, Y., 'An Improved Quasilinearization Method for General Second Order Boundary Value Problems'.

[48] Mohapatra, R.N., Vajravelu, K., and Yin, Y., 'Extension of Quasilinearization Method for Second Order Boundary Value Problems'.

[49] Mohapatra, R.N., Vajravelu, K., and Yin, Y., 'Extension of the Method of Quasilinearization and Rapid Convergence', JOTA, to appear.

[50] Mohapatra, R.N., Vajravelu, K., and Yin, Y., 'Generalized Quasilinearization Method and Rapid Convergence for First Order Initial Value Problems', JMAA, to appear.

[51] Mohapatra, R.N., Vajravelu, K., and Yin, Y., 'Generalized Quasilinearization Method for Second Order Boundary Value Problem', Nonlinear Analysis, to appear.

[52] Moore, Ray E., 'Methods and Applications of Interval Analysis', *SIAM Studies in Appl. Math.* (1979).

[53] Nieto, Juan, J., 'Generalized Quasilinearization Method for a Second Order Differential Equation with Direchlet Boundary Conditions', *Proc. Amer. Math. Soc.*, to appear.

[54] Nieto, Juan, J., 'Quadratic Approximation of Solution for Ordinary Differential Equations', *Bull. Aust. Math. Soc.*, to appear.

[55] Sandqvist, A., and Anderson, K.M., 'Applications of the Generalized Bellman-Kalaba solution formula on Lienard's equation', *JMAA* **186** (1994), 66-90.

[56] Sandqvist, A. and Anderson, K.M., 'A Generalized Bellman-Kalaba Solution Formula for First Order Differential Equations', *JMAA* **185** (1995), 545-569.

[57] Shahzad, N., 'Extension of the Method of Quasilinearization in Abstract Cones', *Applicable Analysis* **57** (1995), 325-332.

[58] Shahzad, N., and Malek, S., 'Remarks on Generalized Quasilinearization Method for First Order Periodic Boundary Value Problems', *Nonlinear World* **V2** (1995), 247-255.

[59] Shahzad, N., and McRae, F.A., 'Method of Generalized Quasilinearization of Stochastic Initial Value Problems', *Nonlinear Times and Digest* **1** (1994), 263-270.

[60] Shahzad, N., and McRae, F.A., 'Extension of the Method of Quasilinearization for Stochastic Initial Value Problems', *J. Appl. Math. and Stoch. Anal.* **V8** (1995), 69-75.

[61] Shahzad, N., and McRae, F.A., 'Another Extension of the Methods of Quasilinearization for Stochastic Initial Value Problems', *Stoch. Anal. and Appl.* **V14** (1996), 191-200.

[62] Shahzad, N., and Sivasundaram, S., 'Extended Quasilinearization Method for Boundary Value Problems', *Nonlinear World* **2** (1995), 311-319.

[63] Shahzad, N., Sivasundaram, S., 'Further Generalization of Quasilinearization Method for Boundary Value Problems', *Nonlinear Times and Digest*, Vol. II, No 1 (1995), 59-68.

[64] Shahzad, N., and Vatsala, A.S., 'An Extension of the Method of Generalized Quasilinearization for Second Order Boundary Value Problems', *Applicable Analysis* **58** (1995), 77-83.

[65] Shahzad, N., and Vatsala, A.S., 'Improved Generalized Quasilinearization for Second Order Boundary Value Problems', Vol. 4, No 1 (1995), 79-86.

[66] Stutson, D., and Vatsala, A.S., 'Quadratic and Semi-Quadratic Convergence of IVP', *Neural, Parallel, and Scientific Computations* **3**:2 (1995), 235-248.

[67] Vasundara Devi, J., Chandrakala, G., and Vatsala, A.S., 'Generalized Quasilinearization for Impulsive Differential Equations with Fixed Moments of Impulse', *Dynamics of Continuous, Discrete and Impulsive Systems* **I**, (1995), 91-99.

[68] Vatsala, A.S., 'Recent Advances in the Method of Quasilinearization', *Proceeding's of the Fourth Colloquium on Differential Equations*, Plovdiv, Bulgaria, (1994), 331-337, VSP International Science Publications.

[69] Vatsala, A.S., 'Present Status on the Method of Generalized Quasilinearization', *Proceeding's of the Fifth Colloquium on Differential Equations*, Plovdiv, Bulgaria, (1995).

[70] Vatsala, A.S., Shahzad, N., and Köksal, S., 'The Method of Generalized Quasilinearization in Finite Systems', *Nonlinear World* **V3** No 2(1995), 151-157.

[71] Vatsala, A.S., and Stutson, D., 'Generalized Quasilinearization and First Order Periodic Boundary Value Problem', *Applied Math. and Computation* **V77** (1996), 113-129.

[72] Yin, Y., 'Monotone Iterative Technique and Quasilinearization for Some Anti-Periodic Problems', *Nonlinear World* **3** No 2(1996), 253-266.

[73] Yin, Y., 'Generalized Quasilinearization for Some Anti-Periodic Problems', to appear.

[74] Yin, Y., 'Remarks on First Order Differential Equations with Anti-Periodic Boundary Conditions', *Nonlinear Times and Digest* **V11** No 1 (1995), 83-94.

Index

J

L

M

N

O

P

Q

R

S

T

U

Other *Mathematics and Its Applications* titles of interest:

A.J. Jerri: *Linear Difference Equations with Discrete Transform Methods.* 1996, 462 pp. ISBN 0-7923-3940-1

I. Novikov and E. Semenov: *Haar Series and Linear Operators.* 1997, 234 pp. ISBN 0-7923-4006-X

L. Zhizhiashvili: *Trigonometric Fourier Series and Their Conjugates.* 1996, 312 pp. ISBN 0-7923-4088-4

R.G. Buschman: *Integral Transformation, Operational Calculus, and Generalized Functions.* 1996, 246 pp. ISBN 0-7923-4183-X

V. Lakshmikantham, S. Sivasundaram and B. Kaymakcalan: *Dynamic Systems on Measure Chains.* 1996, 296 pp. ISBN 0-7923-4116-3

D. Guo, V. Lakshmikantham and X. Liu: *Nonlinear Integral Equations in Abstract Spaces.* 1996, 350 pp. ISBN 0-7923-4144-9

Y. Roitberg: *Elliptic Boundary Value Problems in the Spaces of Distributions.* 1996, 427 pp. ISBN 0-7923-4303-4

Y. Komatu: *Distortion Theorems in Relation to Linear Integral Operators.* 1996, 313 pp. ISBN 0-7923-4304-2

A.G. Chentsov: *Asymptotic Attainability.* 1997, 336 pp. ISBN 0-7923-4302-6

S.T. Zavalishchin and A.N. Sesekin: *Dynamic Impulse Systems.* Theory and Applications. 1997, 268 pp. ISBN 0-7923-4394-8

U. Elias: *Oscillation Theory of Two-Term Differential Equations.* 1997, 226 pp. ISBN 0-7923-4447-2

D. O'Regan: *Existence Theory for Nonlinear Ordinary Differential Equations.* 1997, 204 pp. ISBN 0-7923-4511-8

Yu. Mitropolskii, G. Khoma and M. Gromyak: *Asymptotic Methods for Investigating Quasiwave Equations of Hyperbolic Type.* 1997, 418 pp. ISBN 0-7923-4529-0

R.P. Agarwal and P.J.Y. Wong: *Advanced Topics in Difference Equations.* 1997, 518 pp. ISBN 0-7923-4521-5

N.N. Tarkhanov: *The Analysis of Solutions of Elliptic Equations.* 1997, 406 pp. ISBN 0-7923-4531-2

B. Riečan and T. Neubrunn: *Integral, Measure, and Ordering.* 1997, 376 pp. ISBN 0-7923-4566-5

N.L. Gol'dman: *Inverse Stefan Problems.* 1997, 258 pp. ISBN 0-7923-4588-6

Other *Mathematics and Its Applications* titles of interest:

S. Singh, B. Watson and P. Srivastava: *Fixed Point Theory and Best Approximation: The KKM-map Principle*. 1997, 230 pp. ISBN 0-7923-4758-7

A. Pankov: G-*Convergence and Homogenization of Nonlinear Partial Differential Operators*. 1997, 263 pp. ISBN 0-7923-4720-X

S. Hu and N.S. Papageorgiou: *Handbook of Multivalued Analysis*. Volume I: Theory. 1997, 980 pp. ISBN 0-7923-4682-3 (Set of 2 volumes: 0-7923-4683-1)

L.A. Sakhnovich: *Interpolation Theory and Its Applications*. 1997, 216 pp. ISBN 0-7923-4830-0

G.V. Milovanović: *Recent Progress in Inequalities*. 1998, 531 pp. ISBN 0-7923-4845-1

V.V. Filippov: *Basic Topological Structures of Ordinary Differential Equations*. 1998, 530 pp. ISBN 0-7293-4951-2

S. Gong: *Convex and Starlike Mappings in Several Complex Variables*. 1998, 208 pp. ISBN 0-7923-4964-4

A.B. Kharazishvili: *Applications of Point Set Theory in Real Analysis*. 1998, 244 pp. ISBN 0-7923-4979-2

R.P. Agarwal: *Focal Boundary Value Problems for Differential and Difference Equations*. 1998, 300 pp. ISBN 0-7923-4978-4

D. Przeworska-Rolewicz: *Logarithms and Antilogarithms*. An Algebraic Analysis Approach. 1998, 358 pp. ISBN 0-7923-4974-1

Yu. M. Berezansky and A.A. Kalyuzhnyi: *Harmonic Analysis in Hypercomplex Systems*. 1998, 493 pp. ISBN 0-7923-5029-4

V. Lakshmikantham and A.S. Vatsala: *Generalized Quasilinearization for Nonlinear Problems*. 1998, 286 pp. ISBN 0-7923-5038-3

GPSR Compliance

The European Union's (EU) General Product Safety Regulation (GPSR) is a set of rules that requires consumer products to be safe and our obligations to ensure this.

If you have any concerns about our products, you can contact us on ProductSafety@springernature.com

In case Publisher is established outside the EU, the EU authorized representative is:

Springer Nature Customer Service Center GmbH
Europaplatz 3
69115 Heidelberg, Germany

Batch number: 10755292

Printed by Printforce, the Netherlands